OWEN THE POET

OWEN THE POET

Dominic Hibberd

A boy, I guessed that the fullest, largest liveable life was that of a Poet.
(5 March 1915)

I was introduced as 'Mr Owen, Poet' or even 'Owen, the poet'.
(26 January 1918)

MACMILLAN
PRESS

First published 1986 by
THE MACMILLAN PRESS LTD
Houndmills, Basingstoke, Hampshire RG21 2XS
and London
Companies and representatives
throughout the world

ISBN 0–333–38448–2 hardcover
ISBN 0–333–49104–1 paperback

A catalogue record for this book is available
from the British Library.

Reprinted 1989, 1994

Printed in Great Britain by
Antony Rowe Ltd
Chippenham, Wiltshire

Contents

List of Plates

Preface

This book is a study of what Wilfred Owen called his 'poethood'. It examines his origins and growth as a poet, his understanding of his poetic role, and the unity of his imaginative life, and it assesses his achievement. The publication of almost all his verse and letters, and the release of his surviving books and papers for research and quotation, have at last made it possible to consider his work as a whole. My subject is the poet and his work, rather than the man, so my principal source has been his manuscripts; but biography is included when his poetry can be illuminated by material not fully recorded elsewhere. The poet who emerges is not the impersonal voice of pity presented by the 1920 edition, nor the archetypal soldier poet of Blunden's memoir (1931), nor the champion of pacifism and social reconstruction that some modern critics have described; he is far from the unworldly idealist portrayed by Harold Owen and not quite the straightforward, innocent character shown in Jon Stallworthy's biography. Some commentators have come close to splitting Owen into two people, one before the trenches and one after, but Professor Stallworthy rightly gives him a single identity. There is an unbroken continuity from the lay assistant defying Revival in 1912, through the Aesthete admired by Tailhade in 1914 and the soldier in training and at war in 1916–17, to the poet of 1918. For that reason his early verse is worth more attention than critics have usually given it; unpromising though some of it may seem, it was the foundation for his mature work. His 1918 poetry, written in emotion deliberately 'recollected in tranquillity' as he contemplated 'the inwardness of war' in his cottage room at Ripon – perhaps the last Romantic undertaking in English poetry – deserves more recognition in the history of literature than it has so far received. By temperament and training he was exactly suited to becoming 'the poet of the war', and he has no rival for that title. For one year, beginning in October 1917, no one, soldier or civilian, wrote English poetry more significant than his.

Owen holds a transitional place between the nineteenth century and Modernism, inheriting the aspirations and moral urgency of the Romantics and Victorians but seeing the need for Modernist 'insensibility'. His poethood was shaped by Romantic dedication, Victorian energy, Evangelical fervour, the French and English Decadence, and Georgian innovation. By nature receptive and venerating, with a quick ear for voices, he became, as Murry said of him, 'the splendid borrower who lends a new significance to that which he takes', making original poetry out of material eagerly gathered from wherever he could find it. Passionately bookish but lacking the education that might have dampened his enthusiasm, he found himself in a period which brought culture into unusually violent contact with current events, so that he became a figure in history whom history books will always quote; but, just as the war grew out of the nineteenth-century culture which preceded it, so Owen's war poems were in embryo long before he ever guessed at the nightmare of the trenches. And they speak to our world still, as they were meant to do.

I have not attempted in this book to tackle the large and difficult question of Owen-as-spokesman and whether his view of the war was representative, realistic, useful or in the interests of stable peace. Some historians (John Terraine, for example) have suggested that it may not have been all of those things. The immense shift in general perceptions of the Great War that has developed since about 1930 has undoubtedly distorted the modern image of Owen, but perhaps all shades of opinion could agree that he has done as much as anyone to prevent the reading public from being persuaded ever again that death in battle is 'sweet and decorous'. *Poetry of the Great War: An Anthology*, edited by John Onions and myself (1986), gives some of the context of the war and its mass of verse in which – and against which – Owen's poems were written. I have not attempted here, either, to offer close 'evaluative' criticism of most of his poetry; critical studies by other hands are available in my *Casebook* (1981) and elsewhere, but the best close reading so far is that by Desmond Graham in *The Truth of War* (1984). I have tried not to repeat detailed research when it has already been published, referring instead to the notes and articles in which it first appeared. Over the years one's findings multiply and one's opinions alter; in cases of difference between this book and earlier publications of mine, I hope the book gets it right.

Quotations from Owen's verse are from *The Complete Poems and Fragments* (*CPF*), edited by Jon Stallworthy (1983), except that some passages from drafts which Owen never fully revised are my own readings from the originals, sometimes simplified to show what I take to be his intentions; a few phrases not in *CPF* are quoted from manuscript; and a few amendments to *CPF* are made, usually silently, as a result of recent checks against manuscripts. Owen's spelling is usually corrected. Words cancelled by him are in square brackets. See also Appendix B. Other sources are given in the Appendixes and Bibliography.

My principal debt is to Professor Stallworthy, not only for his labours as editor and biographer but also for his unstinted generosity in making manuscripts available and sharing his knowledge of Owen. It is in the nature of research that one has to record one's disagreements with other people's published conclusions more often than one's agreements, but despite the quibbles in my notes I accept far more than I doubt of his portrait of Owen and his text of the poems. I am also most grateful to Mr Leslie Gunston for his help and friendship and for so generously giving his Owen treasures to the collection at Oxford. Mr John Bell's editing of both the letters and Harold Owen's memoirs has been of inestimable value to Owen studies. I am indebted to the pioneering research of Professor Dennis Welland and the more recent work of Dr Sven Bäckman. Many people have kindly given me information, answered letters and sent me articles, including Mr Sydney Brock and other members of Dr Brock's family; Dr Philip Boardman; Dr Jennifer Breen; M Jean Loisy; Mrs Jean Mitchell and Mrs K. A. Jackson of North Queensferry; Mr F. J. Nicholson; Mr Martin Seymour-Smith; Mr M. M. Stuart; and Dr Simon Wormleighton. Professor Sir William Trethowan has been good enough to read my shellshock chapters. Dr Kathleen McKilligan, Dr Roger Pooley and Mr Tom Coulthard have read other chapters, and Mr James McLaverty and Dr John Onions have gallantly read and savaged the whole typescript; the book, though faulty still, is much the better for their wise advice. I have received tireless assistance from the staff of the English Faculty Library, Oxford, particularly from Miss Margaret Weedon, who established the Owen Collection there with the utmost skill and efficiency. Other librarians and archivists have been unfailingly helpful at the British Library (Manuscript Room and Newspaper Library); the Birmingham Reference Library (War Poetry Collection); King's College Library, Cambridge; the Imperial War Museum; the University Library, Keele (Sociological Society papers); the Brotherton Library, University of Leeds (Gosse Collection); the Berg Collection, New York Public Library (Marsh correspondence); the Bodleian Library, Oxford (Western manuscripts); the National Library of Scotland (Geddes papers); Sheffield Central Library (Carpenter Collection); the Shrewsbury Reference Library; and the Humanities Research Center, University of Texas at Austin. I am grateful to the British Academy for the grant that enabled me to visit Austin.

Acknowledgements are due to the following for permission to quote copyright material: the Owen Estate and Chatto & Windus Ltd, London, and W. W. Norton & Co. Inc., New York, for Wilfred Owen's poems, mainly from the *Collected Poems and Fragments*; the Owen Estate and Oxford University Press for Owen's letters, mainly from the *Collected Letters*, edited by Harold Owen and John Bell; Sir Rupert Hart-Davis for extracts from Siegfried Sassoon's unpublished letters; Mr George Sassoon and Viking Penguin Inc., New York, for Sassoon's published poems; Mr Robert Graves and the Brotherton Library, Leeds, for an extract from an unpublished 1917 letter; Mrs E. Scott Moncrieff for an extract from an unpublished letter by Charles Scott Moncrieff; the Rupert Brooke Estate for an unpublished passage from a 1907 letter; and the Harry Ransom Humanities Research Center, University of Texas at Austin, for extracts from various manuscripts.

I am grateful to Mr John Killham, Mr Leslie Gunston, Mr Sydney Brock and Mrs Gwen Hampshire for their help in obtaining photographs. Illustrations are reproduced by permission of John Killham (1), the Southampton Art Gallery (2), the Owen Estate (3, 4, 5, 15), Leslie Gunston (6, 7), Sydney Brock (8), the British Library (11, 14), the Royal National Institute for the Blind (12) and the Imperial War Museum (13).

1 The Origins of a Poethood

In the decade before the First World War the country town of Shrewsbury was remote from centres of culture, too far for a lower-middle-class schoolboy there to discover much about new developments in literature. It was still important to be earnest, at any rate in a piously Evangelical household. In school, poetry was still Romantic poetry (the respectable sort), its theory laid down by Wordsworth, its emotions and language exemplified in Tennyson and (rather to excess) in Keats. Wilfred Owen worked hard, writing neat essays on the standard authors. He later took pleasure in believing that his poetic vocation came from his native landscape, where he was a worshipper among the hills in Wordsworthian fashion as well as a properly Ruskinian botanist, geologist and archaeologist, but the beginnings of what he later called his 'poethood' had as much to do with books and religion as with nature. In his early years he was very bookish and devout, belonging entirely to the nineteenth century. His seriousness was as typical of that age as it is alien to ours; it is easy to laugh at him, but not very useful.

His artistic ambitions probably began to take shape while he was at school. He was certainly a keen student of literature, as can be seen from some 1907–8 exercise books[1] which show him working on the *Faerie Queene*, at least ten Shakespeare plays and many other texts that were to be of use to him later. His teacher seems to have had a particular interest in the Romantics, although she wrote 'Too strong!' against a 1908 comment that the eighteenth century had lacked 'imagination and moral nature'. She did not argue when he said in the same essay that Wordsworth and his followers, directly inspired by nature, had brought to literature 'a new sympathy with man especially the poor', a statement which gives a clue to the origins of what he described at the end of 1917 as his lifelong 'sympathy for the oppressed'.

1

Love of nature, belief in the imagination, and sympathy with
suffering were characteristics which the would-be poet would
have to encourage in himself. In his schoolboy thoughts about
poetry and its function, Owen was a Wordsworthian, a position
he never fully abandoned.

His only attempt to write a poem about the start of his poetic
career seems to have been jotted down in the summer of 1914,
while he was in Bagnères. The principal event for him in that
momentous August was not the war but his being introduced to
Laurent Tailhade, whose own commitment to poetry had begun
long before in that part of France. Owen had already noticed
that one of the local hills resembled one in Shropshire.[2] Going
for a solitary walk under the moon and wondering what the war
was going to mean for a poet, he took stock of his past; his
memory went back to a Cheshire holiday ten years earlier,
where 'at Broxton, by the Hill . . . was born / [Out of a dark and
disobedient] moon so sweet and [so forlorn] . . . my [poethood]'.[3]
There was probably some myth-making in this, but no doubt
Broxton (Plate 1) had provided his first strong imaginative
stimulus. He referred to its bluebells and heather in later letters.
The hills of Cheshire and Shropshire, steep grassy ridges rising
abruptly from the plain to flat tops covered in 'herb and heather',
were always where his imagination felt most at home. His attic
room in Shrewsbury looked towards Haughmond Hill and the
Wrekin, and beyond them to Caer Caradoc and the Long Mynd,
those 'landscapes whereupon my windows lean' which feature
from time to time in his verse and letters and which seem to be
reflected in the setting of 'Spring Offensive'.

Those 'blue remembered hills', as Housman had called them,
were Welsh border country. Owen was Welsh by descent on both
sides, a fact which he valued because he held the Victorian view
that imagination and strong, often melancholy emotions were
Celtic endowments. The elegiac strain and musical elaboration
of his mature work are qualities typical not only of Tailhade's
ballades élégiaques but also of ancient Welsh verse. Even during
the action in which he won the Military Cross in 1918 he
remembered those earlier soldier poets, 'my forefathers the agile
Welshmen of the Mountains'. Doing some more myth-making
while he was in France in 1914, he said his strength of feeling
was a sign of an artistic temperament and suggested it was an
inherited quality that could also be seen in his mother's brother
(who had taken to drink) and an Owen aunt (who 'literally

palpitates with physical sensation'). Welsh blood was even more worth having if it could be shown to be poisoned; he may have been reading about Baudelaire and other *poètes maudits* who had considered that a tainted ancestry and a capacity for exquisite 'physical sensation' were necessary attributes for a poet.[4]

Whether or not his poethood began at Broxton, there is no evidence that he wrote any serious verse much before April 1911, when, having left school and being 'in a ferment' of hopes and fears about his future, he fell predictably 'in love' with Keats.[5] Like many Victorians before him, he started his juvenilia by imitating Keats in a spate of poems, most of which can be paired with their originals: 'Before reading a Biography of Keats for the first time' ('On First Looking into Chapman's Homer'), 'Sonnet written at Teignmouth, on a Pilgrimage to Keats's House' ('Sonnet Written in the Cottage where Burns was Born'), 'On seeing a lock of Keats's Hair' ('On Seeing a Lock of Milton's Hair'), 'To Poesy' (a youthful declaration of faith like those in 'Sleep and Poetry' and *Endymion*), and others.[6] There are also several prose notes for sonnets that never got written, including this: 'Down by the outlet of the Teign, it is a fascinating sight to watch the brave current in its last moments quenching the breakers that incessantly fight to roll it up. It is like a new dating on his doom.'[7] The allusion to *The Fall of Hyperion* ('Thou hast dated on / Thy doom') shows the young devotee's interest in a poem which was eventually to be echoed in 'Strange Meeting'. The note is also an early example of his habit of sharing in or imitating the experiences of an admired poet. He marked personal details in Colvin's biography of Keats, particularly when they seemed to coincide with his own, noticing that Keats's mind was 'naturally unapt for dogma', that Keats and Hunt were given to 'luxuriating' over 'deliciousness', and that Reynolds came from Shrewsbury and 'lacked health and energy'. He involved himself similarly in the poems. *Endymion* and 'Lamia' kept his pencil especially busy as he underlined the rich vocabulary and marked the lush descriptions, including that of the sleeping Adonis. A bookmarker in *Endymion*, embroidered with the text 'Create in me a clean heart O God', seems to have prayed in vain among sensuous passages in which he evidently delighted, but perhaps guilt overcame him after reading 'Lamia' because four pages of erotic description have been carefully stuck together.[8]

However, too much has been made of Owen's passion for Keats. Critics give it prominence almost as a matter of routine, without paying enough attention to other influences. Mentions of Keats are common in Owen's letters until 1915 but after that they decrease and by 1918 disappear. Even in 1914 he looked back with some embarrassment on the days when he had supposed 'all Poetry' to be contained 'in J. Keats', blaming his ignorance on the narrowness of his life in Shrewsbury.[9] The pattern of his 1911 enthusiasm was repeated in other years: a biography, some pilgrimages, reverential study, and then compositions in the style of the new master. There had been a Coleridge pilgrimage in 1910 and there were probably several Shelley ones in 1912. But 1911 was the year for Keats. Owen explored Teignmouth for Keats associations in April, and in September went to Hampstead, where he presumably saw the lock of hair in the local library, gazed at the outside of Keats's house and walked over the heath.[10] He felt the '[dead's breath]' when he saw the lock, but on the heath heard the voices not only of Keats but also of other 'men long dead'.[11] The plural corresponds with his wish in 'To Poesy' to commune with the 'bards of old', a dream of speech with the dead that was to recur in his poetry until the conversation with the dead poet–soldier in 'Strange Meeting', that 'familiar compound ghost'[12] who embodies not just Keats but many of the poets Owen loved. By 1912 Keats was already losing his supremacy, Owen reluctantly admitting that Shelley was 'the brightest genius of his time, (yes, tho' *I* say it)'.[13]

His books, over three hundred of which are preserved as he left them in 1918, show the range – and limitations – of his interests at school and later.[14] Shakespeare, Scott, Keats and Dickens predominate, but he also worked on Milton, several eighteenth-century authors, and some Elizabethan and late-medieval poets. About two-thirds of his library can be classified as 'English literature', including biographies of at least twenty authors. (There was little demand for criticism in those days before the emergence of English as a major academic subject, so that books about poets tended to be 'lives' rather than critical studies.) There are also nearly fifty books in or about French, a high proportion for someone of Owen's respectable but ordinary educational background. The rest are mostly botany, history and classics. The imprints are often those of the popular 'libraries' of the time – Everyman's Library, the People's Books, the Home University Library, Penny Poets, – cheap editions aimed at the growing market of young people like himself who were keen on self-improvement.

Keats had seen the necessity for 'application study and thought'. Owen's earnest reading was accompanied by equally determined fieldwork in the local countryside, for as an admirer of Ruskin – 'my King *John* the Second' (1912) – he saw the arts and sciences as interrelated. Ruskin would have approved of his helping to found an 'Astronomical, Geological and Botanical Society' in 1907 and a field club in 1917. He seems to have called on Ruskin's biographer, W. G. Collingwood, in the Lake District in 1912, and afterwards, 'a little drunk on Ruskin', to have gone for another of his nocturnal rambles in the hills.[15] In the country or at his desk, he was storing up material for later writing, including his characteristic vocabulary ('fronds', 'granites', 'as quick as lilac shoots'). His approach to his studies was marked by awe and humility to a degree hard to imagine today, for his attitude was essentially religious and the central textbook in his early education was the Bible.

The first books in his library are Bibles. The largest is his mother's, who perhaps put it there. Brought up as a devout Evangelical herself, she reared him in her faith; he fully shared it at first, reading a Bible passage every day with the aid of Scripture Union notes and piously including texts and sermon topics in his early letters. This training was an all-pervasive influence on his approach to life and literature, its effects persisting long after he abandoned orthodoxy in 1912–13. The Evangelical movement was still strong and Owen would have been taught its simple, unchanging messages at prayer-meetings and Bible classes of the kind that he was later to conduct himself at Dunsden. An Evangelical bases his religion on a personal relationship with Christ, maintaining it by means of a daily period of private prayer and scripture-reading in which the Word of God is studied for promises and guidance. At the heart of Evangelical practice there is this literary activity, a discipline of intensive reading which is more imaginative than critical, a daily opening of the mind to a book. A high value is placed on the memorising of texts, so that the believer can hear God 'speaking' at any time and be aware of Christ as a living, ever-present companion. Mrs Owen was not a deeply spiritual woman but her faith was founded on this kind of devotion.[16]

Owen tackled literature in the way that he had been taught to read the Bible. Admiring poetry meant worshipping 'the bards of old', hearing them speak and making pilgrimages to places which their presence had made holy. He memorised texts,

applying them to his own life in an effort to imitate the lives of
the poets. 'Turn [ye] to Adonais; his great spirit seek. / O hear
him; he will speak!':[17] the biblical language ('Seek ye the Lord')
is unmistakable. His repeated wish to know the poets of the past
as companions parallels the Evangelical's desire to 'walk' with
Jesus. As Evangelicalism tends to be more concerned with personal
life and conduct than with abstract or social theory, so Owen was
more attracted to authors' lives than to their ideas. Throughout his
letters there are signs of his taking the behaviour of great writers
as guides to his own: he and his sister must keep their letters,
because Keats and his sister kept theirs; he ought to write home
every day, like Ruskin; he was glad to visit the poor, because
Shelley did.[18] Above all, he read poetry as scripture, looking for
such guidance, exhortation and assurance as might be appropriate
to one who wanted to set his feet on the path.

He was trained to read like this by his mother, whose methods
can be glimpsed in the annotations in her Bible. From 1888
onwards she marked and often dated any verses that 'spoke' to
her, including rules of conduct, calls to righteousness and the
texts which Evangelicals prize as 'God's promises' (a term sourly
echoed in the phrase 'death's promises' in 'S. I. W.'). Several of
her marginal dates record a passage chosen for a New Year
meditation, a custom which may lie behind her son's habit of
reviewing the past year in late-December letters. In 1903 she
pursued an especially serious course of Bible study with him,
apparently in answer to some trouble that he was passing through.
The carefully underlined verses include:

> If we suffer, we shall also reign with him.
> (2 Timothy 2:12, 'Wilfred   Sunday Feb. 8th 1903')

> Satan hath desired to have you . . . But I have prayed for
> thee.   (Luke 22:31–2, 'Wilfred')

> As one whom his mother comforteth, so will I comfort
> you.   (Isaiah 66:13, 'Wilfred   Feb. 16th 03   Promise text').

Approaching his tenth birthday, Wilfred was being shown that
passive endurance was the way to deal with unhappiness, suffering
and temptation. He should rely on God to provide a divine
version of the maternal comfort that was abundantly available
at home. This religion of unquestioning 'self-sacrifice' (a term Mrs

Owen was 'inordinately fond of using'[19]) had many attractions for a boy who was naturally inclined to passivity, but it had to be resisted if he were ever to gain his independence. He was beginning to doubt it by 1911, if not earlier, and by 1913 had rejected it after a hard struggle, but its attractions remained. In 1917 he said at first that Christ was 'literally in no man's land' preaching 'Passivity at any price!', but that was not a view he was able to hold for long in the face of war experience. Mrs Owen had stopped marking her Bible by then. She had listed some texts in the back on his twenty-first birthday (March 1914) and slipped in a newspaper-cutting announcing his commission in 1916, but if she ever used the book after that she left no evidence. The greatest effort of her life was over. In a way she had failed, for her son had lost his faith and not become a clergyman despite her hopes and prayers. Nevertheless, she always believed that he had remained a Christian at heart; 'his Christian ideals . . . were really very deep – too deep to speak of', she wrote to Edmund Blunden after the war, as usual preferring emotional conviction to reasoned statement.[20] She was not altogether wrong, for although Owen may not have been a Christian poet he was certainly a religious one in a general sense.

His relationship with his mother was the closest of his life, but it caused him many difficulties. His literary interests must always have been a mystery to her, although she admired them, for her own reading scarcely extended beyond light novels and the pious, naïve verse of John Oxenham. It has been suggested that she read Keats to her son at Broxton, but that seems highly improbable (even in 1911 she had apparently not heard of the 'Ode to a Nightingale').[21] She thought all modern poetry was likely to be 'rubbish'.[22] Her simple, crudely punctuated letters to Blunden and others after the war show her to have been emotional, affectionate and good-hearted but not remotely intellectual or bookish. Owen's move away from Evangelicalism towards poetry was in part an escape from her into a world which she could not enter. It was not an easy move, for the bond between them was suffocatingly intense; there are connections between it and the many images of (s)mothering and drowning in his poems. Part of him longed for her because she provided rest and safety ('ease / For ever from the vain untravelled leagues'). She encouraged his dependence, sharing little jokes against his father and laughing him out of the few awkward friendships he tried to make with girls. She used her hypochondria – for one must suspect that her

incessant minor ailments were often more imaginary than real –
as a means of exacting sympathetic enquiries, and in return she
encouraged his own similar tendencies so that she could nurse
him as much as possible. Her religion was another way of holding
him. As he freed himself, his health improved and so did his poetry.
He later thought that his first mature poem was 'Happiness', the
first draft of which laments that he is no longer 'a Mother's boy';
there is no mention at all of war in it, but it was written
immediately after his first tour of the trenches. If the completion
of his independence is recorded in that early 1917 sonnet, the
beginning of it seems to be hinted at in the words 'dark',
'disobedient' and 'secret' in his unfinished poem about the Broxton
holiday (and perhaps in Mrs Owen's anxious markings in her
Bible a year before Broxton, when she may have seen signs already
that her son was likely to go his own way). Many years later,
Harold Owen said that his brother had been nursed into poetry
at Broxton in sunlight and 'in the safety and understanding love
that my mother wrapped about him', but that seems to be merely
a pleasant fiction.[23] The fragment tells a quite different story of
a solitary walk in the dark, no doubt a forbidden excursion for
an eleven-year-old. Owen thought of his poethood as having been
'born', 'Nursed' and 'Suckled' in secret solitude on the hills. It
was still a mother–son relationship, but the mother was nature
and Mrs Owen had to be disobeyed.

His inner life was hidden so that even his mother and brother
never perceived it. In his bedroom at Shrewsbury, there were
papers in a locked cupboard which his mother was never allowed
to see. Despite his complaints in 'To Poesy' and other early poems
that he was lonely, he was a Romantic and needed solitude.
Darkness was always strangely congenial, his verse and letters
often associating it with poetic inspiration.[24] One of his most
ambitious early poems, an ode on the ruined city of Uriconium
at the foot of the Wrekin, invites the reader to 'lift the gloomy
curtain of Time Past' in order to see through the 'riven ground'
the 'secret things that Hades hath', mostly traces of war. His
mature poems search the obscurity of his buried experience,
recording a journey through the riven ground into 'the sorrowful
dark of hell'. He found companionship there but that too was
secret.

2 The First Crisis: Religion

Owen began to emerge from his first Keatsian phase towards the end of 1911, under pressure of new influences and painful experience. 1912 was a crucial year, recorded in great detail in his letters because he was away from home for the first time and had much to tell. His verse was still unoriginal (he seems not to have experimented with sound-effects, for example, until he met Tailhade in 1914), but it records the formation of concerns and images that were to be developed in later poems. The commitment to literature which he had expressed in 'To Poesy' was now put to a strenuous test as it came into conflict with the demands of Evangelical orthodoxy, but by 1913 he had found his freedom to 'be a meteor, fast, eccentric, lone, lawless . . .'.[1]

When he left school in 1911 he wanted to try for a university scholarship and his mother wanted him to enter the Church, so it was arranged that he should become an unpaid assistant to the Vicar of Dunsden, near Reading, in return for tuition. Mrs Owen's sister, Emma Gunston, lived nearby, and her own vicar had been able to recommend Mr Wigan as having suitably Evangelical views.[2] After a brief period in which Owen was impressed by Wigan and the unaccustomed elegance of life in the Vicarage, frustration began to set in. There seems to have been little or no tuition and much tedious, formal conversation. Less than two months after his arrival, Owen spoke bitterly of 'the Wasted Hours', a phrase that reappears in a sketch for a poem about his pounding heartbeats;[3] his Dunsden troubles were already beginning to find their way out in physical symptoms and secret verse-writing. As a lay assistant, he was obliged to keep up a Christian front in public and to observe the kind of life that would be his if he were to take orders, so that his own faith came under strain. All literary activity was suspect. Although Wigan was a man of apparent culture, he thought books were an 'alternative to life, or an artificialized life'.[4] Owen found that the more he saw of theoretical Evangelicalism, as opposed to the

9

kindly, unthinking sort he had known at home, the more it
seemed to deny everything he most valued. Even the beauties of
language were potentially sinful because they were 'of this world'
in their appeal to the senses.[5] He was in a dilemma familiar to
many Victorians, caught between a sternly Pauline form of
Christianity which condemned all pleasures of 'the flesh' and a
concept of poetry that was still largely Keatsian.

Conscientiously, he tried his hand at hymn-writing, producing
two hackneyed pieces which seem to be the only verse in which
he ever expressed conventional religious sentiments, but his
devotion to poetry could not be so easily diverted.[6] No doubt
remembering Keats's belief that a long poem was a useful
discipline, he composed two lengthy verse renderings of Hans
Andersen stories, one of them quite un-Christian and the other,
'The Little Mermaid', full of Keatsian descriptions which
sometimes foreshadow the war poems in language ('bugles floated
in far citadels') and in details of pain and horror. He knew from
Colvin that Keats had also trained himself by joining Shelley and
Leigh Hunt in writing poems on chosen subjects. Since his cousin,
Leslie Gunston, was beginning to share his interest in 'Books and
their Makers', they set themselves the task of each composing on
'The Swift', the swifts flying round the Vicarage being convenient
symbols of an enviable liberty and aspiration. There were to be
more poems on agreed subjects in later years.[7]

But these were only poetic exercises. Poetry seemed to make
far higher demands, and they were apparently irreconcilable with
those of religion. In an effort to resolve the issue, Owen turned
to his third main interest, the earth sciences, doing his earnest
but unscholarly best to tackle the Victorian debate between
science and religion. He was soon 'reading, analysing, collecting,
sifting, and classifying Evidence' and 'grappling as I never did
before with the problem of Evolution'. He read a statement of
the Christian answer to Darwinism but contemptuously wrote
'Shallow!' against its discussion of art.[8] His conclusion was
probably summed up in a comment he had marked in Keats's
letters, 'nothing in this world is proveable'; when he met these
words again in W. M. Rossetti's life of Keats, he added, 'at least
proved W. O.'.[9] Proof certainly seemed lacking for personal
immortality, an essential belief of traditional Evangelicalism:
'Science has looked,' one Dunsden poem begins, 'and sees no life
but this'. 'The Dread of Falling into Naught' (18 September
1912) is a typical piece from this period, concealing under its
affected, Tennysonian rhetoric a genuine ache at human mortality

that was to find clearer expression less than six years later in war elegies such as 'Futility'.

However, the crisis was not brought to a head by his own questionings, as might seem the case from existing biographical accounts, but by challenges which were put to him. When he first arrived in the parish, he found that the Vicar had 'the hope of a Revival in the place much on his mind, during these times'.[10] That was ominous. When the phenomenon of Revival sweeps through an Evangelical church at national or local level, believers gain a deeper faith, while doubters and backsliders find strength to make the full sacrifice of their lives to Christ. 'We yearn for times of Revival', said the *Christian* on 25 July 1912, beginning its lengthy reports of the annual Keswick Convention but perhaps not speaking for Owen, who attended the Convention for two or three weeks that summer. He heard the famous preachers of the day repeat their now all too familiar messages. 'The secret of Keswick', said Dr Griffith Thomas, 'is handing over your individuality to Christ.' Some people heard the appeal ('no day was without its trophy of grace') but the *Christian* said gloomily on 1 August that for some the Convention was 'a dark, hidden tragedy, secret sin strangling the soul'. Owen returned to Dunsden in considerable distress but hid his feelings as well as he could.[11] He might have got away with it, but the even pace of parish life was quickening, building up to a fine village drama in November–December. Whether in answer to the Vicar's prayers or as a result of the soul-winning zeal of Clyde Black, a new lay assistant, Revival came at last.[12]

Parishioners began to be converted or converted anew at meeting after meeting. Black was busy 'cornering' people and asking them to accept Christ. The excitement in the Vicarage must have been immense, making it impossible for Owen to keep his opinions to himself or to escape the general expectation that he would join Black in doing the Lord's work. Privately he composed one of his harshest poems ('Unto what pinnacles', dated 6 November), a fierce denunciation of 'good men' who commit 'soul-suicide' in their quest for holiness, becoming obsessed with their own spiritual lives:

And their sole mission is to drag, entice
And push mankind to those same cloudy crags
Where they first breathed the madness-giving air
That made them feel as angels, that are less than men.

The unvarnished, angry style is similar to that of Shelley's tirades against religion in *Queen Mab*, a poem which Owen may well have been reading. His exasperation at the activity around him and at his own bewildered conscience may have been one cause of a Vicarage 'furor' in December in which he was in some way involved.[13] A Christmas present of a new book about Mrs Browning 'from his friend Clyde Black' was perhaps a peace-offering to show that Black, too, liked poetry, provided it was Christian. But the Vicar was stern: if Owen was still considering ordination, he must recognise that it meant complete obedience to God's will and a renunciation of all wordly pleasures, including 'verse-making'.[14]

Owen's letters do not record what happened to the Revival but the outcome of his own crisis was no longer in doubt, although discussions with Wigan continued into 1913. He left the Vicarage on 7 February, not long before his twentieth birthday. One of his manuscripts leaves a vivid record of his state of mind.[15] On one side there is a scrawled first draft of 'The Unreturning' – a sonnet about the finality of death – together with quotations from Keats and a few clumsy lines about romance. 'The Unreturning' did not reach its final version for some years, but without this early draft one would never have guessed that such a forceful poem could have been started as early as 1913. 1913 it must be, however, for on the other side there is a scrappy outline for a letter:

> To Vicar – solely on the ground of affection
>
> I was a boy when I first came to you and [boyishly] held you in the [doubtful] mischievous esteem that a boy has for his Headmaster. It is also true that I was an old man when I left, that is like [an   ol   ] I have a senile & a stupefaction for
>
> The Christian life is – affords no imagination, physical sensation, aesthetic philosophy –
> There is but *one* dimension in the X^n relg the strait line [upwards] to the Zenith – whereas I cannot conceive of less than 3
> But all these considerations are Nothing to the conviction that the philosophy of the whole system as a religion is but a religion and therefore one Interpretation of Life & Scheme of Living among a hundred – and that not the [best] most convenient.

That is all that survives of what was perhaps Owen's only attempt to write down a full statement of his objections to 'the Christian life' (a favourite Evangelical phrase). His mention of 'imagination, physical sensation, aesthetic philosophy' suggests that his 1912 reading had included some recent Aesthetic work, a field he was to explore more thoroughly in the next few years.

Owen's letters from Dunsden are a possibly unique record of a young man's struggle with religion, respectability and growing-up in provincial England before the Great War, but their importance for a study of his poethood is that a knowledge of what happened at Dunsden contributes to an understanding of the later poems. The public role to which he aspired in 1918 was that of an evangelising preacher against hypocrisy, false creeds, oppression and lack of pity. His anger against Wigan, Black and the Keswick speakers in 1912 prepared him for his later indignation at the support which the churches gave to the war, and at the way in which individuals were treated as cannon fodder to be 'dragged, enticed and pushed' into fighting. No longer convinced by promises of salvation and immortality, he grieved over the loss of hope and life which war inflicted in this world, scorning the glib pieties of civilians who said that men killed in battle were saved by God 'even before they fell'.[16] Like the Victorian writers who broke away from Evangelicalism he retained a set of Christian standards, referring his reader to them by means of biblical language and allusion, but he did not recover his early faith; the Christian views which he expressed in some of his 1917 letters would not have satisfied Wigan. He gained much at Dunsden but also lost much, and the loss was permanent. And, in his secret inner life, the visions, torments and delights of his peculiar imagination began to manifest themselves in images which were to reappear later as images of war.

There were other factors in Owen's loss of faith than difficulties over doctrine and personal evangelism. If Wigan was an example, 'the Christian life' did not seem to include much concern for social inequality, nor much allowance for the strong feelings of early manhood. The lay assistant was expected to visit the sick, so that he soon became aware of the contrast between the comforts of the Vicarage and the stark poverty of some of the villagers. He was also expected to teach the local children in Sunday school and choir practices, and the pleasure which he derived from observing youthful beauty was disturbingly inconsistent with the attitudes that he was supposed to be inculcating. As the strain grew, his health suffered and terrors filled his dreams.

Having concluded before he went to Dunsden that the poets he most admired had brought into poetry 'a new sympathy with man especially the poor', he was struck by a comment in a history of literature that the ideals of the French Revolution had 'inspired the British School of revolt and reconstruction in Burns, Shelley, Byron, Wordsworth, Coleridge and Tennyson, till its fires have died down in our own day'. 'Have they!' Owen exclaimed when he read this in April 1912.

> They may have in the bosoms of the muses, but not in my breast. I am increasingly liberalising and liberating my thought, spite of the Vicar's strong Conservatism . . . . From what I hear straight from the tight-pursed lips of wolfish ploughmen in their cottages, I might say there is material ready for another revolution . . . . Am I for or against upheaval? I know not; I am not happy in these thoughts; yet they press upon me.

In the same paragraph he mentions Tennyson and Dickens.[17] Literary history and poetic ambition played a large part in the 'liberalising and liberating' of his opinions. He became especially interested in Shelley, among the poets of 'revolt and reconstruction', noting in a verse letter about the literary associations of the Dunsden area that Shelley had written poetry, music that could still be heard, among 'beechen solitudes' on the nearby Thames. The 'music' which he heard must have been that of *The Revolt of Islam*, for he discovered in January 1912 from a biography of Shelley that *The Revolt* had been composed in a boat 'under the beech-groves' not far away.[18] This poem was to remain in his mind for the rest of his life, providing him with the theme and title of 'Strange Meeting' in 1918. Shelley's passionate defence of freedom against tyranny would have strengthened the lay assistant's rebellion in 1912, just as his advocacy of non-violent resistance affected the young subaltern's attitude to war only a few years later.

Another, much less predictable influence on Owen's thinking at Dunsden and later began in October 1911 when he happened to buy a book of new poems by 'a modern aspirant (unknown to me) . . . . I am idly-busy in trying to discover the talent of our

own days, and the requirements of the public'. This book was
undoubtedly *Before Dawn: Poems and Impressions* by Harold Monro.
Owen read it carefully and could still quote from it two months
later.[19] It was thus at this early stage in his career that he
encountered Monro's work and not, as might otherwise be
supposed, in 1915, when his letters first record a conversation
with Monro in person. *Before Dawn* was a chance find but an
important one, for Monro was soon to become proprietor of the
Poetry Bookshop and publisher of *Georgian Poetry*. There is a clear
line of development from that random purchase in 1911 to the
proud claim six years later: 'I am held peer by the Georgians'.

Just as Owen had imitated Keats and Shelley, so he imitated
Monro, drafting a satirical verse portrait (perhaps of Wigan) on
the back of a respectable hymn he had been writing.[20] Critics
generally assume that he learned to write satire from Sassoon in
1917, but it is time that Monro's 'Impressions' had their due. In
fact Monro's influence can be seen even in some of Owen's
'Sassoonish' 1917 work. One of the 'Impressions', for example, is
a description of a rich man dining in a London club who picks
up the evening paper and begins

> to read it and to carve
> A shilling strawberry. 'Twas about the strike –
> A hundred, in the cause, had sworn to starve.
> He put it down, and muttered: 'Let them starve!'

This ending may be compared not only with the last lines of 'The
Dead-Beat' but also with those of another 1917 piece, an early
sketch for 'The Sentry', in which Owen attempted to cap a grim
record of a trench experience with a picture of civilians in a club:

> turning down his cards.
> The Evening news was [brought into] the Clubbers
> They glanced at Haig's dispatch between the rubbers
> [Nothing! – Advanced   A paltry]
> said: 'Only fifty yards?'[21]

He would have been interested to see that the revolutionary fires
which had burned in Romantic poetry were still alight in the
work of the first modern 'aspirant' he had so far encountered,
and that Monro was hostile to clergymen and religion (some

sceptical lines by Monro are unmistakably echoed in the 1913
version of 'The Unreturning').[22]

Before Dawn lays stress on another theme which was increasingly
alive to Owen at Dunsden, the value of 'physical sensation' and
beauty. Monro's creed is a fervent humanism, symbolised by the
titan of the coming dawn who will be man himself, beautiful in
body as in soul, liberated from sexual as well as social restraints
and able to lead a full physical life. Like other young writers of
the period, including D. H. Lawrence, Monro had been strongly
influenced by the progressive views of Edward Carpenter,[23] whose
advocacy of sexual liberation and fulfilment is implicit in such
Before Dawn poems as 'The Virgin':

> I
> Am made of flesh, and I have tingling nerves:
> My blood is always hot, and I desire
> The touch of gentle hands upon my face
> To cool it, as the moonlight cools the earth.

Owen's verse was to echo this passage several times in the next
few years. His own 'tingling nerves' were troubling him. In
November, with the Revival gathering strength around him, he
gave his mother a glimpse of his new 'aesthetic philosophy':

> *my* philosophy teaches that those mortals who have nerves
> exquisitely responsive to painful sensation, have a perfect right
> to use them . . . *to respond equally keenly to enjoyment.* I know I
> have a tingling capacity for pleasure . . . and if such be the
> operation of a tense nerve, then must I content me with nerves'
> foolish ado when things offend and lacerate them. . . . I am
> willing to pay this price, to purchase the delight to the full . . . .[24]

Like his fragmentary letter to the Vicar, this suggests he had been
thinking about the principles on which his poethood should be
based. Perhaps he had already looked into some literature of the
French Decadence; at any rate, he seems to be aware that sensitive
nerves were important in artistic theory. Late Romanticism held
that the exclusive goal of art was beauty, and that since beauty
could only be apprehended through physical sensation it was
necessary for the artist to have 'exquisitely' responsive nerves, to
observe their workings closely and to react with as much intensity
to pain as to pleasure.

It would be easy but inaccurate to dismiss Owen's many references to his nervous condition and general health in 1912 as signs of nothing more than adolescent hypochondria. He told his mother at intervals that he was suffering from weak sight, insomnia, indigestion, giddy fits, palpitations, and – towards the end of the year – painful breathing. Some of these afflictions, even the eye trouble, may have been caused by nothing more than Vicarage meals; the only picture of him in spectacles is his cartoon of himself undergoing a dizzy 'vertigo' in the dining-room after a heavy lunch (Plate 3). However, the attack occurred in mid November, when the Revival was beginning. Similarly his breathing-difficulty, which culminated in a severe attack of 'congestion of the lungs' after he left Dunsden, was both a genuine symptom and an expression of psychological stress; perhaps the *Christian* had been right in saying that some people had felt 'strangled' at Keswick, and Owen may not have been exaggerating when he said in December that his nerves were in a 'shocking state' and his chest 'continually "too full"'. Just as if one had been over-long in a putrid atmosphere, and had got to the advanced stage of being painfully conscious of it.'[25] His physical troubles represented not only mental distress but also resistance to external pressure (even the vertigo saved him from attending a religious meeting), and they were encouraging proof that he had a poet's temperament. He recognised their psychological origins with lively interest, showing something of that inner resilience that was to keep him sane during his shellshock in 1917.

His collapses at the Vicarage were often rather comical but his illness afterwards seems to have been serious, although little is recorded about it because he was at home and wrote no letters. Looking back on it later, as he often did, he said it had been accompanied by 'phantasies', 'horrors' or 'phantasms'.[26] These seem to have been nightmares or even hallucinations, an acute form of the vivid and unpleasant dreams from which he suffered throughout his life. Their content was probably not unlike that of a 'horror' which haunted his first months at the Vicarage, as recorded in the 'Supposed Confessions of a Secondrate Sensitive Mind in Dejection', an absurdly literary piece which nevertheless records in metaphor a real experience:

> think not, if your life-blood still is warm,
> That ye have looked upon Despondency.
> Ye have but seen her in another's eye,

> As Perseus fearfully beheld the form
> Of Gorgon, mirrored in the stilly well.
> There may ye guess the beauty of that Head,
> The pallor and the mystery – but the Dread
> Ye feel not, nor the horror, nor the spell.
>
> But, face to face, she fixed on me her stare:
> Woe, woe, my blood has never moved since then;
> Down-dragged like corpse in sucking, slimy fen,
> I sank to feel the breath of that Despair.
> With autumn mists, and hand in hand with Night,
> She came to me. But at the break of day,
> Went not again, but stayed, and yet doth stay.
> '— O Horror, doth not Pain take note of light
> And darkness, – doth he not hold off betimes,
> And yield his victim for an hour to Sleep?
> Then why dost thou, O Curst, the long night steep
> In bloodiness and stains of shadowy crimes?'

The violence of these nocturnal ordeals ('bloodiness and stains of shadowy crimes') implies that they were sado-masochistic 'phantasies', comparable in kind if not in degree with his shellshock nightmares in 1917. The poem goes on to say that Despondency will never leave him and that he will eventually be transported to hell. There was a kind of truth in this, for the image of the deathly face always obsessed him, emerging in his 1917–18 dreams and in the poems inspired by them such as 'Dulce et Decorum Est', 'The Sentry' and 'Strange Meeting'. The face was a projection of his own imagination and unspoken urges, arousing guilt, fear and helplessness, but even in its first appearance in his verse he was able to give it a literary form. The allusion to Perseus (who also first appears in his verse here) and the Gorgon's head may be based on a Pre-Raphaelite source such as Burne-Jones's painting *The Baleful Head* (Plate 2), or D. G. Rossetti's poem 'Aspecta Medusa', but the more general image of a fatally beautiful face with eyes that turn men to stone is recurrent throughout Romanticism (it will be discussed under 'Perseus' in the next chapter).[27]

That he continued to be susceptible to 'horrors' in 1912 is evident from his curiously strong reaction to Borrow's descriptions in *Lavengro* of daylight 'fits of the horrors' and from his bookish but attentive account of a near-faint after a fall from his bicycle:

sudden twilight seemed to fall upon the world, an horror of great darkness closed around me – strange noises and a sensation of swimming under water overtook me, and in fact I fell into a regular syncope. I did not fall down however, nor yet lose all consciousness; but the semi-blindness, and the chill were frightful.

The language in this bit of self-dramatising is lifted from Genesis, *Pilgrim's Progress*, and Clarence's dream in *Richard III*, but it points to several images in Owen's later poems, including 'sudden evening' ('Conscious'), 'Suddenly night crushed out the day' (the final version of 'The Unreturning'), or that submarine imagery in 'Dulce et Decorum Est' and 'The Sentry' which marks two rare references to his war dreams. Another Dunsden poem, 'Written on a June Night. (1911)' (apparently written in 1912, despite the title), again uses imagery of burning eyes and violent crime in describing summer sleeplessness. Owen seems to have been accurate in later telling his mother that his nights had been 'terrible to be borne' (and in implying that his sufferings had been a result of suppressed sexual feelings for which he had never been prepared by his parents).[28]

There is persistent imagery in all this of passivity, of smothering, drowning, petrifaction – a nightmare parallel to the maternal affection of his 'idle, protected, loving, wholesome, hidden, intimate, sequestered' home life.[29] When he wrote about beauty, even in 1917 work such as 'The Fates' and 'My Shy Hand', he often gave it qualities of smoothness, embracing, pillowing, sheltering, sleep, languor, timelessness and escape from action. He wrote some verse on his nineteenth birthday about the pleasure of recovering from pain (another bout of indigestion at the Vicarage), saying that he preferred 'the placid plains' of convalescence and being mothered to the exposed heights of health in 'the dangerous air where actual Bliss doth thrill'.[30] After Dunsden he collapsed at home, his 'congestion of the lungs' corresponding not only to the pressure of recent events but also to Mrs Owen's anxious nursing, but he managed to get on the move again after a few months because there was no hope of becoming a poet on those plains. His instinct was to seek shelter and reversion to childhood, but in his first 'mature' poem ('Happiness', 1917) he recognised that the old happiness of childhood was unreturning because he had gone beyond 'the scope / Of mother-arms'. In 'Spring Offensive', his last poem, the soldiers push past the

'sorrowing arms' of brambles, advance up the hill and at the top
meet the 'even rapture of bullets' as grown men facing experience
in the 'dangerous air'.[31] That much at least could be achieved,
but beyond it there seemed to be passivity again, the tormented
helplessness of death which Despondency's visitations had
foreshadowed since 1911 or even earlier. Beauty, the poet's goal,
had two personifications: in one she was a protector and healer,
a bringer of rest, whose eyes were the 'secret gate' by which her
devotee could escape from time, as in 'The Fates'; in the other
she was a 'phantasm' whose terrible stare also brought timelessness
but no rest ('no rest for thee, O Slave of mine', she promises in
the 'Confessions'), a state suffered by the 'encumbered', groaning
sleepers in 'Strange Meeting'. For Owen, even in 1912 before he
knew anything of war, the second personification was the true
one; his imagination allowed him no choice. Painfully discovering
his own nature in the muddle of adolescence, he began to see
that what he might have to say as a poet would not be quite the
same as anything that other poets had said, comforting though
their words had once been. The 'pressure of Problems' was forcing
him to voice his own 'dim reveries', the frightened 'croonings of
a motherless child, in gloom', but perhaps his own verse might
one day bring comfort to someone in need.[32]

Although one cause of his distress in 1912 was religion, another
seems to have been sexual difficulties. There is an erotic element
implicit in much of what he wrote at that time, often suggesting
a deep-seated disorientation. He was still devoted to Keats, saying
even in 1913 that 'I fear domestic criticism when I am in love
with a real live woman. What now I am in love with a youth,
and a dead 'un'!'[33] Reading W. M. Rossetti's biography in 1912,
he was overcome by its account of Keats's death: 'Rossetti guided
my groping hand right into the wound, and I touched, for one
moment the incandescent Heart of Keats.'[34] The language is that
of religious ecstasy, with its allusion to St Thomas, but the
emotion seems to be partly sexual. These statements suggest the
extent of Owen's confusion; his mother ('domestic criticism'),
Keats, the dead, a youth, a woman ('I fear . . .'), sado-masochism,
religion – love seems to have many objects. The fear of criticism
from home was a particularly severe hindrance to his sorting
himself out. Looking back in 1918 on his younger days, he urged
his mother not to deny his youngest brother 'the thing he craves,
as I was denied; for I was denied, and the appeal which, if you
watched, you must have seen in my eyes, you ignored . . . . And

my nights were terrible to be borne.'[35] He was not so clear-sighted in 1912 but gave in to his mother's unspoken demands: 'Oh how do I stand (yes and sit, lie, kneel and walk, too,) in need of some tangible caress from you. . . . my affections are physical as well as abstract – intensely so – and confound 'em for that, it shouldn't be so.'[36] In the absence of his mother he developed a streak of narcissism which was to remain with him, but his feelings also attached themselves to the parish children.

The lay assistant was expected to take an interest in the local youngsters, but Owen was reckless in his friendships with them. He was moved by the innocence and suffering of childhood, finding 'the blanch of sickness, and the dark-cirqued eye' strangely beautiful.[37] In 'Impromptu' he implores a little girl to comfort him, since he is 'in pain for human sin' and 'heart-ease and rest' can no longer be had from 'Mother and Brethren, Teachers, Holy Guides':

> Oh, now, unless my face hath set too granite-hard
> And hurt thy tender hands to stroke it o'er,
>
> Unless the fires that ever rage behind my eyes,
> Hot-sear thy lips in pressing kisses there,
>
> I crave thee, place thy two soft hands upon my cheeks,
> So shall long-treasured tears be loosed at last.

This image of a youth guilty of some unnamed sin, with the stony fixedness of a statue but incessantly burning eyes, was to recur again and again in later poems. One source is Monro's virgin, with her longing for gentle hands to cool her hot face, and another is the legend of Medusa, for Owen had looked on the face of Despondency and felt himself turned to stone, 'too granite-hard' to be restored by a girl of any age.

The child who is the subject of a number of these Dunsden poems was probably Milly Montague, his favourite among the small girls who came to infants' meetings, but his closest friendship was with a thirteen-year-old boy, Vivian Rampton, about whom little is recorded except that he had 'melancholy brown eyes' and was intelligent, fond of music and books, and within a year or two of secondary school. Owen had Rampton to a secret tea at the Vicarage, probably gave him piano lessons at his home, and on at least one occasion 'secretly met with Vivian . . . and went a delicious ramble; lay in hawthorn glades. . . . He read to me, and I told him tales.' He kept all this hidden from the Vicar, apparently regarding Rampton as a private discovery whose intellectual abilities he could encourage without getting tied up with religion. When Revival came, did Black 'corner' the boy and try to convert him (as he had certainly cornered Milly's brother)? Owen would have been furious, insisting that his protégé should be left alone, and the secret would have been out. The Vicar would have been far from pleased to find that his senior lay assistant had been conducting this clandestine friendship for many months and was now even refusing to help save the boy's soul. That seems a possible explanation for a mysterious passage in a January 1913 letter which refers to a recent 'furor' which has 'now abated in the Vicarage, thank Mnemosyne; but I hope that I, who "discovered" him something over a year ago, may' – and here the paper has been cut away in one of Harold Owen's most thorough attempts at censorship. Rampton was to haunt later poems. The friendship probably began when Owen first went to Dunsden, in which case it would have lasted for a year and three months, a span roughly corresponding to 'something over a year' in the letter, and 'a year', 'two years' and 'two short years', three inconsistent phrases in a later fragment about a friendship. In this sketch for part of an autobiographical poem that never got written, Owen remembers how his unnamed friend had undergone the 'inalterable change from boy to man' during the year (or two) that they knew each other, and how 'many of my thoughts were given to him / And many of his hours were given to me'. Then, very hesitantly, he asks whether it would have been better if he had not seen the boy's face, heard his voice or touched his hand.[38] He wrote this in Bagnères in 1914, at the same time as his piece about the birth of his poethood at Broxton; even allowing for myth-making, both fragments are about crucial stages in his growth as a poet. The memory of Rampton persisted, becoming an idealised figure with deep, sad eyes, a hand that could sometimes be touched and a beauty that was for ever unattainable. The relationship was perhaps Owen's first love affair, although he may well have been much less aware of its sexual implications at the time than Wigan probably was. Rampton is present in many poems after 1913, but if he appears anywhere in Dunsden verse it is perhaps only as the brown-eyed

child in 'The Two Reflections', a sonnet evidently based on
recollections of summertime rambles among the pines and beeches
of the Chilterns in 1912.[39]
When Owen finally reached the point of deciding whether or
not to become 'a religious devotee', the Vicar stipulated that he
would have to give up 'verse-making' and friendships with village
children ('all my pretty chickens / At one fell swoop?').[40] Whether
he remained at Dunsden or left, the relationship with Rampton
was over. The 'melancholy brown eyes' and shy hand of his young
friend were to become symbols of pain and loss as well as of love.
He worked out a kind of solution in 1913. Two sonnets which
belong stylistically to 1916–17 are relevant here. In 'To the Bitter
Sweet-Heart: A Dream', Eros takes the poet by the hand,
promising to 'fill with Yours my other hand'; but in the end the
god takes the other hand himself and the human lover is not
found again. In other words, the idea of love is compensation for
a lost relationship (which included hand contact). In the sequel
sonnet, 'To Eros', the poet has been abandoned by the god,
despite sacrificing 'fair fame', 'Old peaceful lives; frail flowers;
firm friends; and Christ'.[41] Sacrificing Christ must refer to
Dunsden, while 'fair fame' may represent Owen's good name in
the Vicarage. Breaking the relationship was made tolerable at
the time by Owen's imagining that he would dedicate himself to
the spirit of love through poetry, an ideal stated in another sonnet,
'Stunned by their life's explosion into love', in which most men
are said to be forced by that 'explosion' into lust or the 'bitter
chastity' of religion without seeing that poetry could slake their
thirst. It was not until later that he realised how unsatisfactory
a solution this was. Eros left him, caring nothing for the sacrifices
of a worshipper who had tried to know Love in solitude.

If this reconstruction of the Vicarage 'furor' is accurate, it sums
up the nature of the Dunsden crisis. Rampton stood for Owen's
private world of music, nature, books, strong feeling, mystery and
delight, as well as of guilt and sin; the Vicarage represented the
smothering prison of piety, respectability, self-denial and sham.
Evangelicalism taught that anyone who had heard the Gospel
and rejected it was certain to be damned. Hell fire was still a
reality, a terror to many Victorian and Edwardian children and
perfectly credible to Owen, with his nights steeped at times in
'bloodiness and stains of shadowy crimes'. 'I can believe in Hell',
he wrote in autumn 1912, despite his doubts about heaven.
Wigan's sermons were 'horrifyingly dismal'.[42] Owen was doubly

damned, having rejected the Christian message and touched a boy's hand. The 'fires that ever rage behind my eyes' were burning dreams, a foretaste of eternal torment. He may have been disgusted at himself, for there are several strong expressions of disgust against humanity in general in the Dunsden poems. That phase passed, but he never quite shook off the fear that his secret desires and his break from religion were sins which would not be forgiven, so that his poetry from 1912 onwards returns obsessively to images of guilt, fire, hell and everlasting pain.

One of the most promising (or at any rate least unpromising) of his 1912-13 poems in which these concerns emerge is 'Deep under turfy grass and heavy clay', in which the funeral of a mother and child prompts a protest against the folly of having children:

> So I rebelled, scorning and mocking such
> As had the ignorant callousness to wed
> On altar steps long frozen by the touch
> Of stretcher after stretcher of our dead.
> Love's blindness is too terrible, I said:
> I will go counsel men, and show what bin
> The harvest of their homes is gathered in.

The first four lines here are perhaps the earliest in which Owen's imagination can be felt as authentic and original, showing the beginnings of his later technical control. The rhyming of 'wed' and 'dead' stresses the dual function of the altar steps, while the word 'touch' at the climax of the stanza is applied to the coldness of death rather than the warmth of marriage. The sensation of chill continues in alliteration into the repeated 'stretcher' of the next line, as generations of coffin-bearers rest their burdens gratingly on stone that seems to have drawn its coldness from the dead. Stretchers more usually carry living people, a meaning which reinforces the unexpected significance of 'touch'; life and death are closer than is realised by those who wed on the altar steps. In the next stanza the grass and clay in the churchyard are seen to be half-burying the wreaths, making the village children wonder 'what might mean / Rich-odoured flowers so whelmed in fetid earth'. The rich flowers and rotting earth take up the contrast between warm life and chilling death; like the dead mother and child, the flowers have grown out of clay and are returning to it. The poet's rebellion is overcome for the moment by the sight of a child 'whose pale brows / Wore beauty like our mother Eve's', but in later years his protest was redoubled when he asked, 'Was it for this the clay grew tall?' ('Futility'), or showed that soldiers understood the deathly significance of the flowers given to them by women ('The Send-Off'). The pale brows of the child made him accept the mortality of beauty, as Eve had made Adam eat the bitter fruit, but five years later the 'pallor of girls' brows' ('Anthem for Doomed Youth') became an emblem of death, like the pallor of Medusa, for the girls had been responsible for sending their men off to war. The child at the Dunsden graveside was to grow into a *femme fatale* in later work.

Owen's religious background left him with the urge to 'go counsel men'. The poet, able to explore farther than other people, had the task of reporting back on what he had seen. This role is expressed in 'O World of many worlds' in the ornate metaphor of the poet as meteor, pursuing a 'lawless' way through space. There the meteor will meet other poets, whose combined light will rival the sun's (as the soldiers challenge the sun in 'Spring Offensive'). He will warn 'the earth of wider ways unknown' (as in 'Storm'), not following the crowd's 'fixed' course ('the march of this retreating world'). Such ideas would have appalled the Vicar, who may have liked to quote St Jude's denunciation of 'filthy dreamers' who creep secretly into the Church and defile 'the flesh', 'wandering stars, to whom is reserved the blackness of darkness for ever'. Owen incorporated this text into his poem, declaring his poetic manifesto in defiance of the biblical orthodoxy which had tried so hard to deny him his freedom.[43]

In discussing the dangers of poetic dreams which might defile the flesh, Wigan may have pointed to St Paul's assertion that 'they that are Christ's have crucified the flesh with the affections and the lusts'. Owen worked this into the closing lines of an extraordinary poem, 'The time was aeon', introducing St Paul in person as a 'small Jew' leading a crowd of (no doubt Evangelical) 'railers'.[44] The poem presents a vision, not a dream but a 'true resumption of experienced things' (here as elsewhere Owen insists that he deals in experience, not mere dreams), in which the world's ugliness is made lovely by a spirit with an evil name, 'the Flesh':

It bore the naked likeness of a {maid / boy}
Flawlessly moulded, fine exceedingly, . . .
His outline changed, from beauty unto beauty,

> As change the contours of slim, sleeping clouds.
> His skin, too, glowed, pale scarlet; like the clouds
> Lit from the eastern underworld; which thing
> Bewondered me the more. But I remember
> The statue of his body standing so
> Against the huge disorder of the place
> Resembled a strong music; . . .
>
> Then watched I how there ran towards that way
> A multitude of railers, hot with hate
> And maddened by the voice of a small Jew
> Who cried [with a loud voice]: saying 'Away!
> Away with him!' and 'Crucify him! Him,
> [With] the affections and the lusts thereof.'

Like the text from St Jude, the Pauline statement is turned back
upon its author, so that crucifying the flesh becomes a crime
similar to crucifying Christ. In this remarkable contradiction of
Evangelical belief, Owen seems to be taking his cue from the
opening poem in Harold Monro's *Before Dawn* ('Two Visions'),
in which the baseness of contemporary life is contrasted with the
beauty of the coming dawn of freedom. Monro describes the
'Titan of the dawn – Humanity':

> His visionary eyes looked out afar
>   Beyond the transient semblances of death.
> No sound of supplication came to mar
>
> The rhythm of his calmly-taken breath.
>   No ripple of a thin or faint delight
> Moved round his crimson lips; and underneath
>
> His bright skin aureoled by the rose twilight
>   Rolled the vast torrent of majestic thews.
> Master of his strong passion . . . .

The homosexual quality of this description was probably not
accidental, since Monro's current guru, Edward Carpenter, was
almost the only person at the time who was prepared to write
publicly in defence of 'Uranian' or homosexual love. Like many
young men of the time, including Sassoon and Graves, Owen owed

his eventual freedom in part to Carpenter's liberalising and liberating views. He may have been completely unaware of both Carpenter and the implications of Monro's poem when he wrote 'The time was aeon', but he certainly first imagined 'the Flesh' as male, adding the alternative 'maid' only as an afterthought. The Vicar's challenge may have forced him towards such honesty.

The 'statue' of the Flesh, with its 'pale scarlet' dawn colour and constantly changing silhouette, is one of a series of similar figures in Owen's poetry. Representing his desires as both man and poet, its various manifestations are often associated with pain and its triumph is never complete. Its earliest appearance is in a passage which he added to 'The Little Mermaid' in September 1912. The mermaid finds on the sea bed

> a marble statue, – some boy-king's,
> Or youthful hero's. Its cold face in vain
> She gazed at, kissed, and tried with sighs to thaw
> For still the wide eyes stared, and nothing saw.
>
> Thereby she set a weeping willow-tree
> To droop and mourn. Full dolefully it clung
> About the form, and moved continually,
> As if it sighed; as if it sometimes wrung
> Convulsive fingers in sad reverie.
> And ever o'er the light blue sand it hung
> A purple shade, which hour by hour the same,
> Burnt softly on, like lambent sulphur-flame.

Despite the traces here of 'Isabella' and *Endymion*, the imagery in the description can now, I hope, be recognised as Owen's own. It contains elements from his Dunsden dreams, experiences and poems, worked into an impersonal pattern. The statue is of a boy with sad eyes. The 'wide eyes' gaze for ever and see nothing, like Moneta's in *The Fall of Hyperion*. Flame burns unchanging, colouring the stone flesh purple, while the convulsive fingers of the shadow make an ever-changing outline. The stony face – for the image is of Owen as well as of Rampton – cannot be softened by a girl's caresses. The submarine setting is a smothering underworld from which there seems to be no release. The material for 'Mental Cases' and 'Strange Meeting' is already beginning to take shape.

Owen had made his rebellion and chosen his course through darkness. He left Dunsden behind him, with the friends and enemies which he had made there. He had found some of the subject matter of his future poetry; now he had to strengthen his 'aesthetic philosophy', develop his own style and establish his freedom, tasks which might be more easily accomplished abroad than at home.

3 Aesthete in France

TAILHADE AND OTHERS

By mid September 1913, Owen was established as a teacher in the Berlitz School of Languages in Bordeaux. He began his two years in France by telling his new acquaintances that he was the son of a knight, a ruse which made him acceptable in higher social levels than had been open to him at home;[1] it was in France that he acquired the social and literary graces which were to make him a welcome member of the circle round Robert Ross, Oscar Wilde's champion, in 1918. The sickly lay assistant of 1912, his languors and ardours derived from Tennyson, Keats and Shelley, was replaced by a dapper, sun-tanned Aesthete, with a *chic* little moustache and a lively knowledge of recent French culture. His dedication to poetry grew stronger than ever, warmed by the southern sun and Laurent Tailhade's encouragement. Owen's debt to Tailhade and French literature has been generally overlooked (partly because fewer poems and letters survive from 1913–15 than from any other period of his adult life); unlike most British poets of his generation, he encountered late Romanticism in France rather than at home. Whereas much of the early verse of Sassoon or Rupert Brooke is modelled on Swinburne, Wilde and the poets of English nineties, Owen's writing shows little trace of Swinburne until 1916 or of Wilde until 1917–18. Although he seems to have been familiar with some English Pre-Raphaelite work, his knowledge of Decadent literature came from French originals, not from imitators such as Wilde. He went to France already interested, as is shown by his reference to 'imagination, physical sensation, aesthetic philosophy' in his 1913 letter to Wigan, as well as by his fondness for purple, the chosen colour of the Aesthetes, and by his sense of belonging to a secret order of poets.[2] He made his first attempt at writing verse in the French style ('The Imbecile') as soon as he reached Bordeaux.

The dominant creed in French art was still that known broadly as Aestheticism,[3] the late-Romantic belief that the artist's sole aim was to hunt and capture beauty. The first French Aesthetes, Gautier ('L'art pour l'art') and Baudelaire in the mid century, and the novelists Flaubert and Huysmans, had led the withdrawal from bourgeois reality into an autonomous, imaginative world where art was supreme. In the eighties, a group of young Aesthetes in Paris declared themselves 'Decadents', maintaining that civilisation had reached a state of ripeness indistinguishable from decay, a cultural autumn. Modern man, with his morbid sensibility and exhaustive knowledge, had grown beyond ideals and morals; left only with his senses and their objects, his highest activity was to develop sensation to its utmost refinements in the pursuit of beauty. Thus the Decadents had at least one idea in common with their contemporaries, the Impressionists and the Naturalists, that the function of art was to explore and embody physical sensation. Art had nothing to do with morality; the work of art concentrated sensation into a perfect form, and the form was all that mattered. When an anarchist threw a bomb into the French parliament in 1893, one of the leading Decadents, Laurent Tailhade, had remarked in a notorious phrase, 'Qu'importent les victimes si le geste est beau?' In meeting Tailhade in 1914, Owen was meeting a poet who had been at the centre of Parisian artistic society in its most famous period. Like Owen, Tailhade had been destined for a career in the Church but had broken away in order to devote himself to art for art's sake. He was famous for his satires against middle-class respectability and for his elegant lyrics. He had been a close friend of Verlaine, a regular attender at Mallarmé's *mardis* (where he had met Wilde), an anarchist sympathiser, a dabbler in the occult and a celebrated dandy. No one could have told Owen more about French literature of the later nineteenth century.[4]

The relevance of Aesthetic, and more particularly Decadent, conventions to Owen the future war poet is not difficult to see. For example, Aestheticism's search for exquisite sensation and its rejection of orthodox moral constraints led to *ennui* and disgust at contemporary society (that was one reason why Brooke and others welcomed war in 1914). Artists were attracted not only to strange cults and religious mysteries but also to violence. There is a strong sado-masochistic element in the literature of the period, in Flaubert's *Salammbô*, for example, or the heady, pagan verse of Swinburne. Women became *femmes fatales*, as in Rossetti's portraits, their brows white with pain and their lips red with the blood of lovers. Blood flowed freely. One very common image was that of a bleeding sunset to denote catastrophe, killing or erotic passion; there are sunsets of this kind in Baudelaire, Flaubert, d'Annunzio, Wilde, Hardy and even late Dickens.[5] The image can be associated with the larger one of the last sunset, the dusk of the nations in which civilisation would finally perish.[6] The great collapse would bring a moment of strange, sweet sensation and then darkness. Meanwhile, an outburst of violence would at least reduce the boredom of living. Swinburne complained that the world had grown 'grey' from Christ's breath.[7] 'The earth has grown too grey and peaceful', exclaims a character in a 1905 novel which Owen read in 1917. 'We need colour – good red splashes of it – good wholesome bloodshed.'[8] Brooke, who was an assiduously Decadent poet in his early years, wrote from Antwerp in 1907 to a friend with similar tastes, 'I am going to drag my tired body out . . . in the faint hope of finding a riot. The sight of fire and street-fighting and men hurt might soothe my seared soul.' In the same letter, he describes the sunset: 'In England of an evening the sun-god used to be crucified in beautiful agony on the red places of the west: here only a suppurating sore is opened afresh . . . .'[9] That is an elegant pose, but Owen's image of the bleeding sun in 'Mental Cases' is in the same tradition:

Sunlight seems a blood-smear; night comes blood-black;
Dawn breaks open like a wound that bleeds afresh.

The purpose is different; the imagery and language are the same.

The blood and pain which pervade Decadent art are usually associated with passive suffering and (preferably illicit) sexual pleasure (even 'Mental Cases' is a version of an earlier erotic poem). Artists thought of themselves as passive victims, often comparing themselves to the original Decadents in ancient Rome, refining their perception of beauty even as the barbarians broke down the gates (a subject for several paintings). 'I love this word decadence,' Verlaine wrote in a well-known passage,

all shimmering in purple and gold. It suggests the subtle thoughts of ultimate civilisation, a high literary culture, a soul capable of intense pleasures. It throws off bursts of fire and the sparkle of precious stones. It is redolent of . . . the consuming in flames of races exhausted by their capacity for sensation, as the trump of an invading enemy sounds.[10]

In political terms the barbarians were the Prussians, who had
marched through Paris in triumph in 1870 and might do so again.
There were many prophecies of international disaster. Death was
strangely attractive as the most intense of all experiences, and
martyrdom was a subject of obsessive interest. Self-sacrifice might
have a religious value or it might be futile; as Tailhade would
have said, what mattered was that the gesture should be beautiful.
St Sebastian, for example, the naked youth passively receiving
the arrows, was a recurrent image of what Wilde called 'all the
pathetic uselessness of martyrdom, all its wasted beauty'.[11] And
the martyr smiles (as in Owen's 'Has your soul sipped') with a
secret joy known only to the few. In their pursuit of sensation
beyond morals, the Decadents were fascinated by the links
between pain and ecstasy, and by all forms of sexual deviance.
The hermaphrodite became a frequent image of tormented desire
and strange, ambiguous beauty ('Perseus', the long poem which
Owen planned in France, was to have included such a figure).
Many poets, including Tailhade, wrote about dangerously beautiful boys, either as classical figures such as Antinous or as tempters
on the modern streets. If such subjects outraged public opinion,
so much the better; one of the artist's duties was to shock the
middle classes (*épater le bourgeois*), especially the older generation.

All this produced a strong sense of exclusiveness. There was a
special vocabulary – 'strange', 'sweet', 'mystery', 'smile', 'secret',
'exquisite', for example, all words which Owen used – and range
of symbols. The near-occult mystery enshrouding French poetry
of the period was particularly apparent in Symbolism, a mode of
writing derived from Gautier and Baudelaire and developed by
Mallarmé and his followers, in which the symbol became an
expression of mystic truth beyond all rational statement, to
be apprehended only through the artistic consciousness. By
meditating on a rose, a waterlily, a jewel, a colour, the artist
emptied his mind of everything that was not art and entered the
secret world of pure form, pure language. Symbolism was
concerned with language alone, with the perfection of words as
music – 'de la musique avant toute chose', Verlaine insisted,
because music was the supreme example of an art form fully
separated from daily reality. But in the hands of lesser writers,
among whom Tailhade and Wilde may be included, the symbols
which Mallarmé wrote about with barely intelligible abstruseness
became repetitive material for charming but shallow lyrics, so that

any young hopeful could adopt both the manner of Symbolism and Mallarmé's (and Tailhade's) ambition to 'purify the dialect of the tribe'.

Later chapters will discuss the effects of French Aestheticism and its English counterpart on Owen's thinking and writing in 1916–18. A few examples here may help to prepare the ground. Dreaming of a book of his own sonnets in 1917, he decided to call it 'Sonatas in Silence' and to have it bound in purple and gold.[12] The quotations given above from Verlaine put these details into place. Music – or, even better, silent music – was the art to which poetry should aspire. Calling a poem a sonata – or rhapsody, impromptu, nocturne, all fashionable titles which Owen duly toyed with – suggested Chopin, who was with Wagner the cult composer of the Decadence. Purple was the cult colour. The Mermaid's willow tree (1912) had cast a 'purple shade', and in the Verlaine-like sonnet 'Purple' (1916) the colour's associations with melancholy and passion are fully, if crudely, set out. The purple wound in 'Disabled' (1917) illustrates the Decadent element in Owen's mature poetry, the colour still carrying its poetic significance and thereby giving new meaning to the bloodthirstiness of war.[13] When he was in the trenches he enjoyed the company of a Captain Sorrell, 'an aesthete' who challenged him to write sonnets. He described Sorrell as 'not virtuous according to English standards' but 'one of the few young men who live up to my principle: that Amusement is never an excuse for "immorality", but that Passion may be so'.[14] It was only in such occasional comments that Owen recorded his Aesthetic creed, but a reader who keeps it in mind can see not only the force of his remark from the trenches, 'extra for me there is the universal pervasion of *Ugliness*',[15] but also the radical change of approach implicit in his 1918 manifesto: 'Above all I am not concerned with Poetry'. Even in 1918, despite that move away from art for art's sake, poems such as 'The Kind Ghosts' and 'Greater Love' show him still writing in the Decadent tradition; and all his war poems, like Sassoon's, had *épater le bourgeois* as a principal motive. That Decadent scorn for middle-class respectability also helped to free his sexual nature and reinforced his sense of being mysteriously set apart ('I had mystery . . . To miss the march of this retreating world'), an isolation that was to be crucial to his 1918 work.

It was in France, too, that he learned the importance of

technique. The French poets, with their high esteem for formal
elegance, were bold experimenters with sound-effects; a 'Sonnet'
by Tailhade includes the lines

> O Lune pâle qui délie,
> Liliale en le soir berceur,
> Ta lueur d'opâle appâlie
> A la douceur d'une alme soeur.

Sequences such as 'pâle . . . -ale . . . opâle appâlie . . . al-' were
the kind of patterning that Owen was to use in his later poems.
The first traces of his famous pararhyme ('mode meed mood')
appear on Bordeaux manuscripts;[16] it is one of the ironies of
literature that a device invented for musical effect in Decadent
love lyrics should have become a means of expressing the harsh
realities of war experience.

1914 was perhaps the happiest year of his life. It started badly,
with illness reminiscent of Dunsden troubles, but there was no
one to fuss over him and no point in languishing. Free at
last from home and piety, he could no longer avoid mature
relationships. In his letters home, he gently rebuked his mother
for having kept him ignorant of sex; nevertheless, he assured
her that all women, 'without exception, *annoy* me, and the
mercenaries . . . I utterly detest'. If Mrs Owen had no 'revelations
to make to me, at 14, I shall have no confessions now I am 21. /
At least none such as must make me blush and weep and you
grow pale.'[17] In April, a friend took him to visit the Poitou family
in the country; he was shaken by the beauty of the daughter,
Henriette, and by finding himself 'like an Egyptian piece of
Statuary' in her presence. Walking with her in the woods may
have reminded him of Rampton; at any rate, in his letters from
France spring flowers and woodlands became recurrent symbols
for physical and psychological states. The Poitous saw him off at
the station, giving him masses of lilac, 'and the odour was carried
by the inblowing breeze into my brain, to be an everlasting
memory. Then those faces and those places fell away from me.'
Back in his lodgings, he wrote '50 lines of poetry in as many
minutes! (thing not attempted for "years")'.[18] The Henriette
episode seems to be referred to in 'The One Remains', in which

the poet seeks perfect beauty 'once for ever, in one face' and remembers

> the secret traces
> Left in my heart, of countenances seen,
> And lost as soon as seen, – but which mine eye
> Remembers as my old home, or the lie
> Of landscapes whereupon my windows lean.

The phrasing suggests that this sonnet was written in France, not only with nostalgia for Shrewsbury but also with a sharper 'everlasting memory' of 'those faces and those places' seen on the Poitou visit. He was troubled at his inability to respond to women; in August, he said he was 'too old to be in love' and in November he complained that he lacked 'any touch of tenderness. I ache in soul, as my bones might ache after a night spent on a cold, stone floor.'[19] Imagery of stone, statuary and age persisted from his Dunsden verse. Then came a change.

In July, one of his pupils, a Mme Léger,[20] invited him to join her as tutor on a family holiday near Bagnères-de-Bigorre at the foot of the Pyrenees. Owen's two months there were to be a milestone in his poetic career. He arrived on 30 July, on the eve of war, delighted to be in fine country. The Légers had rented a villa in the secluded valley of La Gailleste. This was the '[fair] house, secret, ['mid] hills' of the first draft of 'The Sleeping Beauty', and the 'Beauty' was Albine (Nénette), the Légers' only daughter, who immediately adored and captivated Owen although she was too young, as the sonnet records (with regret or perhaps relief), to be kissed 'to the world of Consciousness'. Mme Léger was an elegant, successful businesswoman, her husband a talented man of the theatre. Owen was always to remember his stay with these intelligent, cultured people as a time of 'amazing pleasure'; he described it three years later in 'From my Diary, July 1914', tactfully changing the date to commemorate the last month of peace. The poem incorporates memories of his bathes in a mountain stream, his helping M Léger to get in the hay crop because the farmhands had been mobilised, and the excitement of conducting two flirtations at once, one with Nénette and one, more seriously, with her mother. Quite how seriously is difficult to guess, but Madame was certainly in earnest, telling her tutor flatteringly that she did not love her husband 'excessively' but that she liked handsome people.[21] Then

she asked Owen to go with her on her forthcoming business trip
to Canada. In the event, she embarked alone and he remained
on friendly terms with M Léger, calling on him at least twice in
Bordeaux, but Mrs Owen was alarmed in England and even M
Léger's old friend Tailhade avoided mentioning Owen to him in
1915 for fear of being indiscreet.[22] It was all a welcome change
from the stuffiness of Dunsden and Shrewsbury, anyhow, despite
the distant background of war. When news of the war came,
Owen made the thoroughly Aesthetic gesture of playing Chopin's
Marche funèbre, suffering 'torture' because Nénette forced him to
repeat it but also delight because of her expression. Through the
window he could see a hill 'exactly like' one at home.[23] Pleasure
and pain, Romantic music, the beauty of a child's face and of
the Shropshire landscape thus mingled in the first 'physical
sensation' of war.

At the end of August, Laurent Tailhade came to Bagnères to
deliver some public lectures at the Casino. M Léger gave
supporting recitations, while Owen and Madame sat rather close
together in the audience (Plate 4). Over the next few weeks, the
two poets saw a good deal of each other, Tailhade becoming
very much attached to the young Englishman. They were
photographed together in the garden, Tailhade affectionately
showing him a book (Plate 5). The picture marks a significant
moment in Owen's poethood, as the first famous author he has
ever met introduces him to French literature – for the book is
almost certainly the copy of Flaubert's *La Tentation de Saint-Antoine*
which Tailhade gave him in September.[24] The old poet may not
have had any spare copies of his own books to hand and had in
any case ceased to publish verse, but he must have lent him one
of his two volumes of collected poems because Owen soon started
a translation of a *ballade élégiaque* from it.[25] Later on, Tailhade
presented him with both volumes, inscribing a copy of the lyrical
Poèmes élégiaques (1907) for him when they met again in Paris in
May 1915, 'en souvenir de nos belles causeries et des beaux soirs
à La Gailleste', and at some stage sending him the satirical *Poèmes
aristophanesques*.[26]

Since Tailhade was a master of the *causerie*, it was probably his
conversation rather than his poetry that fascinated Owen at first.
Two principal topics must have been war and literature. Dennis
Welland (1960) has suggested parallels between Owen's war
poems and Tailhade's essays, *Pour la paix* and *Lettre aux conscrits*,
but subsequent critics have echoed this to such an extent that

Tailhade's importance in Owen's life has come to be imagined
as exclusively that of a pacifist. Certainly *Lettre aux conscrits* (1903)
is a powerful appeal, burning with the anarchist fury that had
earned its author a prison sentence in 1901. He urges conscripts
never to kill, arguing that war is a conspiracy by the rich, the
old and the aristocracy. The conscript is warned that he will be
ordered to kill young workers like himself:

> Ils ont, là-bas, des compagnes, leur mère et des amis dont les
> yeux se mouillèrent au départ. Ils aiment, eux aussi, la clarté
> du soleil et le parfum des bois. Ils portent dans leurs veines le
> sang pourpré de la jeunesse. Ils s'avancent, comme toi, pleins
> de la vie, à la conquête du bonheur. Ne tue pas!

This appeal may be distantly echoed in 'Strange Meeting'
('Whatever hope is yours, / Was my life also'). Another passage
suggests some of the imagery in 'Spring Offensive' and 'The Kind
Ghosts':

> L'offrande à Moloch du printemps sacré, de vos vingt ans, ô
> jeunes hommes! laisse indifférentes et soumises, crédules, peut-
> être, à la hideuse fiction du patriotisme, celles mêmes dont les
> entrailles vous ont portés.

Nevertheless, it is as well to be cautious in claiming Tailhade
as a source of Owen's later loathing of war. As a veteran duellist,
the old poet was not opposed to all fighting. The two essays had
been written in protest at a French alliance with Tsarism for
foreign wars, but an invasion of France herself, directed at his
beloved Paris by the same Prussian war-machine that had invaded
it in his youth, seem to have produced a different reaction; he
was 'shouldering a rifle' by November, which made Owen wonder
uncomfortably whether he ought not to be doing the same thing
himself. Tailhade's first comments on the war may have been
more *élégiaques* than *aristophanesques*, nearer the mood of the *Marche
funèbre* than of *Lettre aux conscrits*. He claimed in his 1914 volume
of essays that 'La douleur s'affirme comme le principe de toute
poésie', a statement reminiscent of Wilde's *De Profundis*, Verlaine's
poems and other Decadent works. An essay on masochism, though
deploring 'la folie algolagnique', refers at length to works he
deeply admired, including the *Tentation*.[27] One of his *ballades
élégiaques*, 'Les fleurs d'Ophélie', which Owen liked and probably

translated, contains imagery of a distinctly masochistic, Decadent
and warlike nature:

> Amour! Amour! et sur leurs fronts que tu courbas
> Fais ruisseler la pourpre extatique des roses,
> Pareille au sang joyeux versé dans les combats.

He perhaps discoursed at the villa about youth's noble sacrifice,
the melancholy beauty of death and the purple 'sang joyeux' that
was being spilt. It was that strain, at any rate, which was to be
paramount in Owen's writing until 1917. But anger and satire
were expressed, albeit in an Aesthetic way that is hardly to the
modern taste:

> While it is true that the guns will effect a little useful weeding,
> I am furious with chagrin to think that the Minds which were
> to have excelled the civilisation of ten thousand years, are
> being annihilated – and bodies, the products of aeons of
> Natural Selection, melted down to pay for political statues.[28]

Owen made these Darwinist remarks soon after a conversation
with Tailhade in the privacy of the latter's hotel room. The older
man found him interesting as a poet, being always ready to
encourage the efforts of the select few who were 'tourmentés, à
l'égal de lui-même, par le pur amour de l'art et de la beauté',[29]
but this hardly explains the warmth of his affection. 'He received
me like a lover', Owen reported. 'To use an expression of the
Rev. H. Wigan's, he quite slobbered over me. I know not how
many times he squeezed my hand; and, sitting me down on a
sofa, pressed my head against his shoulder.' At about the same
time, he said that Tailhade 'calls my eyes "So very lovely!!!" etc
and my neck "The neck of a statue!!!!" etc" – because he is a poet,
and unconsciously appreciates in me, *not* the appearance of beauty
but the Spirit and temperament of beauty, Tailhade says he is
going to write a Sonnet on me'. Mrs Owen was sufficiently
convinced by her son's assurances to feel able to quote this passage
(the punctuation is probably hers) to Blunden in 1930, but the
original letter was subsequently 'lost'.[30] There are a great many
sonnets to beautiful youths by Decadent poets, and they are by
no means always concerned only with the 'Spirit' of beauty, as
Owen may have been well aware. To judge from various poems
and anecdotes, Tailhade was not unused to treating young men

as lovers. He taught Owen more things than pacifism, if indeed he taught him pacifism at all.

Owen discovered that expressions of passion between men (including hand contact) were not only permissible but also a subject for poetry. In 'Maundy Thursday', he describes how he kissed the warm hand of the acolyte instead of the 'very dead' crucifix; as in 'The time was aeon', the flesh is set against religion, but the irony and witty sexual suggestion of the sonnet is far from the Monro-inspired portentousness of the earlier poem. 'Maundy Thursday' is perhaps Owen's best effort in Tailhade's *aristophanesque* vein (a style which took a while to master, because the sonnet is clearly a memory of Easter 1915).[31] Tailhade's high regard for formal elegance may explain why there is a deftness in Owen's sonnets from 1915 onwards that is lacking in his earlier work. But in 1914 Tailhade's example aroused an altogether deeper memory of a hand, recorded in a much less polished piece of verse; on the back of his translation from the *ballade élégiaque* Owen roughed out the lines about a boy which have been referred to in the previous chapter (see above, p.22). At about the same time, he wrote his fragmentary record of the birth of his 'poethood', reminded of the Broxton landscape by the hills at Bagnères where Tailhade's own poethood had begun. These manuscripts suggest the depth to which he was influenced by the French poet in 1914, showing him not only translating Tailhade's verse but also, perhaps even on his new friend's advice, exploring the roots of his own poetic imagination by recollecting two symbolic first encounters, one with love and the other with nature.[32]

It seems clear that Tailhade encouraged Owen to start reading intensively, although the clues are sparse. French literature was suspect in England, so that Owen did not say too much about it in his letters home and some of what he did say was later destroyed. On 7 September, 'son vieil ami' gave him the *Tentation* and Renan's *Souvenirs d'enfance et de jeunesse* (an account of yet another writer who broke away from religion in his youth). After returning to Bordeaux with the Légers later in the month, Owen met another poet, Carlos Larronde, who gave him a book of his poems with the hope that 'ces vers lui en feront aimer d'autres – plus beaux – des Poètes que j'aime'. On the thirtieth, he said he was 'studying French literature' and in October he was still busy with 'my interesting books'.[33] A few of those books can be identified.

In his copy of Vigny's *Chatterton*, he marked the sentence, 'En

toi, la rêverie continuelle a tué l'action', and in Renan he marked
a comment that the Celts know how to plunge their hands into
a man's entrails and bring out secrets of the infinite. What he
always thought of as his Celtic strain would have been fascinated
by *La Tentation de Saint-Antoine*, in which Flaubert meticulously
describes the saint's visions of strange and dreadful beings. Owen
read the book with care, underlining frequently. Tailhade had
also marked it, writing 'crétin!' against a criticism by the editor
of the novel's 'grands défauts'. Evidently agreeing with Tailhade,
Owen went on to read at least two more of Flaubert's novels,
Madame Bovary and *Salammbô*. 'Flaubert has my vote for novel-
writing!', he exclaimed to Gunston in July 1915, and he told his
mother that he was reading *Salammbô* 'with more interest than
the Communiqués'.[34] This book is a minutely detailed, gruesome
story of a war in ancient Carthage, told in Flaubert's exquisitely
finished prose. George Steiner has said that 'its scarcely governed
ache for savagery' stems from Flaubert's frustration in the society
of his time and that 'this frenetic yet congealed narrative of blood-
lust, barbaric warfare and orgiastic pain, takes us to the heart'
of the question of war and violence in our century.[35] That Owen
was reading *Salammbô*, and at the same time recording that he
was now keen to enlist (but in the Italian cavalry 'for reasons
both aesthetic and practical')[36] is, one might say, appropriate.
The current of nineteenth-century culture drove him, as it drove
a continent, towards war.

The most obvious result of his first readings of Flaubert is the
curious poem beginning 'Long ages past'.[37] The only manuscript
is dated 31 October 1914, which has led some of its few
commentators to draw what seems to be the groundless conclusion
that it is about the war. It is in fact a rhapsody about a lover
(presumably imaginary), written as thoroughly in the Decadent
style as Owen could manage. Conventional details include
references to the ancient world, human sacrifice, jewels, music,
frenzy, opium, the Gorgon's head, 'wild desire', 'pain' and
'bitter pleasure'. Flaubert's descriptions are full of such material,
providing the model for later works such as Wilde's *Salomé*. True
to form, Owen's subject has the red lips of a vampire and the
pallor of a *femme fatale*, but the gender is left ambiguous, as is
often the case in Decadent poems of this kind; initiates could be
left to assume that the subject was a boy if they wished, but the
attributes of the fatal lover could belong to either sex:

> Thou slewest women and thy pining lovers,
> And on thy lips the stain of crimson blood,
> And on thy brow the pallor of their death.

The 'pale brows' of the child in 'Deep under turfy grass' and the 'pallor' of Despondency–Medusa in the 'Supposed Confessions' were similarly marks of fatal beauty, and the echo of 'Long ages past' in 'Anthem for Doomed Youth' (1917), where dead soldiers are commemorated by the 'pallor of girls' brows', implies that wartime girls have the same deadly quality. The image of the bloodstained mouth also recurs in the war poems, as will be seen. Such images have their origins in the Decadence and, as later chapters will suggest, in those dreams of 'bloodiness and stains of shadowy crimes' which had plagued Owen's nights at least since 1911.

He is likely to have read a good deal of French verse as well as prose during the winter of 1914–15; there are several relevant books in his library, including a few marked anthologies, and a 1914 transcription of Verlaine's sonnet 'Mon rêve familier'.[38] Verlaine's dream of an unknown, unreached lover, with a gaze like that of statues and a calm, grave voice like the loved voices of the dead, left traces in 'The Unreturning', 'Autumnal', 'On a Dream' and other sonnets which Owen probably composed or redrafted in Bordeaux. His first exemplars in the form had been Keats and Shakespeare, but most of his sonnets can be associated with his two years in France, where Verlaine was an obvious model to follow (no doubt Tailhade recommended him); the erotic melancholy and quasi-Symbolist imagery of these poems clearly show the influence of the master and of his sometime disciple, Tailhade. In the end, Owen abandoned the form altogether after 'Hospital Barge' (December 1917), but between 1914 and early 1917 it was his favourite. When he wanted to impress a new literary friend, such as Monro or Sassoon, it was his sonnets that he produced, since they were the only finished poems he had to show for his labours on larger projects such as 'Perseus'. A number of them seem to have been quarried out of 'Perseus' material, a process which had it origins in his 1911 practice of shaping sonnets out of prose drafts. While few of them are work of real distinction, they were at least a useful discipline which was to bear fruit later in the organisation of such war poems as 'Conscious' and 'Inspection', both of which were developed from quatrains into what are in effect sixteen-line sonnets, or 'Dulce et Decorum Est', which begins with a block of eight lines, then one of six, or the long stanzas or verse paragraphs of 'Spring Offensive' and other 1918 poems.

By February 1915 he felt ready to start some writing of his own, having sated his appetite for novels and everything else except 'pure strong Poetry', which was in short supply now that he had read all that he could find. 'All that novelists have to tell has been told me, and by the best of them. I have even found out more. And that *more* which I had not been told I feel I ought to tell.' The stress on realism here suggests that he had been reading Zola, a writer whom Tailhade had known and revered. The experiences which Owen had in mind for writing may have been those of Dunsden, for the phrase 'In Dunsden' is just legible in the manuscript of another February letter before the sentence, 'I have made soundings in deep waters, and I have looked out from many observation-towers: and I found the deep waters terrible, and nearly lost my breath there.' Nevertheless there were also observations and adventures in 1914–15, some of them hinted at in letters and verse but in the end never 'told' in full. It is impossible to know now whether Owen in 1915 really had any '*more*' to tell than the novelists or whether he was merely being self-important, but his feeling of having escaped drowning ('congestion of the lungs'?) and having come back from the dead with new things to tell was to be reflected in the 'Perseus' fragments, with their mysterious tales from the underworld, and eventually in the 'truth untold' of 'Strange Meeting'. The voice of the future war poet reporting from hell is also heard in a detailed description of war wounds, this time not borrowed from Flaubert but seen in a French hospital in September 1914: 'I deliberately tell you all this to educate you to the actualities of the war.' Owen's later urge to 'educate' his readers owes something to his study of French novelists, just as his delight in beauty, form and 'physical sensation' derives in part from his knowledge of French poets. Much of his debt to French literature of the later nineteenth century stemmed from his friendship with Tailhade, and its results began to emerge in the strange fragments of 'Perseus'.[39]

'PERSEUS'

'Perseus', the most ambitious and irrecoverable of all Owen's poetic endeavours, was begun in France and became a central expression of his poethood. It survives only as a few scribbled leaves, understandably ignored by critics as incoherent 'juvenilia', fragments of raw, unfinished verse which seem hopelessly devoid of literary merit despite some odd parallels with the famous poems of 1917–18. But through all this chaotic material Owen seems to be feeling his way towards a myth of his own life and identity, giving it shape in a pattern that is strangely close to that of his war poetry, so that the 'Perseus' manuscripts, crude stuff in themselves, are a key to his eventual achievement.

When he left the Léger household in the autumn of 1914, he took up freelance teaching in Bordeaux for some months, corresponding occasionally with Tailhade, who had returned to Paris, and becoming something of a man about town. Rumour has always had it that it was in this period that his first sexual encounters occurred;[40] certainly, his involuntary coolness towards Henriette and Mme Léger, as well as Tailhade's demonstrativeness and the liberty of a great French city, would have given him all the clues he needed. He described several youths with incautious admiration in his letters, while the homosexual element in his verse became conspicuous in 1915. Whatever the case, he enjoyed his freedom, despite repeated assurances to his mother that he would like to return home. In December 1914 he accepted a post at Mérignac on the outskirts of the city as tutor to some Anglo-French boys. The local woods and moors in the spring provided imagery for his 1915 work. He said little in his letters about his poetry but referred mysteriously to 'certain writings' which he wanted to finish. In response to British recruiting-propaganda (there was little in France because conscription was in force from the start) and parental worries about his future, he felt more strongly than ever that his career had long been chosen. 'A boy, I guessed that the fullest, largest liveable life was that of a Poet. I *know* it now . . . .' Universities and the Church had lost their attraction. 'There is *one* title I prize, one clear call audible, one Sphere where I may influence for Truth, one workshop whence I may send forth Beauty, one mode of living entirely congenial to me' (March). He quoted Keats, adding that 'as trees in Spring produce a new ring of tissue, so does every poet put forth a fresh, and lasting outlay of stuff at the same season' (April). The purple heather on the moors in August reminded him of Broxton. As for enlisting, he maintained at first that a poet was more useful to his country alive than dead but later he began to think that military life might be good for his writing.[41] These statements of literary ambition, fired by his recent explorations of French literature, were accompanied by work on the new vehicle for 'Truth' and 'Beauty', a long poem or verse drama under the title of 'Perseus'.

A full account of Owen's plans for 'Perseus' can never be given, since much of the poem has perished or was never written down. He was probably unsure about them himself, for he kept returning to the manuscripts with additions and alterations. The earliest drafts may date from 1914, at least three belong to 1915, one to 1917, and even in 1918 he was still intending to finish the project. In the account which follows I have given each fragment a number (P1, P2, etc.) for ease of reference and have omitted many small details, in an attempt to disentangle what is left of an undertaking that occupied him intermittently over several years.[42]

The longest fragment, P1 (*CPF*, 467–70), consists of a little more than fifty lines of blank verse monologue, scribbled hurriedly on a large double sheet. It is marked 'For Perseus' and has a few added mentions of Pluto and Persephone, but there are no references to classical myth in the main body of the verse except two to Eros, who also appears in several of Owen's sonnets. The original speaker of the monologue seems to have been Owen himself, not some classical character. It is possible, in fact, that this is the '50 lines of poetry' which he poured out 'in as many minutes' after meeting Henriette in April 1914; there is certainly no other fragment surviving among his papers that fits this description. By 1915 he seems to have decided that he could make the material presentable by incorporating it into a long work on Perseus, its vaguely mythical quality making it suitable for such treatment. The subject of Perseus was probably selected in consultation with Gunston, for the two cousins had apparently agreed to compete once more, each choosing a classical hero associated with flying, just as they had agreed to write on 'The Swift', another flying subject, in 1914. In the summer of 1915 Gunston wrote ninety lines on Icarus, his own subject, churning them out at a line a minute, a speed his cousin had no doubt recommended as a way of getting started.[43] But P1 was never altered much, so that it has nothing to do with the Perseus legends though plenty to do with Owen's 'Perseus'.

P1 opens with a heavily cancelled line, 'What have I done, O God, what have I done?' If this and what follows are part of Owen's response to his helplessness in the presence of Henriette, when he had felt like 'an Egyptian piece of Statuary', the fragment shows him starting very soon after the event to turn experience into something like myth. The speaker describes a night when Eros descended to him with a sound of fire, heralding the perfection of love: 'I knew the thing was come at last'. But then, mysteriously, this first and last chance is lost (Owen's wording loses all coherence at this point) because the speaker pauses to 'renounce the world', using biblical language reminiscent of a lapsed Evangelical's training and guilt:

> Let not the sun drop sweetness on me more
> Nor let the moon be comforting by night . . .

The speaker turns away 'great friends' and 'all youth', for he has 'touched the god; [and] touch me not'.

> On all that had been sweet to me, I laughed.
> I trampled on the flowers I so much loved
> . . .
> [And the dear books I blotted out and tore]
> Let music [be my terror] a jar [to all] my peace
> I would no more remember my old home
> Nor the old days or people or the places
> And turned adrift even my helpless hopes
> And bade the flame of this love burn them down.[44]

There are echoes here of 'The One Remains' ('my old home') and of the Henriette excursion ('those faces and those places'), while the last lines contain a rough version of the burnt offering in 'To Eros':

> In that I loved you, Love, I worshipped you.
> In that I worshipped well, I sacrificed.
> All of most worth I bound and burnt and slew:
> Old peaceful lives; frail flowers; firms friends; and Christ.[45]

It is evident that the two sonnets 'To the Bitter Sweet-Heart: A Dream' and 'To Eros' maintain the sequence of P1; like other sonnets, they may have been quarried from draft work for the long poem. The autobiographical content seems clear: when Owen left Dunsden, he left religion ('Christ'), children ('all youth'), the old people he visited in the parish ('Old peaceful lives'), botanical studies at Reading ('flowers'), musical activities,

and his hopes for a degree ('the dear books'?). Then he left
Shrewsbury ('my old home'), family and friends. The reasons
for the sacrifice can only be guessed at but it was made to Eros,
the treacherous god of love. 'What have I done, O God . . ?' The
question seems to be answered: 'I have touched the god', thereby
losing the power to respond to an ordinary lover. Was Owen
remembering touching a boy's hand in a hawthorn glade in 1912?
Poets have agonised over smaller matters.

There is considerable confusion over genders in P1. The speaker
could be of either sex, but the voice is so personal that one assumes
it to be the poet's own. The speaker's lover was originally male,
either Eros or a human equivalent, but Owen later made some
half-hearted attempts, as in 'The time was aeon', to change this
representative of beauty into a female figure, in this case the
goddess Persephone. He also tried to turn the speaker into
Persephone's husband, Pluto, who was to have been shown
miserable and alone in hell, lamenting his queen's annual absence
in the upper world. This would have disguised the homosexual
and personal implications of the original draft, while perhaps
allowing Owen secretly to imagine Pluto as himself alone in the
city, and Persephone as Henriette, goddess of the spring woods.
Since several of the 'Perseus' fragments refer to the seasons, it
may be that he intended to base his scheme on the seasonal cycle,
a structure which he used for rather too many of his shorter
poems, but there does not seem to be any ancient Perseus legend
which allows for this or for a descent into the underworld. He
would have had to have taken as many liberties with classical
myth as Keats had done in writing about Endymion or Hyperion.
Whatever his plan, he never got round to amending P1 consist-
ently, and the lover remains neither Persephone nor Henriette
but Eros or an idealised youth, with cloudy brows and snowy
feet:

> Between the summits of his shining shoulders
> Are there not valleys for my rest for ever
> Is not the flower of all flesh opened me?
> Shall I not have enough of joy [with him]?

The sexual confusions of P1 and other poems end in disaster.
Although the poet accepts Eros in 'To the Bitter Sweet-Heart' as
a substitute for his partner and reaches the happy valleys ('My
face fell deeply in his shoulder'), in 'To Eros' he is deserted by

the god. Similarly, P1 ends with 'my love' rejecting the speaker, who finds himself suddenly corpse-like, with a skull smile and empty eye sockets.

> I entered hell's low sorrowful secrecy
> My arms were wrenched with inexplicable weakness
> Methought my heart would loosen out of me
> Most like it was what men tell of the first
> Terrific minute of the hour of death . . . .

After this death-like agony, he meets a 'haggled crone', Age, who promises to be his new lover in hell. An excessive devotion to young male beauty seems to have brought painful retribution.

Some of the leading images in Owen's later poems can be seen in all this, including the skull smile and descent alive into hell ('Mental Cases', 'Strange Meeting'), the rejection of the sun's blessing ('Spring Offensive'), the climax of agonised death and the prohibition of touch ('Greater Love'), the figure of Beauty that offers 'rest for ever' (or 'ease / For ever' in 'My Shy Hand'), imagery of the male body as cloud and mountains ('The Wrestlers'), and the unreachable glory of youth. Material from or similar to P1 occurs frequently in Owen's verse from 1915 onwards, as will be seen.

P2 (*CPF*, 465–6) consists of some unfinished rhymed stanzas headed 'The cultivated Rose (From the Greek)'. Since the marginal note 'Danae speaks' is the only link with the Perseus story, this fragment also seems to have begun as an independent piece. Like P1, it seems impossible to date with any certainty, but the subject is a little like that of 'Long ages past' (October 1914), being a beautiful youth racked with desire. He has eyes dark 'with gloom of death and hell', a mouth red 'like fresh, uncovered blood' and beauty 'Dangerously exquisite, from toe to throat; / Contused with sap of life like a bursting bud':

> As shakes a high-trilled note
> His fingers ever trembled, with his eyes
> While hot thoughts ever loosened from his hair
> Perfume; and pearls of the fierce, vast desire.

The fragment ends with an attempt to explain this figure as a hermaphrodite tortured in the hell of unquenchable lust, but this and the title may have been afterthoughts designed to conceal

any personal elements in the poem by placing it even more
thoroughly within Decadent convention.[46] The core of P2 remains
one of Owen's obsessive images, a boy in the grip of helpless
sexual frustration, a hellish parallel to the triumphant 'Flesh' in
'The time was aeon'. The Rose's ever-trembling fingers are
like those of the Mermaid's willow-tree, while several of his
characteristics reappear in 'Purgatorial passions', a 1916 fragment
about sufferers in hell which was to be used for 'Mental Cases'.
His tormented eyes come from the most persistent of Owen's
nightmares, that of the staring face. The Rose is thus an
embodiment of Owen's personal difficulties and of the sort of
figure that he found desirable; here as elsewhere the poet and his
object seem one and the same.

P3 (*CPF*, 464) is one of a group of three fragments apparently
written in the spring of 1915. By this stage Owen has decided on
Perseus as a subject from the start, for Danae and Perseus are
both mentioned in the verse. According to the Greek myth, Danae
was the daughter of a king who locked her up in a tower to keep
her away from men, a vain precaution because Zeus himself fell in
love with her and entered the prison in the form of a shower of
gold or, in versions by William Morris and others, a shaft of
sunlight. Owen made a few notes for this event and its result:
'And then the whole sky fell / In vortices of flashings and
showering gold', and 'Of all this sprang the infant Perseus. / Fair
from the moment of his birth'.[47] However, most of this brief
fragment is about Danae, alone in her cell, suffering from desire
too intense to be satisfied by any human and watching 'trembling'
human lovers who seem to be compared to trees in a storm
('Shook and were bowed before embracing winds'). The similarity
between her torments and those of the Rose in P2 and the speaker
in P1 suggests that Owen is beginning to adapt existing material
to fit the Perseus legend, giving Danae the sufferings of those two
characters and Zeus the fiery descent which had been attributed
to Eros in P1. The imagery here was to remain in his poetry even
as far as his last poem, 'Spring Offensive', where climactic fire
again falls from 'the whole sky'. There is a particularly clear
parallel between P3 and the mysterious sonnet 'Storm' (1916),
in which someone (Eros? a youth?) is like destructive lightning
fire:

> His face was charged with beauty as a cloud
>   With glimmering lightning. When it shadowed me,

I *shook*, and was uneasy *as a tree*
That draws the brilliant danger, *tremulous, bowed.*

(My italics.) If 'Storm' is compared with the Danae legend it becomes a little more intelligible than it has been until now, although such a reading still leaves questions to be answered. In both the sonnet and the fragment, the descent of fire is a sexual assault which results in the birth of beauty; stated in those general terms, the metaphor can be seen to represent the descent of inspiration on the poet and the resulting birth of the truth and beauty of poetry. Owen may have had some idea at this stage of developing 'Perseus'into that archetypal Romantic poem, a myth about the nature and origins of poetry itself.

P4 (*CPF*, 449–50) is another spring 1915 manuscript. I take it to be further work for 'Perseus', since it includes phrases about the sunlit air being 'thick with gold' or 'fire'. Furthermore, one of Perseus's flying-exploits was the rescue of Andromeda from a sea beast, and this folio refers to wartime Germany as a 'vast Beast / Whose throat is iron and whose teeth are steel'. There are also lines about the arrogant, shameless 'Emperor' (Kaiser), 'our' (French?) ministers smiling 'ministerially', and a Zeppelin ('horrible absurdity') over the Thames, all of which show that Owen took the popular Franco-British view of Germany in 1915. So 'Perseus' briefly and unpredictably emerges as a poem about the war. One of Owen's 1916 letters gives some clues as to what he had in mind; it discusses his short-lived ambition to become a pilot in the Royal Flying Corps and to battle against 'Zeppelin, the giant dragon', wearing the wings of Hermes as his badge and being in the same fellowship as 'Perseus and Icarus, whom I loved'.[48] The modern Perseus-pilot, born of Zeus (sunlight) and endowed with the power of Hermes, the winged messenger of the gods, would rescue Andromeda–France from the German dragon.

This romantic–heroic view of the war matched Owen's thoughts about enlisting in 1915. His Mérignac letters refer several times to 'the power of the midday sun'; taking pleasure in every 'nature-shock' from sun and wind was a 'sure sign of being Right with Nature'. He was conscious of being much fitter than he had been at Dunsden, where he had been supposed to have had a weak heart. 'This aft. I ran, with the boys, almost a mile – significant!' – he was strong enough for war. France was strong enough, too: the grapes were ripening in the sun and 'the muscle of France keeps resisted, – while her heart scarcely beats faster – The

Emperor yet frowns imperially; and our ministers yet wear a
ministerial smile' (August 1915, quoting from P4).[49] His exulting
runs in the hot sun were preparation for what he described in
1917 as the 'extraordinary exultation' of advancing across No
Man's Land under fire.[50] That 1917 advance was actually at
walking pace, but when the experience was reworked for 'Spring
Offensive' in 1918 his imagination returned to running, moors
and trees; the men in the poem stand like trees, then race across
a moor in sunlight until 'the whole sky' burns against them, the
final 'nature-shock' in his poetry.

   P5 (*CPF*, 448), the third spring 1915 manuscript, contains a
few draft images which were probably more work for 'Perseus'.
A gloomy winter forest is contrasted with a summer landscape.
Like the trees in P3, the forest has a psychological connotation,
seeming like the 'horrors of an obscene mind . . . afraid of self'.

   P6 (*CPF*, 470) is a monologue marked 'Speech for King' and
'Perseus'. It contains some familiar material about frustration
and finding ideal love but it is remarkably late, being written on
a type of paper which Owen used at Craiglockhart in 1917.
Overleaf is the line, 'O piteous mistake: O wrong wrong word'.[51]

   These six fragments, and a few others which may have been
intended for 'Perseus', reveal little about the shape that Owen
planned for his project. Perhaps he had in mind a Romantic epic
or drama, like *Endymion* or *Atalanta in Calydon*, but he seems to
have begun P1 and P2 as separate poems, only later deciding
that he might be able to incorporate them into his new project.
How he imagined he could include Danae, Perseus, Pluto,
Persephone and Eros in a poem about the Great War, heaven
knows; P6, the 1917 fragment, makes no reference to the war, so
by then he had perhaps abandoned that aspect of the scheme.[52]
Nevertheless, whatever the formal plan for 'Perseus' was, the
fragments and some of the poems related to them do seem to
contain a shadowy but insistent pattern, a secret myth which he
seems to have been making out of his own experiences. It is in
this pattern that the importance of 'Perseus' lies for any study of
his poethood, since it can be related to some of his most successful
1918 work as the following outline may indicate.

   The pattern starts with the hope for an ideally beautiful but
human and usually female lover ('The One Remains', 'On a
Dream', 'Mon rêve familier', P6). If the ideal is found, she is
sexually unawakened and the poet cannot or will not 'kiss her to
the world of Consciousness' ('The Sleeping Beauty', Henriette,

several poems about Dunsden children). Frustrated desire grows 'vast' and is seen in symptoms such as severe trembling (Danae, the Rose).

Then Eros or Zeus descends in sunlight, fire or lightning (P1, P3, 'Storm', 'Spring Offensive'). This is the moment of awakening, when the poet has the opportunity of sexual fulfilment and of generating beauty–poetry. There is a parallel here with the sun at Mérignac, a fiery challenge to enlist which Owen met with answering excitement, believing that enlisting might benefit his poetry. At this point the poet sees a beautiful youth of ambiguous status: a living statue, whose shifting outline sometimes makes him appear androgynous; a god, with human flesh; a spiritual symbol of 'physical sensation', embodying the beauties of music, landscape, seasons, jewels (the Mermaid's statue, 'The time was aeon', 'Long ages past', P1, 'My Shy Hand', 'Spells and Incantation').

The poet now sacrifices to Eros, initially hoping that he will be able to love his earthly lover more fully ('To the Bitter Sweet-Heart'). Other loves – friends, family, reputation, creed – are offered up (P1, 'A Palinode', 'To Eros'). 'I have touched the god': hand contact seems the crucial action (the 1914 fragment about touching a boy's hand, and much hand imagery in other poems).

The new relationship suddenly collapses, the poet being rejected and/or abandoned by the lover or the god (P1, 'A Palinode', 'To Eros'). Something he has said or done seems to be a cause of this disaster ('What have I done, O God . . . ?', 'O wrong, wrong word'). The sun ceases to 'drop sweetness' on him and is angry ('Reunion', 'Spring Offensive') or withdraws its power or light (P1, 'Futility', 'Exposure' and the numerous 1917–18 poems set in darkness).

The abandoned devotee finds himself entering hell–Hades, alive but like a corpse, with sensations of paralysis, drowning, being turned into stone ('Supposed Confessions' and other Dunsden poems, P1, 'Dulce et Decorum Est', 'The Sentry', 'Mental Cases', 'Strange Meeting', 'Spring Offensive').

He meets some of the inmates of hell, who show him his own possible fate (the Rose, Age, 'Purgatorial passions', 'Mental Cases', 'Strange Meeting'). These descriptions have plenty of literary parallels, of which Dante's *Inferno* is the most obvious (Owen owned a copy of the Cary translation). Lewis Morris's *The Epic of Hades* and Harold Monro's 'The Swamp' (in *Before*

Dawn) are other examples which he certainly knew.

Those seem to be the main stages in the obsessive sequence which underlies 'Perseus' and many of Owen's subsequent poems. It had little to do with Perseus originally, but the Perseus story itself had several features which Owen would have found attractive once the subject had been agreed upon with Gunston. First, many comments in his letters show that he was fascinated by flying, both as a modern skill and as an image of poetic creation. References to the 'physical sensations' of flying, falling and drowning (the fate of Icarus) can be found throughout his verse. Secondly, traditional representations of the Greek myths, including some in his schoolbooks, allowed for descriptions of the male nude. Mr Gunston still has a little bronze figure of Hermes which Owen seems to have bought in Bordeaux (Plate 6).[53] Beauty is often represented in Owen's verse as a naked statuesque figure (Tailhade had even told him that he had the beauty of a statue himself). Thirdly, the most celebrated object in the Perseus legends is the head of Medusa the Gorgon, with its terrible stare that turned men into statues. He had used that image already in 'Long ages past' and in the 'Supposed Confessions', where Despondency–Medusa is blamed for his adolescent 'horrors'.

The Gorgon's head has two different forms in artistic tradition: the original, archaic image is of a devilish face with glaring eyes and lips drawn back in a death grin, but in later representations the face becomes one of agonised loveliness. The archaic version has been associated by many psychiatrists with the terrifying faces that are a common feature of dreams. It is the face given to shellshock nightmares by a cartoonist in the Craiglockhart Hospital magazine in 1917 (Plate 10). Dr William Brown, who was in charge of the Casualty Clearing Station where Owen was a patient, recorded a case of shellshock in which the victim was rigid and trembling continually, especially in the arms ('My arms were wrenched with inexplicable weakness'): 'his face wore a fixed stare of horror, with starting eyes, dilated nostrils, the mouth slightly open . . . . This was the petrifaction of fright, such as the Greeks knew and portrayed in the Gorgon myth.'[54] The 'fixed eyes' of the dead poet in 'Strange Meeting' or the tormented faces of the damned madmen in 'Mental Cases' are combinations of artistic convention, dream and wartime experience. The later, beautiful version of Medusa was known to Owen through nineteenth-century literature and painting. As Frank Kermode (1957) has shown, 'the face' is a pervasive image in Romantic art, notably in *The Fall of Hyperion*, where Keats is frozen into a timeless agony by the eyes and deathly beauty of Moneta, reaching his poetic vision through this torment. To see and bow down before the face of death-in-life is the poet's privilege and the reward for his suffering. More generally, staring faces and fixed eyes are frequently encountered in Gothic novels and other Romantic narratives. Mario Praz (1970) has studied the psychological significance of 'the face', beginning with Shelley's poem on Medusa's head and showing how this image of fatal beauty developed into that of Fatal Woman. The *femme fatale* (of whom Salammbô is one of the first examples) is, according to Praz, a sado-masochistic, Decadent symbol by which the male artist allows himself to be tortured and consumed. Death and erotic beauty are inseparably linked, sometimes barely distinguishable; the male's role is one of passive suffering (hence the frequency of the hermaphrodite as a symbol). Much of this would have made immediate sense to Owen by 1915. He never got as far as writing about the Gorgon for 'Perseus', but her paralysing stare and beautiful or terrible face were to be manifested in some of the most forceful images in his poetry.

I have tried to outline some of the reasons why the verse that Owen wrote and might have written on his chosen title of 'Perseus' is worth more attention than critics have so far given it. As a work of art, the finished piece would probably have been old-fashioned and cumbersome, so that one need not regret his failure to complete it, but as an expression of the principal drives in his imaginative life it remains a fascinating if enigmatic key to the rest of his poems, especially to those which describe his later visions of the underworld. It is the connecting link between his Dunsden torments, his explorations of life and literature in France, his trench experiences and his 1917–18 war poems.

The rest of the 'certain writings' which he wanted to finish at Mérignac are mostly not known. He redrafted 'The Swift' and 'The time was aeon', both relevant to 'Perseus', and composed or copied out afresh two little poems, 'Nocturne' and 'A Contemplation', which show something of the social conscience which critics have tried to find in his early verse. 'Impromptu', a lyric about a child's hand and eyes, is worth noticing for its evidence of a new interest in sound-effects of the kind used by the French poets ('And kiss across it, as the sea the sand').[55] Material probably intended for verse can be seen in several letters, for instance in a description of yet another springtime ramble with a child, this time a small boy who was heir to a large estate:

> Into his woods we go in the hot afternoons; into the woods we
> go; and the floor of the coppice surges with verdure; the
> meadows heave with new grass as the sea with a great tide;
> the violets are not shy as in England but push openly and
> thickly; the anemones are dense like weeds, and the primroses
> like the yellow sands of the sea. And there the nightingale ever
> sings; but I had rather she did not, for she sings a minor key.
> And the idyll becomes an ode-elegiac.[56]

(Or a *ballade élégiaque?*) Owen expresses his own concealed
excitement and the energy of the growing boy in characteristic
imagery of promiscuous flowers and foliage. With passages such
as this in mind one can make some sense of the 'Perseus' drafts
and of that curious sequence of three 1914 fragments: the Tailhade
ballad (which is about going into 'virginal' April woods), the
lines about touching a boy's hand, and the attempt at a poem
about the birth of Owen's poethood at Broxton.

He returned to England in May 1915 for a month on, of all
things, a trade mission for a scent-manufacturer. On the way, he
paused for a night or possibly more in Paris, where he was given
a taste of Parisian social life by Tailhade. He spent his last two
evenings in England in the East End of London, wandering
through the Jewish quarter in Whitechapel. 'I never saw so much
beauty, in two hours.'[57] The 'beauty' seems to be explained in a
1917 poem, 'Lines to a Beauty seen in Limehouse' (also entitled
'A Vision in Whitechapel'), a Decadent portrait of a godlike
youth whom he had admired but not dared to approach. He may
well have walked from Whitechapel on that 1915 evening through
the dock areas of Limehouse and Shadwell (seeing a Zeppelin
over the river) and emerged at Tower Bridge, 'sonorous under
rapid wheels'. The Cockney youth and the places associated with
him were to be material for several later poems, including
'Disabled' (1917).

Traces of these experiences and of 'Perseus' are evident in 'A
Palinode', a riddling poem apparently written in October 1915.
By then Owen had left Bordeaux for good after a busy summer
and was joining the British Army.

4 Preparing for War

The modern notion that the Great War came as a complete break in continuity, a shattering surprise to Europe, is a product of post-war mythology rather than of history. The same notion dictates that critics should be astonished at the phrase 'stunning guns' in Owen's 'Little Mermaid' (1912) because it seems to foreshadow his war poems, or that they should accept Sassoon's suggestion that 'Exposure' was written a month after its author came out of the trenches, as though front-line experience alone had created Owen's mature style. The world which Owen grew up in was rife with military activity and prophecies of coming disaster. There had been either a war or a war scare nearly every year since 1895. As a child, he had been photographed in uniform with a home-made rifle and had played what a friend later remembered as battles of 'attrition' with toy soldiers. Large-scale Army exercises were reported at length in the newspapers while he was at Dunsden. A bishop said at Keswick in 1912 that the 'growth of armaments still remains a standing threat to our peace and that of the world'.[1] In art and philosophy, as in many of the political groupings of both right and left, there had been a hunger for action and violence. Rupert Brooke was true to his antecedents as a young Decadent and Socialist when he greeted the war as a glorious chance not only for martyrdom but also for radical social change. If Owen was never as enthusiastic, that was more because of his Evangelical, conservative background and his absence from England during the war fever of 1914 than because he was out of sympathy with Brooke's feelings. Joining the Army in 1915 did not significantly alter his poetic direction, for his training in Romanticism had attuned his imagination to war and given him a language and imagery with which to tackle its strange conditions. What changed his style was not experience of the trenches but meeting Sassoon, yet even that had been prepared for by his meetings with Harold Monro at the Poetry Bookshop in 1915–16. If there is one theme in his 1916 verse, it is that of

55

growing-up, a process which involves both change and continuity.
It has been usual to argue that Owen tried to get rid of the
Romantic elements in his war poetry and that what remains of
them is there only for the purpose of bitter contrast. This
seems too simple. He went into war equipped in two ways by
Romanticism. On the one hand, the humanitarian values of
Shelley, Wordsworth and others gave him a standard by which
to judge his eventual knowledge of the trenches; thus measured,
war seemed an evil opposed to everything literature stood for. So
his 1917–18 poems contain many deliberate allusions to the early
Romantics. On the other hand, the later Romantics' obsession
with pain, martyrdom and physical sensation made war a
profoundly interesting experience. It is no coincidence that Owen
seems to have started reading Swinburne in earnest in 1916.
When he returned to the front in 1918, knowing that he would
kill and probably be killed, he took volumes of both Shelley and
Swinburne with him, but after he had been in action he sent the
Shelley back to Shrewsbury, keeping only Swinburne's *Poems and
Ballads*, the one book of poetry still in his kit at his death. It was
Swinburne's gloomy underworld, not Shelley's bright heaven,
that seemed his home at last, but he never ceased to be a disciple
of Romanticism.

Owen is interesting not only for the intrinsic worth of his late
poetry but also as a representative figure of his time. Like the
men in 'Exposure', many of the war generation believed that
they had been born to die in war. The culture and fashions of
their youth had predisposed them to such thoughts. Of the
innumerable prophetic parallels between late Romanticism and
the war, one bizarre example may be enough here: Max
Beerbohm's parody of Decadent conventions, *Zuleika Dobson*
(1911). In this urbane and entertaining novel, the undergraduates
of Oxford drown themselves *en masse* for love of the heroine, a
femme fatale. Only the poorest specimens are left alive and the
dons dine in silent halls. In 1914, three years later, the President
of Magdalen said of the real Oxford that 'emptiness, silence reigns
everywhere . . . . The remnant of undergraduates, the invalid, the
crippled, the neutrals, make absolutely no show at all.'[2] The only
men who could live and die in the waste land of the trenches
were of necessity fit and young; they left behind them the old
men, who praised their valour without understanding their
suffering, and womenfolk, like those in 'Disabled' and 'The Send-
Off', who encouraged them to go. The attitude of the early

volunteers themselves was often one of sacrificial joy as they turned from the world like 'swimmers into cleanness leaping', a famous simile from one of Brooke's sonnets that was, as it happened, drawn from river bathing at Cambridge. As Beerbohm's young idealists leap into the river at Oxford, an old don falls in by mistake:

> He whimpered as he sought foot-hold in the slime. It was ill to be down in that abominable sink of death.
>   Abominable, yes, to them who discerned there death only; but sacramental and sweet enough to the men who were dying there for love. Any face that rose was smiling.[3]

This was to become a standard language for war poetry. Although 'love' was reinterpreted as love of country or even more commonly as the 'greater love' of Christ's saying before the Crucifixion ('Greater love hath no man than this, that a man lay down his life for his friends'), the term did not altogether lose its sexual meaning. Dying for love in war was seen as sweet, decorous and sacramental,[4] a sacrifice comparable to that of the Mass yet described in terms that were sometimes as much erotic as religious. Civilian war poems of this kind eventually irritated the soldier poets, who responded by stressing the exclusiveness of martyrdom: death in battle had a beauty unintelligible to old men and beyond the beauty of women. Only a soldier could understand a soldier's smile. The best-known poems in this idiom are Owen's 'Greater Love' and 'Apologia pro Poemate Meo', which several reviewers of the 1920 edition of his poems, including experienced soldiers such as Blunden, singled out as the truest and most beautiful in the book. The attitudes of the generation of 1914 were rooted in Romanticism, which is one reason why Owen can be seen as, in John Middleton Murry's phrase, 'the poet of the war'.[5]

Owen's early responses to the crisis were characteristic of the period, with little trace of either originality or scepticism. His sonnet '1914', perhaps written in that year, sees the late nineteenth century as a European autumn, as the Decadents had seen it, and the war as an inevitable winter during which young men's blood would be the 'seed' for a new spring. Like other people, he was not blind to the probability of heavy casualties. British heroics in 1914–15 are misleading to a modern observer if they all seem to be ways of suppressing reality rather of coping with it. Before he enlisted, Owen read *Salammbô* and the *Song of Roland*, visited

a medieval battlefield (scene of a major English defeat by the
French), dreamed of joining the Italian cavalry and began turning
'Perseus' into a war allegory, but what made him finally decide
to volunteer was the simple conviction that Germany had to be
resisted and that calls for reinforcements could not be ignored
after the Gallipoli losses. He had seen war wounds by then and
knew what he was risking, but there seemed no alternative in the
face of what virtually everyone in France and Britain saw as
ruthless German aggression. P4 (1915) describes the Kaiser as
shameless and 'arrogant', a view repeated in the 'Artillery' sonnet
(early 1917?), where one of 'our' guns is urged to hurl 'Huge
imprecations' at enemy 'Arrogance', and even in preliminary
work for 'Anthem for Doomed Youth', where 'our' guns are said
to utter 'deep cursing', 'deliberate fury' and 'majestic insults'.[6]
Owen's political understanding of the war was as unquestioning
as most people's until he met Sassoon in 1917.

His emotional attitude begins to emerge in December 1914
when he described soldiers as 'the thousand redeemers by whose
blood my life is being redeemed', an image no doubt picked up
from the newspapers he was receiving from home. Reading the
casualty lists was like reading Severn's account of Keats's death
or Christ's prophecy over Jerusalem.[7] This sentimental, religious–
poetic note continued into 'The Ballad of Peace and War',
subtitled '1914', which was to be his longest-running war poem.
The first drafts date from 1915, if not 1914; another was mentioned
in 1916; three more seem to have been written at Craiglockhart
in 1917, by which stage it was the 'Ballad of Kings and Christs';
and the last version, 'The Women and the Slain', may even be
1918 work.[8] All these drafts describe soldiers as modern King
Arthurs and Christs, at first very confidently but by 1917 with
some hesitation. In the end 'The Women and the Slain' attributes
these heroic statements exclusively to women, who are flatly
contradicted by dead soldiers. The ballad shows how Owen's
views changed but also how he retained a weakness for popular
wartime imagery; even though he decided in 1917 that the poem
was based on false values, he could not bring himself to scrap it
but instead turned it upside down by ascribing its idealism to
ignorant women. Metaphors of Arthur and Christ for soldiers
were perhaps the most common of all wartime images, turning
up somewhere in the work of most 1914–18 poets and extensively
used in pictures and memorials. Owen drew on both, identifying
Tommies as Arthurs even in December 1917 in 'Hospital Barge'

and as Christs later still in 'Greater Love'. The soldier–Christ image implied that a soldier who died in battle was earning salvation for himself and his country, an attractive consolation, but the idea was irreconcilable with Protestant doctrine, as the *Christian* sternly pointed out in 1914, because it denied the unique value of Christ's death.[9] Its occurrence in Owen's war poems is sometimes mistakenly seen as proof that his faith had revived, but the image was more sentimental than Christian; it was suspect, too, because it lent itself to exploitation by propagandists in support of the war. He came to recognise it in 1917 as 'a distorted view to hold in a general way'[10] but even so he went on to give it elaborate expression in 'Greater Love'. The war only deepened his original belief that front-line soldiers were heroes and martyrs, beautiful in death.

The first drafts of his 'Ballad of Peace and War' contain several other themes which were widespread in contemporary verse. The third stanza declares with no hint of irony that it is 'sweet' and 'meet' to die for brothers – the 'old lie' that he was later to denounce so savagely in 'Dulce et Decorum Est'. Ensuing stanzas develop a more interesting idea, touched on in '1914', that men were dying to ensure the return of spring: the 'Sun is sweet on rose and wheat', the 'soil is safe' and 'children's cheeks are ruddy'

> Because the good lads' limbs lie cold,
> And their brave cheeks are bloody.

This was to be hinted at again in 'Exposure' (1918), where there is a distinct echo of the ballad ('Nor ever suns smile true on child, or field, or fruit'). Owen may have been aware of *The Golden Bough*, J. G. Frazer's recent study of pagan vegetation rites in which kings or gods (or, as it might be, Arthurs or Christs) were sacrificed in winter to guarantee a fertile spring. The hope that society might be born anew out of destruction kept many soldiers going, but Owen finally rejected it in 1918, writing its elegy in 'Strange Meeting' and 'Exposure'.

Most of the ideas in the few poems which he wrote directly about the war before 1917 were expressed better by other writers. The most striking effect of his getting into uniform was not any advance in his thoughts about public events but the experience of working as a man with men. His growing-up was almost finished at last. Excitement is evident even in the handwriting of his breathless letters in the winter of 1915–16. His 1916 verse

revolves around themes of social and sexual maturity with such
strange ambiguity that love and war seem indistinguishable. New
images from parade ground and firing-range are used for both
subjects alike. For example, in April he heard a drum-and-fife
band:

> A thrilling affair. The sound . . . has finally dazzled me with
> Military Glory.
> The fifers are worthy to rank with the demented violins that
> make Queen's Hall to spin round as a top, and with the
> Cathedral Choir that pierces thro' the heights of heaven.
> Sweetly sing the fifes as it were great charmed birds in Arabian
> forests. And the drums pulse fearfully-voluptuously, as great
> hearts in death.[11]

The language of late Victorian lyricism is being adapted to a
military subject, and Owen is accumulating a store of poetic
material: 'great hearts in death' reappears as 'hearts made great
with shot' in 'Greater Love', 'demented violins' as 'demented
choirs' in 'Anthem for Doomed Youth', while other phrases turn
up in work for 'The End', 'Music', 'An Imperial Elegy' and other
1916 poems of love and war.

Some of the excitement may not have come from Army life, since
his postings to various training-camps gave him opportunities to
meet civilians. There were spring rambles again, this time with
some Boy Scouts whose 'affection – which has come up swiftly as
the February flowers – seems without bounds and without
restraint',[12] and there were visits to London. The manuscript of
'To——' is marked 'May 10 1916 London', an unusually
precise dating; he revived a pre-war habit on that day by visiting
the annual Academy Exhibition, presumably with an old friend.[13]
The metaphor in the octave of Eros running between the two
boys on the beach may well have come from a painting, since
Academicians were fond of such scenes, but an actual relationship
seems to lie behind the poem. Eros is soon to awaken the pair
from childish affection to adult love. But the last line, 'The sea
is rising . . . and the world is sand', sounds as much like a reference
to the war as a lament for lost innocence.

Another sonnet, written later in the year, begins with an octave
that is entirely about the lost 'heaven' of childhood; not until the
sestet of 'A New Heaven' does one discover that this is a war
poem:

Let's die back to those hearths we died for. Thus
Shall we be gods there. Death shall be no sev'rance.
In dull dim [chancels, flower new shrines for] us.
For us, rough knees of boys shall ache with rev'rence;
For girls' breasts are the clear white Acropole
[Where our own mothers' tears shall heal us whole.][14]

Death in battle is proposed as a way of regaining childhood
happiness. This remained a possible consolation even in
September 1917, when Owen rephrased the idea in the sestet of
'Anthem for Doomed Youth', but neither 'A New Heaven' nor
'Anthem' expresses his maturity as a poet or soldier. In soldiering
as in love there was no way back to childhood and maternal
affection. His 1918 poems were to explore the implications of a
world in which the doors of home were closed for ever. The
soldiers in 'Exposure' go back in dream to 'those hearths we died
for' but they see only the 'innocent mice' rejoicing by the embers
while women and children sleep in forgetfulness, so they return
to their dying in the winter night. Owen battled for years with
his ache for home until accepting, in what he regarded as his first
'mature' poem, that he could never again be 'Harboured in
heaven' as 'a Mother's boy'.[15] In his more truthful moments in
1916 and earlier he knew he had forfeited 'heaven', both at home
and in the next world. Somehow his destiny seemed to be
darkness and torment, either in the 'low sorrowful secrecy' of the
underworld of guilt and desire (P1) or in the 'perishing great
darkness' of war ('1914'). 'The sea is rising': drowning and
underwater imagery are recurrent in his poems of nightmare.
More positively, 'O World of many worlds' had foreseen that
'blackness of darkness' was the realm of meteor-poets, so that the
darkness and pain of maturity were to be welcomed. 'The Poet
in Pain', an undated sonnet, suggests a purpose for the suffering
that seemed an inevitable accompaniment of poetic creation:

Some men sing songs of Pain and scarcely guess
Their import, for they never knew her stress.
And there be other souls that ever lie
Begnawed by seven devils, silent. Aye,
Whose hearts have wept out blood, who not once spake
Of tears. If therefore my remorseless ache
Be needful to proof-test upon my flesh
The thoughts I think, and in words bleeding-fresh

> Teach me for speechless sufferers to plain,
> I would not quench it. Rather by my part
> To write of health with shaking hands, bone-pale,
> Of pleasure, having hell in every vein,
> ' Than chant of care from out a careless heart,
> [And mock with musics man's eternal wail.][16]

Pain was a way of proving truth on the pulses and of learning how to be a spokesman for 'speechless sufferers', an idea expressed first in 'On my Songs' (1913) and eventually in Owen's famous statement in 1918 that he wanted to 'plead' for his voiceless soldiers. There is a similar continuity from his Dunsden poems about the endurance of human suffering to the phrases 'man's eternal wail' in 'The Poet in Pain' and 'whatever moans in man' in 'Insensibility' (1918). The poet could only bring health and light to others if he shared in their misery to the utmost. Owen hoped in 1917 that his best poems would 'light the darkness of the world'.[17] The consistency of metaphor in his statements of poetic intention is a further illustration of the unity of his early and late work. 'The Poet in Pain' was probably written in wartime but it may refer to the pain either of love or of war, for his imaginative responses to both were strangely similar.

Most of the poems which he managed to finish during his fourteen months of training in England were sonnets, drawing some of their style from his reading in France but owing their existence mainly to a new partnership with Gunston and a friend, Olwen Joergens. The trio agreed to write sonnets on at least ten chosen subjects.[18] Six of Gunston's sonnets are in the book which he published in November 1917, two by Joergens survive in manuscript and four more by her are mentioned in Owen's letters. The subjects, with the dates of Owen's contributions, are as follows: 'Purple' (September 1916), 'Music' (final draft dated 'Oct 1916–17'), 'The End' (probably late 1916), 'Golden Hair' and 'Happiness' (both February 1917), 'Sunrise' (May 1917), 'How Do I Love Thee?' (not later than summer 1917), 'Fate' (Owen's 'The Fates' was written early in July 1917), 'Beauty' ('My Shy Hand' was first drafted on 'Aug 29–30, 1917' as 'Sonnet to Beauty') and 'Attar of Roses'. If Owen wrote on this last subject, his completed sonnet has not survived.

The two cousins may have hoped to submit their work to the Little Books of Georgian Verse series which had been launched in June 1915 by the publisher Erskine Macdonald, with support from the editor of the *Poetry Review*, Galloway Kyle. A slim volume of Joergens's poems appeared in the series in April 1916. Kyle, an enthusiastic discoverer of youthful 'talent', used to encourage aspirants to send work to Macdonald, who would then consider their manuscripts on condition that they bought four Little Books and subscribed to the *Review*. Anything that Macdonald published was duly puffed by Kyle in the *Review*. Gunston, who apparently subscribed to the magazine, sent his cousin at least one Little Book and three more are still in Owen's library. Among Owen's papers there is what seems to be a plan for a contribution to the series, to have been entitled 'Certain Sonnets of Wilfred Owen' or 'With Lightning and with Music' (a phrase from *Adonais*) or, inaccurately but with an eye on the nature of the series, 'Minor Poems – in Minor Keys – By a Minor'. An accompanying note declares that his sonnets have simplicity, unity of idea and 'a solemn dignity in the treatment'. Recovering from shellshock in 1917 and meeting someone who claimed to know Macdonald, he sketched out what was probably to have been another Little Book, this time entitled 'Sonnets in Silence' or, preferably, 'Sonatas in Silence'. It was as well that none of this got anywhere, because 'Macdonald' was an unscrupulous publisher of amateur verse, making profits out of his unsuspecting authors; in fact, a court case in 1922 revealed that the name was nothing more than an alias for Galloway Kyle, who had played a double role throughout the war of helpful critic and demanding publisher. One of the few people who suspected Kyle of deception was Harold Monro, who advised Owen against the Little Books in 1916. Nevertheless Owen continued to hope that his sonnets might get into print; it may have been as late as early 1918 that he fair-copied most of them and arranged twenty in a numbered sequence.[19]

Some of the workings for his 1916 sonnets show him starting out with a set subject and an idea to illustrate it, but some are less straightforward. 'The End' was a title which caused particular trouble; he tried it out on what eventually became 'To Eros' and on an unfinished piece that was designed to end a book, before attaching it to some notes he was making for a poem about a sunset. He said that the sonnet which emerged was 'intentionally' in Joergens's style but could have said more accurately that it was in Swinburne's, a poet whom Gunston frequently imitated. Owen's manuscript begins with a stanza from Swinburne's 'Laus Veneris' about man's mortality:

> And lo, between the sundawn and the sun,
> His day's work and his night's work are undone;
> And lo, between the nightfall and the light,
> He is not, and none knoweth of such an one.

Then comes Owen's first attempts at a quatrain:

> There [blew] a blast of sunlight from the east
> A flourish of high clouds; and [men marched hot]
> A mighty noise of drumming rolled, and ceased.
> The bronze west blew retreat; [and they were] not.

Work for further lines includes:

> The fire of day fell blazing from its fount
> Cascaded thro' the leafage; ricoched
> Across the rippling pool.

> The light so smote the forehead of the earth
> She lay in daze as dead.

> The touch of heat was like a too bold lover's [touch]
> Full of offence

> Aloof lone foxgloves in their ferny glooms

> The Pond, like an eternal July night
> [Wetted and cooled and hid his body white]
> [Closed on the youth with cold and dim] delight
> Wetted his body with a dark

> The same pond [which] was a [London] Paris July night[20]

Most of this harks back to France: bathing at Bagnères, the sun at Mérignac ('This kind of blazing heat stuns one pleasantly, like strong music', August 1915[21]), visits to London and Paris (May and June 1915), the rape of Danae in 'Perseus'. However, the metaphors of military music and gunnery belong to 1916, and the fragment is typical of 1916 work both in its echoes from earlier periods and in its oddness. It was to have been something more than a description of evening, for the youth sinking into a 'Paris night' seems to be a captive of love. Hence the link with 'Laus

Veneris', which is about the soldier poet Tannhäuser, the prisoner of Venus in the Mountain of Love.[22] Swinburne dwells on the knight's agonies of unsated desire, which are much the same as those of the Cultivated Rose; the youth sinking into the pond may have been doomed to similar torment.

However, Owen scrapped all of this Swinburnian fantasy except the first four lines. These he added to further lines from a quite different fragment about mortality to form the octave of the final sonnet, completing the poem with a newly composed sestet:

> After the blast of lightning from the east,
>   The flourish of loud clouds, the Chariot Throne;
> After the drums of time have rolled and ceased,
>   And by the bronze west long retreat is blown,
> Shall Life renew these bodies? Of a truth,
>   All death will he annul, all tears assuage?
> Or fill these void veins full again with youth,
>   And wash, with an immortal water, age?
>
> When I do ask white Age, he saith not so:
>   'My head hangs weighed with snow.'
> And when I hearken to the Earth, she saith:
>   'My fiery heart shrinks, aching. It is death.
> Mine ancient scars shall not be glorified,
>   Nor my titanic tears, the seas, be dried.'

The summer day in the first draft has become the Last Day, ending with the last sunset of the Decadents when not one but all youth dies and the great darkness begins. The sonnet may be read as a comment on war, but one could hardly call it a war poem. Its conclusions go back to Owen's loss of belief in immortality as he watched the Dunsden children, its imagery to the 'thrilling' military band and the stunning sunlight at Mérignac. One of the least personal and most memorable of his sonnets, 'The End' owes such force as it has to its being founded on a number of personal but significant experiences. The invisible presence of actual experience in the later poems is similarly a source of their strength.

He was as yet uneasy about using his own experience in verse, feeling the typically late-Romantic need to conceal it under symbols and large statements. Although he read Rupert Brooke in 1916, probably for the first time, he seems to have had reservations about that poet's Georgian egocentricity, for in the only draft of 'An Imperial Elegy' (1916) he contradicted Brooke's most celebrated phrase: 'Not one corner of a foreign field / But a span as wide as Europe . . . I looked and saw'. There was not one grave at the front but a trench of death right across the continent. 'And I heard a voice crying / This is the Path of Glory.' This unpromising piece was never finished but its grand language shows how Owen felt about the war while he was in training. This time the style really is that of Joergens, who was prone to similar prophetic vagueness (the 'Elegy' manuscript actually quotes two lines from her book).[23] The opening lines refer once again to the drum-and-fife band and the draft is subtitled 'Libretto for *Marche Funèbre*', the tune he had played to Nénette in August 1914. He may even have been strumming the march with one hand and writing with the other, since the words are scattered about the page in unusual confusion. The confusion increases when one finds over the page not more of the 'Elegy' but (echoing a change in Chopin's sonata?) an outline for a lyric comparing a beautiful body to jewels and seasons ('diamonds like the diamond dawns of spring', and so on).[24] These comparisons were rescued for use in 'Spells and Incantation' in 1917 but for the time being Owen scratched them out and began a fresh sheet.

And again there is a startling change of subject:

[Purgatorial passions]

> Their teeth leered wicked like the teeth of skulls
> Their necks were bowed like moping foxgloves all
> Aloof, lone foxgloves in their ferny glooms
> And like that melancholy flower's their mouths
>   Bitterly and wide
> Hung drooping and stroke on stroke of pain
> Has gouged blue chasms round their eyes [. . .]
> [Pain weltered] from their feet and from their palms.
> Eternal terrors fingered in their hair.
> Their crimson eyeballs twinged and twitched
> The Wrong that wrung their fingers like a rack.
> The awful falsehood of the smile of skulls
> Their hideousness of set, hilarious teeth
> And the attention of their empty eyes.

On the back of this second sheet there are some notes for a description of an infernal landscape:

It was an evening
   Yet less an evening than a stagnant dawn,
   . . .
[Wherein the sun] for ever smouldering [blood]
   [He found not Cybele
He saw not Proserpine, for he was lost,
   As all their ways [were] lost Who wandered there]

I thought 'My heart has failed,
As I desired, in sleep. And this is Hell.[25]

Much of this material was used again in 1918:

Who are these? Why sit they here in twilight?
Wherefore rock they, purgatorial shadows,
Drooping tongues from jaws that slob their relish,
Baring teeth that leer like skulls' teeth wicked?
Stroke on stroke of pain, – but what slow panic,
Gouged these chasms round their fretted sockets?
Ever from their hair and through their hands' palms
Misery swelters. Surely we have perished
Sleeping, and walk hell; but who these hellish?

That is the first stanza of 'Mental Cases'. In the other stanzas, the shellshocked soldiers are further described: they 'helpless wander', 'their eyeballs shrink tormented', 'Sunlight seems a blood-smear', 'their heads wear this hilarious, hideous, / Awful falseness of set-smiling corpses', 'their hands are plucking at each other'. 'Purgatorial passions' has been thought of as an early draft of 'Mental Cases', but it does not seem to refer to war. It seems more like work for 'Perseus' (although its association with 'An Imperial Elegy' makes that hypothesis an awkward one to prove). The image of the foxgloves, borrowed from the first draft of 'The End', suggests a gloomy setting but not a wartime one. Cybele and Proserpine, goddesses of the underworld, seem irrelevant to the hell of the trenches. Owen is thinking of the caverns wandered in by Endymion (who did see Cybele) and the hells in *fin de siècle* works such as the *City of Dreadful Night*, where bleeding and stationary suns were standard properties. The damned sufferers, racked by 'passions' and a 'Wrong' which they themselves have presumably committed, are in the same state as the sinful Tannhäuser, tormented in the prison of the Venusburg (Swinburne describes him as trembling and feverish), or the

Cultivated Rose. The 'Wrong' is lust not war-making, and its victims now watch the bleeding last sun of the Decadence in an eternal agony of 'physical sensation'. The 'blue chasms' round their eyes are the same as the purple marks 'round a youth's eyes, strained with love looks' in preliminary work for 'Purple'.[26] More broadly, the adjective 'purgatorial', which survives into 'Mental Cases', suggests Dante.[27] The inmates of Dante's hell are there as a consequence of their own sins; the transfer of guilt in 'Mental Cases' to 'us who smote them, brother . . . us who dealt them war' is not implied in 'Purgatorial passions'.

The similarities between this fragment and 'Perseus' are evident. The Cultivated Rose had eyes dark 'with gloom of death and hell', ever-trembling fingers and a sweating scalp, as he writhed in the 'everlasting fire not quenchable' of lust. The speaker of P1 had sought Persephone (Proserpine) in 'hell's low sorrowful secrecy', finding himself smiling 'the wide bright smile of skulls', looking through empty eye sockets and suffering 'inexplicable weakness'. Earlier verse had described the ever-changing outline of 'the Flesh' and the Mermaid's statue standing in the 'purple shade' of the willow tree's 'Convulsive fingers'. In a series of poems from 1912 to 1918, Owen returned to his vision of hell, a place incorporating elements of the Inferno, the Venusburg and the subterranean and submarine regions explored by Endymion. Its inhabitants are versions of himself paralysed and drowning under the gaze of Despondency–Medusa at Dunsden, but their eyes are nightmare versions of the 'melancholy' eyes of Rampton. Like the hermaphrodite, each of these figures seems to be two persons in one. Most of Owen's mature poems have their origins in his earlier verse, not because he liked (as he certainly did) to make use of images and turns of phrase that had pleased him in the past, but because the development from love poetry (if one may call it that) to war poetry was an expression of his whole self, including his knowledge of literature. He took the twilight of the Decadence out of dream and fantasy and related it to history, too grandly in 'The End' but with grimly controlled accuracy in 'Mental Cases' and 'Strange Meeting'; at the same time, his private 'horrors' were fulfilled in his 1918 poems as he looked on the face of death-in-life, the staring eyes and skull-like smiles of war's victims. It was only after he had been in the trenches, repeating in vastly magnified form the crisis he had suffered at Dunsden, that he was able to transform his pre-war self into the poet of 1918; in 1916, although he was in the Army, his verse derived from Dunsden and Bordeaux. Nevertheless the speed with which he moved from 'An Imperial Elegy' to a fragmentary love poem to a description of hell indicates both his peculiar suitability for his future role as a war poet and the usefulness of the literary tradition in which he had trained himself. Late Romanticism had always associated love and horror, so that an image such as the staring face, which had originally been a way of interpreting his own fears and craving for beauty, could develop easily into a metaphor for the 'truth untold' of war.

Owen needed advice. His 1916 verse was moving too far towards generalisation, partly because he was imitating Olwen Joergens. Its tone is sombre, elegiac, with 'a solemn dignity in the treatment', tending as always towards the biblical and hymn-like; its view is wide, sometimes aerial, prophetic (like Isaiah, he hears 'a voice crying'). The best of this was to serve him well in such poems as 'The Show' and 'Insensibility' but the worst was no better than the sort of thing that could be found in any Little Book of Georgian Verse. He was fortunate in getting some help from Harold Monro, whom he met in October 1915 and several times thereafter. Monro's Poetry Bookshop had become widely known as a centre for all serious readers and writers of poetry in London; it stocked a large range of books (except titles published by 'Erskine Macdonald') and callers were welcome to browse. Owen attended several of the famous poetry readings, bought Monro's latest book, *Children of Love*, and became a familiar visitor. He was impressed by the war poems in *Children of Love*, for Monro was one of the few poets who had always regarded the war with sad scepticism, writing about it in a cool, modern style far from the nebulosities of Joergens or the naïve heroics of the 'Ballad of Peace and War'. In March 1916 Owen stayed in one of the rooms at the shop which were available to needy poets, and apprehensively asked Monro to read his sonnets. Monro had received dozens of similar requests but showed his usual kindness and shrewd judgement:

> last night at eleven o'clock, when I had strewn about my goods preparatory to sorting and packing, up comes Monro to my room, with my MSS! So we sit down, and I have the time of my life. For he was 'very struck' with these sonnets. He went over the things in detail and he told me what was fresh and clever, and what was second-hand and banal; and what Keatsian, and what 'modern'.

He summed up their value as far above that of the Little Books of Georgian Verse.[28]

By 'modern' Monro meant something nearer 'Georgian' than 'Modernist', although he would not have used either of those terms. He had considerable doubts about the advanced work of Ezra Pound and the Imagists but was both a contributor to and the publisher of *Georgian Poetry*, the anthology edited by Edward Marsh which had caused a mild sensation with its first two volumes (1912 and 1915). Monro was friendly with most of the younger contributors and had known their original leader, Rupert Brooke. As an example of freshness, cleverness and modernity, Brooke was the obvious choice. Owen bought a May 1916 reprint of Brooke's poems, slipping into it a photograph from a magazine captioned 'A "Corner of a foreign field that is for ever England": Rupert Brooke's grave on Skyros'. There is little sign of Georgian influence in any of his 1916 verse but in the longer term Brooke's clear, anti-Victorian language would have appealed to him, especially in 1917 after he had received advice from two more Georgians, Sassoon and Robert Graves, both of whom were contributors to the third *Georgian Poetry* (1917). Graves's *Over the Brazier* (1916) was published by the Bookshop, where Owen may well have seen it. Monro gave him access to new work that was to be invaluable to him in 1917–18 and may have drawn his attention to several established writers whom he had hitherto neglected (Yeats, Housman and Tagore, for instance, are mentioned in 1916 letters for the first time). But at the end of the year Owen's reading and writing came to a sudden halt. On 30 December he returned to France, not as a tutor this time but as an Army officer prepared for active service.

5 The Second Crisis: Shellshock

CASUALTY

As the strain of events at Dunsden had led to Owen's nervous collapse in 1913, so his trench duty in 1917 ended in 'neurasthenia' or shellshock. He had never guessed at the horrors of the trenches but they were not entirely unfamiliar; waking and sleeping, his mind translated them into images he had known before. There is no need here to give more than a brief outline of his front-line experience since his own letters tell the story vividly, as they were meant to do, and Professor Stallworthy has filled in most of the gaps, but his shellshock and its treatment are worth more attention than they have so far received from his commentators. He reached the war zone in the first week of January, keeping his mother informed of his whereabouts by means of a simple code[1] and assuring her, echoing Sydney Carton, that 'I cannot do a better thing or be in a righter place.'[2] But soon, somewhere near Beaumont Hamel, he found himself obliged to hold a flooded dug-out in No Man's Land where he endured the events described in 'The Sentry'; as his futile match burned out in the darkness, leaving him with the mental picture of his sentry's 'huge-bulged' eyes, his training as an officer and as a poet came to an end. He had finally reached maturity, and the fixed, sightless eyes of nightmare had stared at him 'face to face' at last, a dream made real.

The disciple of beauty found the trenches hard to bear, but his preaching instincts were aroused:

extra for me there is the universal pervasion of *Ugliness*. Hideous landscapes, vile noises, foul language . . . everything *unnatural*, broken, blasted; the distortion of the dead, whose unburiable bodies sit outside the dug-outs all day, all night, the most execrable sights on earth. In poetry we call them the most glorious.[3]

Some literary parallels were still appropriate though, however inadequate: 'Gehenna', 'Hades', 'seventh hell', 'inferno', 'the Slough of Despond', 'Sodom and Gomorrah', 'the eternal place of gnashing of teeth'.[4] He longed for 'an inoffensive sky, that does not shriek all night with flights of shells', adding that he was writing down unpleasant details 'for the sake of future reminders'.[5] This anticipation of the offensive sky in 'Spring Offensive' and the 'shrieking air' in 'The Sentry' is one of so many passages in his trench letters used eventually for poetry that it seems clear he did indeed turn back to these records when he came to write poems months later, but at the time his motive for accurate reporting was simply that he wanted to tell the truth to people at home. Many soldiers chose not to do that on the grounds that it would cause useless distress, but Owen had no such qualms. 'I must not disguise from you the fact that we are at one of the worst parts of the Line', he wrote from Beaumont Hamel, and then, 'I can see no excuse for deceiving you . . . I have suffered seventh hell.'[6] His family were used to being 'educated' in this way: 'Why am I telling you these dreadful things? . . . to let them educate you, as they are educating me, in the Book of Life' (1911); 'I must not conceal from you my malady' (1913); 'I deliberately tell you all this to educate you to the actualities of the war' (1914); 'that *more* which I had not been told I feel I ought to tell' (1915).[7] Just as he had been unable to contain his indignation at Dunsden, so he insisted now that his letters should be typed out and circulated as a small contribution to waking up the nation. 'The people of England . . . must agitate. But they are not yet agitated even.'[8] He particularly wanted his reports to be read by the Gunston brothers, Leslie, who was medically unfit for military service but prone to patriotic sentiments, and Gordon, whom Owen considered to be what the newspapers would have called a 'shirker'.[9]

In the light of this strong response to front-line conditions, it may seem surprising that Owen wrote no poems aimed at the civilian conscience for more than six months after he first saw the trenches, but all the evidence suggests that such was the case. 'Exposure' was once believed to have been written in February 1917 but in fact it was not even started until December, while

'At a Calvary' and 'Le Christianisme', two short pieces formerly ascribed to early 1917, seem likely to belong to the end of the year. There were several reasons for his not writing any significant war poems before Craiglockhart. First, he lacked a theoretical basis for protest. Many of his early 1917 comments were the standard 'grouses' of front-line troops which he was picking up for the first time. Soldiers grumbled endlessly about such topics as the supposedly fit men who had avoided conscription or the ignorant complacency of civilians, especially journalists.[10] Justified though such complaints may have been, they were not a foundation for original poetry, nor did they imply any weakening of the general belief that only war could end the war. Protests against 'shirkers' were protests in support of the war effort, not against it. The intellectual arguments against further fighting were being hammered out mostly by civilians, but Owen knew little or nothing of them until the summer. A second reason for his not working on serious verse until then was simply that conditions were not right. He always needed solitude, quiet and even secrecy, valuing his privacy at Shrewsbury and Dunsden and even renting a room outside camp in 1918 so that he could work undisturbed. Apart from his literary friends, few people who knew him were ever aware that he wrote poetry. Thirdly, the horrors of trench service began to take effect well before his shellshock was diagnosed, so that reflecting deeply on immediate experiences would have been an intolerable ordeal.

What kept him going as a verse-writer in the first half of 1917 was his partnership with his cousin, with whom he remained on the friendliest terms, and Olwen Joergens. Several sonnets resulted but no war poems; he seems to have kept war in one mental compartment and poetry in another, registering surprise when an artillery officer gave him 'a book of Poems to read as if it were the natural thing to do!!' Sent behind the lines for the whole of February, perhaps because his superiors had already seen signs of the strain he had undergone, he was soon 'settling down to a little verse once more, and tonight I want to do Leslie's subject "Golden Hair" and O.A.J.'s "Happiness" ' – extraordinary subjects, it might seem, for a poet who had just spend a month on the Somme. But, despite the fact that he signed both sonnets 'Adolescens' when he rewrote them in the summer, 'Leslie's subject' proved to be a fruitful one. Later, in August, it was from the first draft of 'Happiness' that he quoted what he had come to think of as his first mature lines:

Tennyson, it seems was always a great child.
So should I have been, but for Beaumont Hamel.
(Not before January 1917 did I write the *only lines* of mine
that carry the stamp of maturity: these

> But the old happiness is unreturning.
> Boys have no grief so grievous as youth's yearning;
> Boys have no sadness sadder than our hope.

The theme of growing-up which had been prominent in his 1916
verse is here brought to a conclusion. He had just been reading
A. C. Benson's *Tennyson*, which records Coventry Patmore as
having said that 'Tennnyson is like a great child, simple and very
much self-absorbed.'[11] Owen's comment has frequently been
quoted as a judgement on Tennyson's poetry, but in the context
of Patmore's remark it should be seen more accurately as a
characteristic response to a piece of biographical information, for
Owen had not lost his habit of comparing his own life with those
of 'the bards of old'. Beaumont Hamel seemed in retrospect to
have ended his self-absorption and forced him at last into
maturity.

Even in 'Happiness', however, there is no mention of war. The
old happiness of childhood has been lost because the poet has
gone 'past the scope / Of mother-arms', laughing 'too often since
with Joy' and committing 'sick and sorrowful wrongs'. 'You must
not conclude that I have misbehaved in any way from the tone
of the poem (though you might infer it if you knew the tone of
this Town)', Owen hastily assured his mother from Abbeville in
February, implying that the 'sick' (or in a later revision 'strange')
'wrongs' were sins of the flesh rather than acts of violence in war.
The language is of the evasive, mysterious kind that had been
used by the Decadents when they had confessed to 'strange' (but
unspecified) sins. Another draft concludes that we who have
'played with human passions for our toys, / Know that men suffer
chiefly by their joys'. The 'passions' here are those of 'Purgatorial
passions', that 1916 portrait of smiling sufferers in torment for an
unexplained 'Wrong'. In the final version of 'Happiness' the
maternal arms become, memorably, 'the wide arms of trees', a
metaphor which moves Owen's poetry away from private statement towards the impersonal strength of 'Spring Offensive',
where the embrace of maternal nature was to be represented in
the more specific image of clinging bramble shoots. The break

from his mother had become a break from nature too ('Let not the sun drop sweetness on me more').[12] Everything in war was '*unnatural*, broken, blasted'; the Mérignac sun had been trying to destroy him after all. As that transition developed in his imagination under stress of experience in 1917–18, so also did the evolution of his erotic verse into war poetry. His poethood grew fast but it maintained its continuity from his original 'dark' and 'disobedient' escape from his mother at Broxton in 1904.

Within a few weeks of returning to the trenches in March he was in a Casualty Clearing Station (CCS) with concussion, having fallen into a ruined cellar. A letter describing this incident was seen by Blunden but has since been lost, so that not much is known about what happened. One significant detail in another letter has been overlooked, however: 'I lost count of days in that cellar, & even missed the passing of night & daylight, because my only light was a candle.'[13] The poet of the underworld had been trapped in a dark hole underground for twenty-four hours or more. The dreadfulness of the experience may be guessed at, coming as it did not only after his peacetime horrors of paralysis and drowning but also after his fifty hours in the dug-out, when the roof had threatened to collapse under shellfire and he had been tempted to let himself drown in the slowly rising floodwater. It was not safe to think or feel deeply. He managed to find a cheap exercise book in which to draft sonnets while he was in hospital, completing 'With an Identity Disc', a piece that has some wit and charm but little substance.

'My long rest has shaken my nerve', he admitted on 4 April, back in action after about a fortnight in the CCS. The enemy was retreating to the Hindenburg Line and the Allies were taking the bait; Owen's battalion moved beyond the trenches to fight in a relatively undamaged landscape, advancing in the open to attack Savy Wood on 14 April. He got through the barrage unscathed, feeling an 'exultation' which he was always to remember, but his accounts of events after that are confused and incomplete. It seems that his unit was left fighting without relief for twelve days; his nerve finally broke when a shell just missed him on a railway embankment, after which he spent several days sheltering 'in a hole just big enough to lie in' near a brother officer who had been blown to pieces some weeks earlier. He never said why he remained in the hole for so long. He may have been pinned down by enemy fire or he may have been helpless with shock, perhaps even unconscious for a while because he later spoke of 'coming-to after the Embankment-Shell-Shock'. Whatever the case, he can have been little use to the men he was supposed to be leading. In the end he returned to base safely but was obviously in a bad state. Perhaps word got round; at any rate a week later the Colonel spoke sharply to him, ordering him to report to the Medical Officer. Neurasthenia was diagnosed and Owen found himself back in the CCS.[14]

It is now generally forgotten that the circumstances of Owen's shellshock were a matter for some debate soon after the war. In 1920 Charles Scott Moncrieff said in print that Owen had been officially recorded as having suffered a loss of morale under shell-fire, a remark that was taken up by several reviewers of the newly published poems. Sassoon drew Mrs Owen's attention to the spreading report, remarking acidly that Scott Moncrieff's 'very objectionable' article had been written 'entirely for his own advertisement'. Edith Sitwell, no doubt with Sassoon's encourage-ment, published a counter-attack, quoting Mrs Owen as saying that 'Wilfred was *not* sent home on account of any loss of morale.' The anxiety of Owen's supporters that he should not be presented as a coward helps to explain the insistence on his courage that often appears in early criticisms of his poetry, a insistence which is still sometimes repeated although the argument which caused it has long since faded from view. Scott Moncrieff is unlikely to have been lying, since he worked in the War Office in 1918 and almost certainly saw Owen's file. He replied to Edith Sitwell's protest but did not substantially alter his point. His evidence seems to be reinforced by the less reliable testimony of Robert Graves, who met Owen in 1917 and said in the first (1929) edition of *Goodbye to All That* that Owen 'had had a bad time . . . in France; and, further, it had preyed on his mind that he had been accused of cowardice by his commanding officer. He was in a very shaky condition.' When Blunden came to write his memoir of Owen in 1930 he drafted a denial of this 'callous mis-statement', quoting Sassoon as remembering that 'The "cowardice" was only an indefinite idea which worried O', but the Owen family asked him to omit all reference to Graves and the cowardice question. Blunden therefore deleted the passage, also omitting part of a letter which he had received from Mrs Gray, a friend of Owen's in 1917. Mrs Gray told him that Owen had 'grieved deeply' over his failure to live up to his own standards.

Nevertheless in his most despondent moods he could never be said to have experienced despair. His courage was too indomitable for that, and he never laid down his arms, – even in the grip of his painful delusion, the belief in his commanding officer's regarding him as a coward.

Blunden dropped the words after the dash, although they shed some light on what precedes them. Owen was under no delusion. The reprimand may have been no more than an angry exclamation from an exhausted man, but it was given and it would have been a severe blow in an age when most people did not clearly recognise the distinction between cowardice and involuntary nervous collapse. Valuing the esteem of his fellow soldiers and usually remarkably happy in their company, Owen would have felt himself disgraced. The importance of the incident lies not in whatever truth there may have been in the accusation, since the notion of 'cowardice' on those battlefields has now rightly become meaningless, but in its effect on Owen's subsequent poems and actions. It helps to explain his return to the front in 1918 and the heavy weight of guilt that loads his finest writing.[15]

As Wigan had perhaps contributed to his illness in 1913, so the Colonel's words seem to have precipitated the neurasthenia which had been threatening since March. Owen did not break down completely, any more than he had done in 1913, but he must have been seriously ill. The senior psychiatrist at the CCS, William Brown, reckoned to send seventy per cent of patients back to the trenches after about a fortnight,[16] but Owen was in hospital for six months and unfit for active service for another nine. The visible symptoms were probably not much more than the usual sweating, especially from palms and scalp, and uncontrollable shaking. He may have experienced those before, since they figure in 'The cultivated Rose' and 'Purgatorial passions'. Other common symptoms which would have been familiar to him were a rapid pulse and a sense of suffocation, both of which he had often mentioned in his verse. More severe cases of shellshock were marked by temporary paralysis, acute depression and terrifying dreams. He had described himself as paralysed by the stare of Despondency in 1911–12, and it seems possible that he suffered actual paralysis in that 'hole just big enough to lie in' on the railway embankment, the third and perhaps the worst of his near-burials at the front. Mrs Gray remembered his 'despondent moods' at Craiglockhart, so that depression was probably another of his symptoms, but his letters and poems show that the most painful and enduring effects of his illness were violent nightmares, predictably enough. Even before the 'horrors', 'phantasies' or 'phantasms' of his post-Dunsden illness, Despondency had steeped his nights in 'bloodiness and stains of shadowy crimes'. It was well known among psychiatrists that any weakness that a patient had suffered from before the war was likely to emerge in a more acute form in shellshock.[17]

Although the doctors no doubt forced Owen to remember and describe the events at Savy Wood, since full recall was considered to be an essential step on the road to health, such memories were not yet material for poetry. He began drafting sonnets again in his notebook but none of them refers to war.[18] One at least is on a set subject, 'Sunrise', and some fragments in the book may be attempts at 'Nocturne' and 'A Wind in the Night', subjects which Mr Gunston remembers as having been selected by the trio of poets. There are attempts at Tailhadesque sound-patterns ('Waned . . . wanner . . . was . . . wan . . . worn', for example, within three lines in 'Sunrise') and the usual imagery of jewels, flowers, seasons and times of day, exercises revealed by some flippantly archaic spelling in 'Sunrise' to be self-conscious literary diversions. One of the drafts introduces violence but only as Decadent metaphor: the fire of love having been 'quenched by mine own blood in spurts', the poet seeks for 'Beauty the eternal' in the 'sweet wound' left in his flesh by the flames. A phrase in the same sketch, '[Fastening of feeling] fingers on my wrist', is a connecting link between the hands that had been laid on Owen's arm 'in the night, along the Bordeaux streets' and the erotic menace of the snow in 'Exposure' ('Pale flakes with fingering stealth come feeling for our faces'), illustrating once again the continuity from his Bordeaux experience into his 1917–18 poems, and the sexual unease that underlies some of his most successful late imagery.[19] But there was a long way to go before work such as 'Exposure' could even be imagined. He went on making notes for sonnets, struggling to shut out what he had been through. Perhaps there would be enough for a Little Book of Georgian Verse – 'Sonatas in Silence', purple and gold, with type 'like Before Dawn' . . . but he had forgotten Monro's advice.

Shellshock continued to affect his verse for some months. He was sent home by stages, arriving at Craiglockhart War Hospital for Neurasthenic Officers, just outside Edinburgh, on 26 June.[20] He was soon telling his cousin that a new ballad was 'going strong', but it was not a serious poem. He called its heroine Yolande (she was originally Mildred), a name probably borrowed from Swinburne, whom he seems to have been reading again.[21] He also wrote two more sonnets on agreed subjects, 'The Fates' and 'How Do I Love Thee?', the former setting out what he told Gunston was 'almost my Gospel', the Aesthetic creed that the only way to escape 'the march of lifetime' was to gaze into the eyes of Beauty. Neither sonnet was among his best, but as his health began its slow return he tried some more original writing. His early Craiglockhart manuscripts contain not only his first attempts at pararhymed verse but also two of the strangest statements he ever made of his private imaginings, 'Lines to a Beauty seen in Limehouse' and 'Has your soul sipped'. Whether as a result of shellshock and its aftermath or because the doctors had made him look into his inner mind, two of his most indelible memories emerged with peculiar explicitness; both mental pictures were of smiling youths, one a handsome boy he had seen in the East End and the other a soldier whose death he had witnessed, and both were to appear several times in later poetry.

The 'Beauty' of Limehouse, presumably remembered from Owen's walk there in 1915 or perhaps a more recent visit, is compared to a carved and painted idol receiving human sacrifices:

> I watched thy lips
> Vermilion like a gods; [dyed bright] with dips
> [In my] own blood in dreams that woke me faint.

But the 'half-god' ignores the poet, who knows that the boy is not bound like himself by social convention but is free to 'take thy pleasures with thy kind, / Where love is easy, where I cannot go'. The poem was not worth finishing but it recorded a strongly felt experience: seeing a beautiful youth in a working-class district, Owen had come up against the social and psychological obstacles which prevented him from offering more than an admiring glance, despite dreams of making himself a blood sacrifice at the boy's 'smooth, smooth naked knees'. The fragment revives imagery of sculpture, stoniness, incense, eyes and fingers from earlier letters and verse, including 'Perseus' work. But the special interest of the 'half-god' is his connection with the soldier in 'Disabled', a link suggested by the note '(Thy hands shall be gloved with?)' among rough workings for the Limehouse poem.

A group of quotations may demonstrate the origins of 'Disabled' more clearly:[22]

(a) Now approach the days when the evening light draws out seductively: tempting all such as have movable hearts and movable bodies to stray forth and behold clouds, games, lamps, stars, riversides, swallows, and the daughters of men. (To Colin Owen, aged fifteen and working on a farm, April 1916)

(b) Of the Last Draught that went out, men I had helped to train, some are already fallen. Your tender age is a thing to be valued and gloried in, more than many wounds.

Not only because it puts you among the Elders and the gods, high witnesses of the general slaughter, being one of those for whom every soldier fights, if he knew it; your Youth is to be prized not because your blood will not be drained, but because it *is* blood; and Time dare not yet mix into it his abominable physic . . . .

Let your hands be gloved with the dust of earth, and your neck scarved with the brown scarf of sunshine. (To Colin Owen, August 1916)

(c)   Ah! He was handsome when he used to stand
Each evening on the curb, or by the quays.
His old soft cap slung half-way down his ear;
– Proud of his neck, scarfed with a sun-burn band,
And of his curl, and all his reckless gear,
Down to [the] gloves of sun-brown on his hand.
      (Cancelled stanza for 'Disabled', autumn 1917
      or later)

(d)   Ah! he was looked at when he used to stand
In parks each evening [outside cinemas] himself to
    please
[Who mated with him . . . . 'Oh the *dear*!'
Wealthy old ladies said quite loud, and scanned
Unpleasantly the rose behind his ear . . .
And eyed all up and down his reckless gear]
      (Another attempt at the same stanza)

(e)   Voices of boys were by the riverside
      Sleep mothered them; and left the twilight sad.
      (From 'But I was looking', a fragment
      written at Craiglockhart)

(f)   He sat in a wheeled chair, waiting for dark,
And shivered in his ghastly suit of grey,
Legless, sewn short at elbow. Through the park
Voices of boys rang saddening like a hymn,

Voices of play and pleasure after day,
Till gathering sleep had mothered them from him.

About this time Town used to swing so gay
When glow-lamps budded in the light blue trees,
And girls glanced lovelier as the air grew dim, –
In the old times, before he threw away his knees.
Now he will never feel again how slim
Girls' waists are, or how warm their subtle hands.
All of them touch him like some queer disease.

There was an artist silly for his face,
For it was younger than his youth, last year.
Now, he is old; his back will never brace;
He's lost his colour very far from here,
Poured it down shell-holes till the veins ran dry,
And half his lifetime lapsed in the hot race
And leap of purple spurted from his thigh.
　　　(The first three stanzas of 'Disabled', October 1917)

These interrelated passages imply that Owen had been working
on a poem in 1916 about a young man waiting to pick up a girl,
a subject based on his East End memory and associated with his
awareness that Colin was entering manhood. The 'riversides' in
the first extract would have been the paths along the Severn in
Shrewsbury, perhaps in the Quarry, where games, lamps and
swallows are to be seen in summer, but the setting of the cancelled
stanza seems more urban, 'quays' suggesting the Limehouse
docks. The second letter to Colin introduces the image of the far-
seeing gods, who are detached from human suffering like the
'everliving' in the Yeats epigraph to 'The Show' and the 'half-
god', who had gazed beyond human mortality. Envious of the
Limehouse boy's self-assurance, freedom and detachment, and
attracted by his good looks, Owen brooded on the memory until
it buried itself in his imagination and was eventually (after he
had met Sassoon) released into 'Disabled'. The handsome boy in
'Disabled' is under nineteen, much the same age as the recruits
in 1916 whom Owen had helped to train, and his young blood
has been 'drained' like theirs. Like the 'half-god', he had beautiful
knees ('Someone had said he'd look a god in kilts') but they have
been shot away. He waits again at dusk in the park, this time

sitting impotent in grey hospital uniform instead of standing in
'all his reckless gear', but the women no longer eye him 'up and
down', preferring to avoid looking at him. Vermilion and purple
have been poured away, leaving only grey. Such is one of Owen's
measures for the destructiveness of war. There may be something
of the poet himself in the soldier in 'Disabled'. Memories of Paris
and London evenings may have gone into the making of the
poem, and in October 1917 he described himself as 'an obviously
unmarried young man in a reckless soft-cap', using phrases from
the cancelled stanza.[23] The sunburnt youth with a handsome
neck can be compared with the young poet at Bagnères who was
praised for his beauty by Tailhade in the language of the
Decadence and who dreamed of enlisting for the sheer romance
of it.

The 'half-god' has vermilion lips, dyed with the poet's blood
in dreams, just as the lips of the 'painted idol' in 'Long ages past'
bear 'the stain of crimson blood' of lovers. These details give some
indication of an element in Owen's dreams before 1917, including
the 'bloodiness and stains of shadowy crimes' inflicted on his
nights by Despondency in 1911–12. It appears from two other
poems that the vision took an actual shape in 1917 when Owen
saw a young man dying while bleeding at the mouth. One of
these poems is 'Has your soul sipped', which seems to be one of
the verse exercises he wrote at Craiglockhart before he met
Sassoon (it is also possibly his first composition in pararhyme).
The poet claims to have been 'witness / Of a strange sweetness',
the adjectives 'strange' and 'sweet' introducing a list of such
Decadent properties as nocturnes, nightingales, dying loves and
martyrdom, but sweeter than all these things

> To me was that Smile,
>   Faint as a wan, worn myth,
> Faint and exceeding small,
>   On a boy's murdered mouth.

> Though from his throat
>   The life-tide leaps
> There was no threat
>   On his lips.

> [Is it his mother
>   He feels as he slips

Or girls' hands smoother
And suaver than sleep's?]

But with the bitter blood
And the death-smell
All his life's sweetness bled
Into a smile.

There can be little doubt that this grotesque piece was written in the summer of 1917, not in 1916 as was once supposed,[24] but one's response to it should be tempered by the probability that Owen was reading Swinburne at the time. Its unpleasantness may have been a consciously Decadent attempt to be shocking, the repetition of 'sweet' (nine times in various forms) being comparable to the way in which Swinburne harps on that word in describing how the leprous Yolande was 'sweeter than all sweet' to her lover ('The Leper'). On the other hand, 'The lifetide leaps' (originally 'The crimson leaps') and the smoothness of girls' hands anticipate the 'leap of purple' from the soldier's thigh and the 'subtle hands' of girls in 'Disabled'; if Owen was capable of thus turning some of the affected, hackneyed language of 'Has your soul sipped' into such intensely serious imagery only a few weeks later, it probably had some importance for him from the start. The memory of the boy's dying face seems to have been a real one, appearing again in the little poem 'I saw his round mouth's crimson deepen as it fell', where the bleeding mouth is seen as a 'magnificent' sunset. There are more generalised images of bleeding mouths in later poems, notably 'Greater Love' and 'The Kind Ghosts'.

The Swinburnian metaphor of torn mouths as red roses in 'The Kind Ghosts' (1918) brings together Owen's nightmares and his late-Romantic inheritance in an image which is designed to express something of the mysterious 'truth untold' of war. It horrifies the reader, as Owen intended, but it could not have come from a poet whose understanding of the war had been confined to reasoned reflection on external experience. Owen's war poems are not simply protests or statements of pity. They constantly return to certain obsessive images and to guilt, desire, darkness and blood. He might have gone mad as many of his fellow soldiers did, but instead he got his imagination under control and wrote with an increasingly serene self-discipline. His recovery was partly due to his own strength of character but he was greatly helped by his Craiglockhart doctor, A. J. Brock.

DR BROCK

It had always been assumed that the treatment given to Owen at Craiglockhart was much the same as that given to Sassoon, some commentators even following Robert Graves's inaccurate statement that both patients were under the same doctor, W. H. R. Rivers.[25] Rivers was a famous man, made more famous later by Sassoon's portrait of him in *Sherston's Progress*, but he had little or no contact with Owen, who was treated by Arthur Brock (Plate 8) on a system differing considerably from that used by Rivers. According to Sassoon's letters and memoirs, Rivers worked through 'therapeutic conversations', leaving his patients free for much of the day to play golf or otherwise amuse themselves. Brock, on the other hand, regarded spare-time activities as an essential element in therapy. A serious, talkative man, full of ideas, he brought to his wartime job an extensive range of interests in art, classical scholarship, literature and the new science of sociology. Having trained and practised locally before the war, he had many friends in the city. His enthusiasm for sociology, like his commitment to interdisciplinary enquiry, owed much to the celebrated Edinburgh polymath Patrick Geddes, whose teaching strongly influenced his work at Craiglockhart.[26]

Geddes maintained that the ills of modern living were a result of people's having lost touch with their environment; work and life had become separated, towns had become shapeless, communities had fragmented. His formula 'Place–Work–Folk' expressed his belief that the crucial link between people and place is what they *do*, their work in its widest sense. One way to reforge that link was to encourage 'Folk' to undertake 'Regional Surveys' of their 'Place', studying every natural and civic aspect of their surroundings. Seeing the parallel between Geddes's formula and the biologist's 'organism-function-environment', Brock interpreted shellshock as an extreme form of the social failings which Geddes had analysed: the organism had become violently detached from its environment and could no longer relate to it by means of its usual functions. Treatment had to concentrate on 'function' and 'work' if health was to be restored.

Brock had a ready pupil in Owen, who had been interested in biology for years and found Geddesian thinking immediately congenial.[27] The doctor gave him the task of composing an essay on the Outlook Tower, an old house near Edinburgh Castle which Geddes had established as a centre for sociological studies.[28]

Owen's surviving notes for the essay show that he visited the
Tower, attending carefully to its symbolic organisation; like all
visitors, he would have gone first to the roof, with its view of the
city, and then descended through rooms containing 'Regional
Survey' exhibitions, first of Edinburgh, then of Scotland and
finally of the world as a whole (Plate 9). 'I perceived that this
Tower was a symbol,' he wrote, 'an Allegory, not a historic
structure but a poetic form . . . . The Tower is suggestive of the
great Method of Philosophical Thinking, which is Correlation or
Co-ordination.' The Tower embodied Geddes's aim of enabling
people to realise their full individuality by growing beyond
specialisms, parties and nations towards an understanding of the
interdependence of man and nature as a whole. The Tower's
chief enemy, Owen noted, was 'the Spirit of exclusiveness'. The
outline for his essay contains such typically Geddesian sequences
as 'Individual – Family – Township – the Nation – the Internation'
and Tennyson's line, 'The Parliament of Man, the Federation of
the World'. Later in the year, he was to talk to a school class
about 'the international idea' and to make notes for a play set in
a federated Europe of the future.[29] It seems clear that the Tower
stimulated his political and social thinking, its 'poetic form'
offering a purposeful, imaginative set of ideas which revealed,
among other things, the unnaturalness of conflict between nations.
At the other end of the scale, the 'Individual' made one small
step towards regaining a right relationship with his environment
by the simple act of looking thoughtfully at the Tower.

Brock's theories are set out in his post-war book and in many
articles. Writing to Geddes in 1920, he said that a section of the
book was to be on 'Treatment of the War Mind': 'Away
from Other-worldliness and back to the Here and Now – i.e.
Regionalism'; 'Treatment = Civics beginning by putting each
man on his own survey'.[30] This section was unfortunately never
published, but the Craiglockhart magazine, the *Hydra*, and
Owen's letters show that the doctor's patients were indeed
required to study their social and natural environment as a
central therapeutic activity. Like Geddes, Brock placed a high
value on 'Work', believing that the main symptom of the
dissociation of 'Folk' from 'Place' was 'ergophobia', a fear of
effort, and that the appropriate treatment was, in his own special
word, 'Ergotherapy', a work cure.[31] He expected his patients to
work both physically and morally, requiring them to survey not
only the region but also themselves, facing up to their past

experience and overcoming it. The practical aspects of Ergotherapy came from Geddes but much of its underlying philosophy derived from Bergson, a writer whom Brock much admired, and the Greeks. Bergson insists on the importance of the integrated personality, in whom intuition and imagination make life a process of continuous, creative change. The individual is responsible for his own growth, understanding his past and open to his future, living as spirit and not as machine.[32] Brock saw his task as helping shellshock victims to reconnect themselves with their past, including its terrors, and with their future, as well as with the external world.[33] It followed that he needed to understand their backgrounds, so that in a sense he himself had to make a 'Regional Survey' of each patient, a practice which he found advocated by the greatest doctors of the ancient world, particularly Galen, who believed that an organism could only be understood in terms of its environment.

Brock found that it was common for shellshock sufferers to regress to childhood and long for maternal affection. Owen wrote to an aunt soon after his arrival at Craiglockhart, 'I am not able to settle down here without seeing Mother. I feel a sort of reserve and suspense about everything I do.' He did not 'do' very much, except write a little verse and rashly attend a church service ('It gave me the indigoes, not half'). Mrs Owen hurried up to Edinburgh and was no doubt interviewed by the doctor, who liked to meet patients' families (they were part of the 'environment' which his diagnosis had to take into account). Owen said that his first sight of her was accompanied by the kind of 'exultation' he had felt after getting through the barrage at Savy Wood. He noticed her grey hair, 'the ashes of all your Sacrifices'. This emotional reunion, arousing language associated with his two nervous crises (shellshock and the 'ashes' of his Dunsden sacrifice), seems to have been the starting-point of his cure, for after it Brock was able to start an intensive course of Ergotherapy.

Owen was soon 'full of activities' and enjoying 'a Greek feeling of energy and elemental life' (a phrase which clearly shows the influence of the disciple of Galen and Bergson).[34] Regional Survey began with an afternoon of visits to a factory, a munition works and a brass-foundry, the next morning being devoted, as Geddes would have advised, to some practical metalwork. However, the main elements in the cure were always selected by Brock to suit the special interests of the patient, which meant literature and the earth sciences in Owen's case. On the day after the factory

visits, Owen became a founder member of a new Field Club, with Brock as President, and soon he was also busy editing the *Hydra*. He remained editor until the end of September, seeing six numbers through the press and contributing a good deal of unsigned material, his hand being often evident in reviews and stories. At the same time that he took on the *Hydra*, he finished his essay on the Outlook Tower and began work on a poem which Brock had set him on the subject of Antaeus.

The story of Antaeus was one of Brock's favourite myths, often referred to in his writings. An anonymous article in the *Hydra* for November 1917 encourages patients to join in hospital activities, warning them that it might be 'literally suicidal' not to; they needed

> vital contact with our surroundings (physical, organic, and social), and we shall each of us live over again the experience of the giant Antaeus, who gained fresh springs of life at every fresh contact with his Mother Earth.
>
> The Field Club in a sense aims at coordinating all the other scientific groups. Its immediate object is a regional survey, *i.e.*, a survey of the Craiglockhart region, from all the different aspects (geological, botanical, economic, etc.), which will, at the same time, show the absolute interdependence between these aspects. The field of external nature is one . . . .[35]

The January 1918 number explains the myth more fully:

> Antaeus was a young Libyan giant, whose parents were Gaia and Poseidon, Earth and Sea. In a wrestling combat he could not be overthrown as long as his feet were on his Mother Earth. When he was raised off the earth his strength rapidly failed, only to be renewed again at the first contact with the soil. Finally Hercules, seeing this, lifted him bodily up in the air, and holding him there, crushed him to death in his arms.
>
> Now surely every officer who comes to Craiglockhart recognises that, in a way, he is himself an Antaeus who has been taken from his Mother Earth and well-nigh crushed to death by the war giant or military machine.
>
> . . . Antaeus typifies the occupation cure at Craiglockhart. His story is the justification of our activities.

This article continues that the patient must 'act in relation to his environment' and must 'beware of Art for Art's sake', a warning consistent with Outlook Tower teaching. Geddes believed that art should serve the community, so the paintings and other works which he commissioned for the Tower were designed as contributions to its educational function. Artists or poets who concentrated on their own specialisms were scorned by Brock, who would have strongly disliked several of the poems Owen had recently been drafting, including 'The Fates'. Owen's declaration in his 1918 Preface, 'Above all I am not concerned with Poetry', is a statement of Geddesian principle. He had not fully absorbed these ideas when he wrote on Antaeus, but his stock comparisons between seasons, landscape and the human body began to acquire some underpinning from the myth. He took the chance of including plenty of 'Perseus'-like descriptions of male beauty but they have 'a Greek feeling of energy and elemental life' that is new in his verse. The same energy emerges in a bolder use of alliteration and assonance than he had ever tried before.

> the thews and cordage of his thighs
> Straitened and strained beyond the utmost stretch
> From quivering heel to haunch like sweating hawsers . . .

The pairing of 'thews' and 'thighs' here is a reminder that pararhyme was another of the skills which Owen worked on at Craiglockhart. To end the wrestling-match, Hercules tugs Antaeus upward ('Rooted him up') as he had pulled up oaks in Argos, an image of the fighter as a tree uprooted by 'the war giant or military machine' that points to the Geddesian significance of the comparison between trees and soldiers in 'Spring Offensive'. But 'The Wrestlers' is not a very satisfactory illustration of Brock's theories, because Owen is more interested in Hercules than Antaeus, carefully adding the details that Hercules was 'the son of Perseus' and that he gained his wisdom from his beautiful boy companion, Hylas.

   'The Wrestlers' and the Outlook Tower essay are the only pieces known to have been written at Brock's request, but he probably encouraged other writing and must have been responsible for Owen's becoming *Hydra* editor. Other Ergotherapeutic activities included teaching in a school (all Brock's patients were expected to do something in the local community), as well as joining in the hospital's debating and dramatic groups. Owen lectured to the Field Club on 'Do Plants Think?' on 30 July; since the ideas in this talk were partly based on information he had gleaned in 1911 from Gunston's *Cassell's Popular Science*, they were hardly new or original but they were very much in line with Brock's.[36] Everyone in the hospital knew the doctor's concern with organism and environment (the January 1918 *Hydra* reports facetiously that 'We hear . . . That a certain stoical M.O. has discovered an organism entirely without environment'), so Owen's lively discussion of plants' responses to external stimuli was well received by the Club. The lecturer felt 'exultation' once more when he was applauded; perhaps he wrote the anonymous *Hydra* report, which said that 'the lecture carried us to the farthest point of modern research'. The Field Club also went on expeditions, including one on 10 August when 'Two wanderers from Shropshire saw no small resemblance between the Pentlands and the Longmynd range'. Owen marked the opening of this sentence for his mother's attention; his memories of the Shropshire landscape were as active as they had been in 1914, when he had seen a hill at Bagnères as 'exactly like' the Long Mynd or one of its neighbours. A further phrase in this *Hydra* article, 'steep grassy hills', again points towards the setting of 'Spring Offensive'.

Doctor and patient must have talked at length about their scientific, social and literary interests. Brock obviously soon discovered that Owen wrote poetry, but whether he saw any of it, apart from 'The Wrestlers', remains a matter for conjecture. It is difficult to imagine Owen showing him the Limehouse fragment or 'Has your soul sipped', for example, although those two curiously personal poems were written while the work cure was in progress. However, two topics which might well have come up for discussion were the techniques of verse and the image of the terrifying face, since the first required hard work which could be regarded as part of Ergotherapy and the second may have recurred as often in Owen's reports of his shellshock dreams as it does in his writing.

Owen seems to have taken a special interest in the technicalities of verse composition during his first months at the hospital, trying his hand at a ballad, sonnets, blank-verse narrative, and a variety of stanza forms and sound-effects. He wrote a small lyric called 'Song of Songs', publishing it anonymously in the *Hydra* (1 September) because Sassoon admired it:

> Sing me at morn but only with your laugh:
> Even as Spring that laugheth into leaf;
> Even as Love that laugheth after Life.

He sent a proof of 'Song of Songs' to Gunston with the comment,
'My first printed Poem!', but he could have added that it was
the first published poem in English to use pararhyme as a regular
scheme. A closely related piece, 'From my Diary, July 1914', may
have been written under the stimulus of Sassoon's praise. These
two, with 'Has your soul sipped', seem to be his first complete
works in pararhyme, so it may be that his use of the device was
one of the fruits of Ergotherapy, although the idea had occurred
to him before and probably derived from France. Pararhyme was
thus not invented for war poems, as is sometimes claimed, but
for sensuous, erotic 'songs' in the Decadent style.[37] 'From my
Diary' looks back to the delights of Bagnères, where Tailhade
had encouraged him to begin his enthusiastic study of French
literature and had joined Mme Léger in awakening him to his
own attractiveness and freedom. Despite its being based on
memories of that happy summer, the poem is 'seriously and
shamelessly worked out' (to borrow a phrase from a 1915 letter),[38]
using Owen's well-tried frame of a single day but relating
morning, noon and evening to the energy, passion and completion
of an imagined sexual relationship. Displaying his skills for his
own and possibly his doctor's enjoyment, the craftsman packs in
as much rhyme, pararhyme, alliteration and assonance as the
lines can hold:

> Birds
> Cheerily chirping in the early day.
> Bards
> Singing of summer, scything thro' the hay.

The careful organisation of these Craiglockhart verse experiments
may well reflect Brock's advice.

It was certainly through Brock that Owen made most of his
new friendships in Edinburgh. Miss Wyer, for example, was a
leading member of the Outlook Tower Association; she showed
him round the 'Open Spaces', small plots of land established in
the slums by Association members. Owen also met John Duncan
and Henry Lintott, two artists who had designed murals and
book illustrations for Geddes's many enterprises, and became
friendly with two more families who lived nearby, the Grays and
the Steinthals. Only in the Légers' house in Bordeaux had he
seen a household where art and culture were so much in evidence.
Mrs Gray adored him, taking him slum-visiting (no doubt she

too was an Association member) and lending him books. At about the time that he completed 'Disabled', in which the young soldier is said to have had 'an artist silly for his face', Mrs Steinthal painted his portrait. What testimony it bore to the beauty and horror which he carried with him cannot be known, for Mrs Owen destroyed it after the war. Although his Edinburgh friends did their best to make his days rewarding, they knew that his nights were hard to bear.

'There is this advantage in being "one of the ones" at the Hospital, that nurses cease from troubling the weary who don't want rest.' There is plenty of evidence in the Craiglockhart letters that Owen deliberately stayed up late in order to shorten his sleeping hours, but he was reticent on the subject of his dreams. He reported 'having had some very bellicose dreams of late' in mid August, but by that stage the worst was over. 'The Barrage'd Nights are quite the exception', he wrote only a week later, adding after another week that, 'I still have disastrous dreams, but they are taking on a more civilian character, motor accidents and so on.' There was 'one horrid night' late in September and no doubt others, but they were not experiences that could be written about easily.[39] The word 'Barrage'd' suggests that the barrage at Savy Wood was a recurrent nightmare, while two references in poems show that other frequent subjects were two horribly blinded faces:

> In all my dreams, before my helpless sight,
> He plunges at me, guttering, choking, drowning.
>
> If in some smothering dreams you too could pace
> Behind the wagon that we flung him in,
> And watch the white eyes writhing in his face . . .

> ('Dulce et Decorum Est')

> Eyeballs, huge-bulged like squids',
> Watch my dreams still . . .
> ('The Sentry')

Brock said in his post-war book that shellshock patients often had nightmares of being pursued or rendered helpless by terrifying, devilish apparitions. Claustrophobia was also common, patients fearing to walk 'in long, dark passages' even by day (there were many such passages at Craiglockhart).[40] In 'Strange Meeting', the poet finds himself going down 'some profound dull tunnel' where he meets a dead enemy who holds him with a stare from 'fixed eyes'; this is precisely the kind of 'smothering' dream that was common in shellshock cases and Owen himself had probably experienced it.

It has already been suggested that these horrors were not unlike his pre-war visions. The 'phantasms' of early 1913, the dreams of violence brought by Despondency's stare in 1911–12, the waking tortures described in 'Written on a June Night' – these were the weaknesses which the doctors would have expected to reappear in an exaggerated form in shellshock. The overlap between pre-war and wartime nightmares is evident even in 1918, when Owen's remark that 'my nights were terrible to be borne' refers to his adolescent frustrations, not to 1917.[41] He may also have suffered from claustrophobia before the war; his mother seems to have done so, as his comment on a 1916 incident reveals:

> When I was going up the subway at Liverpool St. . . . I noticed the passages unduly encumbered, and found the outlet just closed, and Liverpool St. in complete darkness. We were corked down in those subways for close on 3 hours. This should appeal to *your* susceptibilities especially. There was just room to move from one Exit to another seeking an escape. After all the Zeppelins never got over the City, though we heard the guns.

This description may be compared with that of the 'fifty hours' in the dug-out, where 'One entrance had been blown in & blocked' and the guns thundered overhead,[42] and with the 'escape' into the dark 'tunnel' of 'Strange Meeting' where '*encumbered* sleepers' groan; it is in such settings that the dead or disabled figures in Owen's visionary poems are often encountered.

Dr Brock had a Geddesian explanation for dreams of this kind. In folklore, a subject in which he was widely read, the countryman who does his work well is rewarded by benevolent spirits such as Robin Goodfellow, while one who fails in his work and is out of step with his environment is plagued by malicious goblins. In a 1924 article, Brock said that the modern world had lost touch with nature on a massive scale: 'If civilisation is to be saved from perishing, it must rapidly, like Antaeus, regain its footing upon Mother Earth.' He described various legendary spirits which represent man's relationship with his environment, including the

Kobold, a sea spirit which is usually benevolent but which torments idle sailors and is fatal if seen:

> When, in fact, the seaman's morale falls so low that he begins to 'see things', he is indeed in a parlous state . . . he has failed to assimilate his environment and suffers in effect from a kind of nautical 'shell-shock'; hence these nightmares that Morpheus the dream-god (literally 'form-maker') sends him; they are essentially the same as those which haunted the pillows of the soldiers morally disabled by the strain of war; in these latter cases the evil Spirit of the Battlefield constantly took shape as some terrible Kobold or ogre of Frightfulness.[43]

The goblins of folklore and the apparitions in war dreams were thus 'true' in the sense that they represented profound human experience. This interpretation of 'imaginary' beings could easily be extended to art, and in pp. 171–2 of *Health and Conduct* (1923) Brock uses Owen's poems as evidence to prove his point:

> The psychology of all sorts of poetic or artistic creation is, of course, closely allied to the production of dreams or legends. Every work of imagination is a projection into sensory form of the artist's deepest personal experiences. Dante's *Vision of Hell* is an imagery of his own moral struggles; so also with Bunyan's *Pilgrim's Progress*, Mrs Shelley's *Frankenstein*, the gruesome paintings of Wiertz, etc. In the powerful war-poems of Wilfred Owen we read the heroic testimony of one who having in the most literal sense 'faced the phantoms of the mind' had *all but* laid them ere the last call came; they still appear in his poetry but he fears them no longer.

Ergotherapy had '*all but*' succeeded. We do not know what part Brock played, if any, in getting Owen to 'face the phantoms of the mind' through writing poetry; he seemed to regard the poems as only the record of a process which had been achieved by other means. Some of the principal poems about 'phantoms', such as 'Strange Meeting' and 'Mental Cases', were written after Owen left Craiglockhart, but 'Dulce et Decorum Est' and 'The Sentry' were begun while he was still under Brock's supervision and it is possible that the doctor encouraged such writing as a therapeutic exercise. In any case, he gives us a way of reading Owen's war poems which is of great interest; his association of the poems with

the *Inferno*, *Pilgrim's Progress* and *Frankenstein* is persuasive.
The mention of Owen in *Health and Conduct* is apparently Brock's
only reference to him in print, but Mrs Owen wrote to Blunden
after the war, 'You will like to see Dr Brock's letters'; these letters
seem to be lost, but they are presumably the source for the
quotation which Blunden makes in his Memoir that Brock
regarded his patient as 'a very outstanding figure, both in intellect
and in character'.[44] Owen's last two and a half months at
Craiglockhart were dominated by his friendship with Sassoon but
Brock's teaching remained with him. At the heart of his reflections
on the nature of war in 1918 lay the conviction that the organism
had to live in harmony with its environment and that destruction
was bound to ensue from its failure to do so. His own means to
wholeness were to face his personal nightmares, recognising them
for what they were. Whereas the dying man in 'Dulce et Decorum
Est' (one of his 'early' war poems) has a face 'like a devil's' and
is blind, the dead man in 'Strange Meeting' is no devil and sees
the poet with 'piteous recognition'. Such self-knowledge could
not end the war, however; despite Ergotherapy, the recognition
in 'Strange Meeting' is the prelude to death among groaning
sleepers in a war dream from which there is no awakening.

6 Sassoon

Siegfried Sassoon arrived at Craiglockhart on 23 July 1917, having made a public protest that the war had become one of 'aggression and conquest'. It was easier for the authorities to treat him as a shellshock case than to risk the publicity of a court martial. Some people believed then and later that his being sent to the hospital was yet another example of the unscrupulousness of politicians, but the facts have yet to be fully established. Robert Graves, his friend and fellow officer, said afterwards that Sassoon really had been suffering from shellshock (and hallucinations),[1] but then Graves had played a major part in getting him classified as neurasthenic, believing that his protest had been made at the bidding of pacifists while he was under severe stress. Sassoon himself confided to Lady Ottoline Morrell in April 1917 that he was 'very near the snapping point' and in November that he had 'taken to war nightmares again'. In the following February he admitted that 'I realise now that I couldn't have stood any more French horrors without breaking down.' Worried about his own mental stability, he found Craiglockhart profoundly depressing: 'My fellow-patients are 160 more or less dotty officers. A great many of them are degenerate-looking. A few are genuine cases of shell-shock etc. One committed suicide three weeks ago.' He had to revisit 'the cursed place' briefly in 1918: the 'drifting patients looked more haunted and "mental" than ever. I noticed the same types there, (though they are of course an entirely new crop)'. He does not seem to have taken part in hospital activities or to have found any of the pleasures which Owen enjoyed in Edinburgh; country walks and golf were his only solace.

It was true that Sassoon's protest had been formulated with the help of several pacifists, notably Bertrand Russell, whom he much admired for a time. He told Lady Ottoline in July that he would 'like above all to know that B. R. is satisfied that I've done something toward destroying the Beast of War'. But Russell at a distance was no match for Dr Rivers, who held an hour's

95

consultation with Sassoon every other day. While not suggesting that his patient's nerves were disordered, Rivers maintained that his attitude was abnormal and tried to talk him into modifying his views.[2] A medical officer's job, after all, was to restore his patients' fitness for general service. Rivers would have quickly identified and worked on Sassoon's two most vulnerable points, his affection for his men and his knowledge that they were safest when well led. Quietly but insistently the doctor suggested that the place for an experienced officer was in the battle line. Sassoon's anger against the war did not abate but he began to doubt the usefulness of his protest. Withdrawn and inward-looking at the best of times, he was now in a torment of self-mistrust, his wretchedness made worse by the news that his closest pre-war friend had been killed on 14 August. It is hardly surprising that he was not very interested in Owen when they first met.

Sassoon had published several slender volumes of Aesthete's verse before the war. Like Owen, he was at heart a late Romantic with a fondness for melancholy lyricism, but his early poems are less energetic and interesting than Owen's. Such verse had soon seemed out of place in the trenches. He read Hardy's poems there, saying in March 1917 that Hardy had 'always been more to me than any other writer, in the times I've spent out here. "A thinker of crooked thoughts upon Life in the sere, And on That which consigns men to night after showing the day to them." '[3] His changing attitude to the war, from sacrificial ardour in 1915 to anger in 1916, protest in 1917 and weary acceptance in 1918, can be traced in the unique record which he left in his poems, diaries, memoirs and numerous unpublished letters. In 1916 he began writing his epigrams, forceful and passionately sincere (albeit often clumsy) little poems designed to force civilians into recognising what the trenches were really like.[4] These satires were admired by intellectuals who were trying to turn public opinion against the war. At a time when Germany seemed inclined to seek peace by diplomacy, while the Allies were reluctant to state their war aims as a preliminary to negotiation, the opinion was growing in some circles that Allied politicians were now more interested in 'aggression and conquest' than in liberating Belgium. This change in intention seemed to be supported by profiteering capitalists and a stridently militarist right-wing press. All too many civilians appeared content to acquiesce; Sassoon used satire, realism and straightforward, outspoken language in his efforts to wake them up. He declared in his 1917 protest that he was speaking on behalf of the troops, implying that the officer poet's duty was to put the sufferings of the inarticulate into words, just as writers in earlier periods had spoken up for the poor. His satires were published in the *Cambridge Magazine*, an organ attacked in Parliament in November as pacifist propaganda (which it was not), and he was mentioned admiringly from time to time in its editorials. Other journals also accepted his work, so that by mid 1917 he was beginning to acquire the status he held between 1918 and the publication of Owen's poems in 1920 as the most impressive new poet of the war.

Sassoon's first substantial volume, *The Old Huntsman and Other Poems*, came out in May 1917. It was dedicated to Hardy, who delighted its author by praising the war poems in it and ignoring the lyrics.[5] A contrary opinion came from John Gambril Nicholson, a leading 'Uranian' poet, who praised the lyrics but was doubtful about the war poems. In reply, Sassoon defended the war poems but expressed his own preference for 'The Death-Bed', 'The Last Meeting' and 'A Letter Home': 'they have the *best* part of me in them, the quest for beauty and compassion and friendship'.[6] These three pieces, occupying the final pages of the 1917 edition, are 'lyrical war poems', not satirical epigrams of the kind for which Sassoon is still most generally known. He never regarded his epigrams as his most valuable work, often grumbling in later years at being represented by them in anthologies. 'Am I *never* to be given credit for having written from my heart?' he lamented to Blunden in 1965, referring to the selection made by Ian Parsons for *Men who March Away*.[7] He would have been pleased when Owen agreed with him in September 1917 that 'The Death-Bed' was the best poem in *The Old Huntsman*.[8]

Owen read the book in mid August: 'I . . . am feeling at a very high pitch of emotion. Nothing like his trench life sketches has ever been written or ever will be written.'[9] About a week later he dared to introduce himself. Sassoon later remembered finding him an 'interesting little chap' (he always referred to him as 'little' because of the considerable difference in their heights) but 'rather ordinary' and 'perceptibly provincial'.[10] Owen, on the other hand, was enthralled, his capacity for veneration aroused more strongly than ever before; he persisted in his visits until by mid September he had won Sassoon's confidence. The relationship was always one-sided, more so than is now usually supposed. In contrast to the many fervent passages about 'the Greatest friend

I have' in Owen's 1917–18 letters, Sassoon's letters of the same period contain very few references to 'little Owen'.[11] For Owen it was the friendship of a lifetime, brought about by sheer coincidence like his discovery of Monro's poems in 1911 or his relationship with Tailhade in 1914–15 but far more important to him than either of those encounters. For Sassoon it was a pleasant companionship that brought relief in a period of intense unhappiness.

The first lesson to be learned from Sassoon's precept and example was that truth to experience was essential to poetry, a maxim which Owen had always half known but never seen so clearly before. He realised with sudden excitement that his knowledge of the trenches was not just material for letters. Sending a copy of *The Old Huntsman* home, he said that 'except in one or two of my letters, (ahem!) you will find nothing so perfectly truthfully descriptive of war'.[12] As usual he was quick to imitate, writing 'something in Sassoon's style' immediately after his second visit. This was the first draft of 'The Dead-Beat', which he sent next day to his cousin marked '*True* – in the incidental', adding 'Those are the very words!' against the doctor's brutal comment in the closing lines.[13] The accurate representation of events evidently seemed a principal characteristic of 'Sassoon's style'; others were colloquial language, deliberately unpoetic imagery, dramatic abruptness and topical references.

> He dropped, more sullenly than wearily,
>   Became a lump of stench, a clot of meat,
>   And none of us could kick him to his feet.
> He blinked at my revolver, blearily.
>
> He didn't seem to know a war was on,
>   Or see or smell the bloody trench at all . . .
>   Perhaps he saw the crowd at Caxton Hall,
> And that is why the fellow's pluck's all gone –
>
> Not that the Kaiser frowns imperially.
>   He sees his wife, how cosily she chats;
>   Not his blue pal there, feeding fifty rats.
> Hotels he sees, improved materially;
>
> Where ministers smile ministerially.
>   Sees Punch still grinning at the Belcher bloke;

> Bairnsfather, enlarging on his little joke,
> While Belloc prophesies of last year, serially.
>
> We sent him down at last, he seemed so bad,
>   Although a strongish chap and quite unhurt.
>   Next day I heard the Doc's fat laugh: 'That dirt
> You sent me down last night's just died. So glad!'

Shown this first attempt at a 'trench life sketch', Sassoon pointed out that 'the facetious bit was out of keeping with the first & last stanzas'. Reluctant to waste the offending section and needing to fill space in the next *Hydra*, Owen hastily revised it; his anonymous editorial (1 September) comments that, when he and his readers had returned from France,

> some of us were not a little wounded by apparent indifference of the public and the press, not indeed to our precious selves, but to the unimagined durances of the fit fellow in the line.
>   We were a little *too* piqued by the piquancy of smart women, and as for the dainty newspaper jokes concerning the men in the mud, we could not see them at all . . .
>   Our reflections, like our reflexes, may have been exaggerated when on first looking round England, we soliloquised thus: –

> Who cares the Kaiser frowns imperially?
> The exempted shriek at Charlie Chaplin's smirk.
> The *Mirror* shows how Tommy smiles at work.
> And if girls sigh, they sigh ethereally,
> And wish the Push would get on less funereally.
> Old Bill enlarges on his little jokes.
> *Punch* is still grinning at the Derby blokes.
> And Belloc prophesies of last year, serially.

The style stumbles over itself as Owen tries too eagerly to copy the new way of writing (and to affect a soldierly colloquialism) without abandoning everything from his previous work. As he admitted to Gunston, the line about ministers with its awkward rhyme word was 'years old!!', originating from a 1915 'Perseus' manuscript, and his annoyance at *Daily Mirror* photographs went back at least as far as February. 'The Dead-Beat' was cobbled up from old material as well as new, but it marks a new phase in his poetry, standing at the beginning of his *annus mirabilis*.

He remained dissatisfied with the first draft and its offshoot, revising the poem many times. The final draft is as follows:

> He dropped, – more sullenly than wearily,
> Lay stupid like a cod, heavy like meat,
> And none of us could kick him to his feet;
> – Just blinked at my revolver, blearily;
> – Didn't appear to know a war was on,
> Or see the blasted trench at which he stared.
> 'I'll do 'em in,' he whined. 'If this hand's spared,
> I'll murder them, I will.'

*　　　　*　　　　*

> A low voice said,
> 'It's Blighty, p'raps, he sees; his pluck's all gone,
> Dreaming of all the valiant, that *aren't* dead:
> Bold uncles, smiling ministerially;
> Maybe his brave young wife, getting her fun
> In some new home, improved materially.
> It's not these stiffs have crazed him; nor the Hun.'

*　　　　*　　　　*

> We sent him down at last, out of the way.
> Unwounded; – stout lad, too, before that strafe.
> Malingering? Stretcher-bearers winked, 'Not half!'

*　　　　*　　　　*

> Next day I heard the Doc's well-whiskied laugh:
> 'That scum you sent last night soon died. Hooray!'

The five quatrains of the first draft, which had been made excessively conspicuous by their heavy rhymes, have been reworked into a more flexible verse paragraph, divided by asterisks where dramatic pauses are needed. Three passages of direct speech have been added, but the doctor's 'very words' of the original have been modified into a single, more forceful line; truth to experience did not necessitate word-for-word reporting. The punctuation throughout has been enlivened (Owen grew fond of the dash). The obscure reference to Caxton Hall gives way to an ambiguous threat from the soldier himself; there is still a momentary puzzle (who is threatened?) but it is worth solving. All the political and topical references are scrapped, because the world of London clubs, ministers, bishops and entertainers, which Sassoon often satirised, was known to Owen only through newspapers (as the *Hydra* verse demonstrates). The 'valiant' civilians in the final draft might be anybody's family, while the 'low voice' which refers to them is that of an unidentified, representative soldier. There is an implied tension, which was to be developed much further in later poems, between Owen-as-officer, using his revolver to enforce discipline, and Owen-as-poet, diagnosing the man's collapse with a sympathy beyond the reach of the war-hardened medical men. The poem has become more thoughtful and dramatic; several other Sassoonish poems went through a comparable process of revision, often spread over many months.

Owen responded to his new friend's first advice, 'Sweat your guts out writing poetry!', with characteristic enthusiasm, making use of everything he could learn from Sassoon's work. One of his Craiglockhart manuscripts shows him making notes for three or four Sassoonish onslaughts against civilian attitudes. He jotted down 'Accidental Death' and 'Killed by accident', presumably titles for what was to become 'S.I.W.', and roughed out a first attempt at 'The Sentry' with an ending in the style of Harold Monro (see above, p.15) as well as of Sassoon. Then he scribbled two fragments, perhaps intended as parts of a single poem, criticising civilians for being more concerned with such things as the sugar shortage than with the scale and significance of the casualty lists.[14] The presence of all this material on one sheet of paper shows how rapidly ideas for poems took shape once he had started to 'sweat his guts out'. It did not take him long to rival and in some respects outdo his new master. Most, possibly all, of his poems 'in Sassoon's style' (except 'Smile, Smile, Smile') seem to have been at least begun at Craiglockhart. Many of them can be matched with poems in *The Old Huntsman*: 'The Dead-Beat' with 'Blighters', for example; 'The Letter'[15] with 'In the Pink'; 'The Chances' with ' "They" '; 'S. I. W.' with 'The Hero'; 'Inspection' with 'Stand-to: Good Friday Morning'. 'The Chances' with its grimly dramatic ending, is as spare and powerful as anything by Sassoon. 'The Dead-Beat', 'S. I. W.', 'Inspection', 'Disabled' and 'Conscious' were all first drafted in quatrains, a form which Sassoon used often, but were then given looser shapes.

'S. I. W.' was further divided up by means of ironic, literary headings, the style of each section being adjusted appropriately so that 'The Action' is terse and broken up by abrupt punctuation, 'The Poem' is elaborately worded, and 'The Epilogue' has a dismissive yet arresting brevity:

> With him they buried the muzzle his teeth had kissed,
> And truthfully wrote the mother, 'Tim died smiling.'

'Truthfully', because the face wore both agony and relief, a macabre joke which prompts the reader to reconsider possible associations between 'Mother', 'kissed', 'smiling', 'muzzle' and the last, loving relationship between Tim and his rifle. One may pause, too, at the unexpected word 'teeth' and notice how Tim has acted out a phrase in the poem's epigraph from Yeats ('that man there has set his teeth to die'). Few of Sassoon's poems offer this kind of interest and concreteness.

Most of Owen's work 'in Sassoon's style' was repeatedly revised, often well into 1918. The multiplicity of versions, and in some cases the absence of first drafts, make it difficult to establish dates of initial composition. 'Conscious', for example, exists in two different versions, either (or neither) of which may represent Owen's final intention. All three drafts of the poem are on types of paper which he used after Craiglockhart, yet it seems more likely than most to be Craiglockhart work in origin because it evidently derives from 'The Death-Bed', the piece which both he and Sassoon considered to be the best in the *Old Huntsman* collection. On the other hand, 'Conscious' does not belong to Owen's first flush of enthusiasm, for there is criticism implicit in his doing in sixteen lines much of what Sassoon does in forty-two. Both poems describe a soldier on the verge of consciousness in hospital. Sassoon uses dreamy imagery of the kind he often introduced into his letters and lyrics:

> He drowsed and was aware of silence heaped
> Round him, unshaken as the steadfast walls;
> Aqueous like floating rays of amber light . . . .

By contrast, 'Conscious' opens with activity ('His fingers wake, and flutter', 'His eyes come open with a pull of will'), and in place of mellifluous generalities there are sharp, isolated details. The man is described in terms of his fingers and eyes, which seem to have wills of their own. His surroundings are perceived as separate, sharply focused objects: yellow flowers, the blind-cord 'drawling' across the window sill, a smooth floor, a rug. Then 'sudden evening blurs and fogs the air'; the things which the patient was beginning to recognise and question blur again as his senses become confused ('Cold, he's cold; yet hot', 'there's no light to see the voices by'). Sassoon's soldier drinks, 'unresisting', but Owen's is aware that 'There seems no time to want a drink of water'. 'Conscious' is full of struggle and effort, whereas the boy in 'The Death-Bed' drifts passively into reverie, almost in the way that Sassoon himself seems to have done in his lyrical moods. Sassoon's dying patient swoons through 'crimson gloom to darkness', night in the ward and his drink of water bringing him dreams of a river, coloured clouds, and rain on roses, 'passionless music', for some nineteen lines. Owen reduces most of this into three:

> Nurse looks so far away. And here and there
> Music and roses burst through crimson slaughter.
> He can't remember where he saw blue sky . . . .

The second line seems to be an attempt to condense Sassoon's rather obvious descriptions into Symbolist imagery. Owen may even have come across the line, 'L'éclat mystérieux des roses et du sang', in *Poèmes élégiaques*, since he had the book with him at Craiglockhart.[16] Tailhade's craftsmanship still had its value, although Symbolism and Georgian realism would have to be blended more subtly. But truth to experience was the rule above all others which Owen needed to learn and 'Conscious' shows him making good use of it, remembering details such as the yellow flowers from his own stay in the CCS and striving to re-create that sick, visionary sensation of being on the border of life and death that he had felt in dreams, and after his 1912 bicycle accident, and perhaps on the railway embankment after Savy Wood.

Sassoon showed him not only how war poems could be written but also how they could be based on a consistent, reasoned opposition to the war. Owen might have been less easily persuaded by a civilian pacifist such as Russell, although Sassoon seems to have lent him Russell's books and introduced him to liberal periodicals such as the *Cambridge Magazine* and the *Nation*. There can have been few officers in 1917 able to expound the war as clearly as Sassoon did; coming from him the case seemed overwhelming. Owen summed it up in 'The Next War', which he published in the *Hydra* on 29 September in order, he said, to 'strike a note': the war was now being waged for 'flags' (or 'money-bags' and 'empire-dreams', as he wrote in a draft).[17] The poem implies that soldiers understand war more clearly than anyone else, having a special knowledge and comradeship. Owen had acquired some of this knowledge for himself before he met Sassoon, which was why he was so moved when he first read *The Old Huntsman*. His letters from the trenches had expressed anger at the attitudes of civilians, especially churchmen. To the former lay assistant, Sassoon's mockery of priests and bishops would have had a familiar ring. Owen admired 'The Redeemer', a statement of the popular soldier–Christ image which he said he had been 'wishing to write every week for the last three years',[18] but Sassoon would have told him that the poem had been written in 1915 and no longer represented its author's view of the war. Shown the weakness of the image, Owen restated it later as a condemnation of organised religion in 'At a Calvary near the Ancre' and 'Le Christianisme'.

Owen and Sassoon discussed ideas more than experiences, avoiding mention of horrors or their own sufferings, with the reticence characteristic of veterans. The element of swagger in 'The Next War' reflects Owen's pride in being a hardened soldier and a comrade of Sassoon (the poem's epigraph consists of the closing lines of *The Old Huntsman*[19]); 'we', says the poem, have known Death as an 'old chum'. To some extent this was an assumed attitude rather than an expression of Owen's deepest inclinations. It may explain why he took a while to write directly of war's brutality; his first attempt at horrific description, 'Dulce et Decorum Est', was not written until October. Quite apart from soldierly conventions, however, conversation about the trenches was bad for Sassoon's nerves and worse for Owen's. When Sassoon read Blunden's memoir of Owen in 1930, he wrote to its author:

> The passages from W's war letters are a revelation to me, and I believe that their effect on people's minds will be tremendous. At Craiglockhart, he and I talked very little about our experiences of the disgusting and terrible, as seen in France. I discouraged him from reviving such memories, knowing they were bad for him. And little Wilfred was such a modest chap that I never fully realised his imaginative grasp of the scene.

Tonight I have felt that he has grown greater than ever, in my mind.[20]

Although Robert Graves was undoubtedly correct in saying that it was meeting Sassoon which started Owen writing war poems,[21] the stimulus came from Sassoon's example and beliefs rather than from discussions of 'the disgusting and terrible'.

Indeed, the two poets probably talked more about literature than anything else. Owen found that they had been following 'parallel trenches all our lives' and had 'more friends in common, authors I mean, than most people can boast of in a lifetime'. By chance, Sassoon was reading a small volume of Keats which Lady Ottoline had sent him. He shared Owen's interest in the late-Victorian poets, including Housman, whose influence is often apparent in his war poems, but Owen was surprised to find that he admired Hardy 'more than anybody living'. No doubt Sassoon persuaded him to start reading Hardy's poems. In return, Owen showed him Tailhade's book and asked Gunston and Mrs Owen to send all the manuscripts in their possession, Mrs Owen even being instructed to break into a locked cupboard and take out three folders of papers without reading a word. Looking over Owen's manuscripts, Sassoon censured 'the over-luscious writing in his immature pieces'.

> Some of my old Sonnets didn't please him at all. But the 'Antaeus' he applauded fervently; and ['Song of Songs'] he pronounced perfect work, absolutely charming, etc. etc. and begged that I would copy it out for him, to show the powers that be.

Continuing to comment patiently on all he was shown, the new critic 'condemned some . . . amended others, and rejoiced over a few'. Some of the amendments can be seen on surviving manuscripts. 'Song of Songs' met with favourable comment when it appeared (unsigned) in the *Hydra*:

> I have been doubtful whether to [?confirm] the 'Song of Songs' as mine. But now I find it well received by the public and praised by Sassoon with no patronizing manner but as a musical achievement not possible to him. He is sending copies of the *Hydra* to Personages!

Sassoon had clearly been struck by Owen's use of pararhyme, although his reason for sending copies of the magazine to several friends – Robert Ross, Lady Ottoline, probably Roderick Meiklejohn and Edward Marsh, and perhaps others – was that it also contained his own poem 'Dreamers'. Under 'Song of Songs' in Lady Ottoline's copy he wrote, 'The man who wrote this brings me quantities & I have to say kind things. He will improve, I think!'[22]

Having to 'say kind things' helped to draw Sassoon out of his introspection and he was glad of Owen's answering interest in his own work. He had a notebook with him in which he was assembling the poems for his next volume, *Counter-Attack* (1918), a counter-attack against civilian complacency; he read some of them on 7 September to Owen, who thought them 'superb beyond anything' in *The Old Huntsman*. One, written the night before, seemed 'the most exquisitely painful war poem of any language or time'; this was perhaps the last stanza of 'Prelude: The Troops', which is added in pencil in the notebook and dated 'September 1917'. Owen must eventually have heard or read the rest of the notebook's contents, including two pieces which were omitted from the 1918 volume because they were too brutal. The distinction of final place in the book was originally given to 'The Triumph', dated 'Oct. 1917', a poem which Sassoon later suppressed, perhaps feeling that it displayed more of himself than he wished to reveal. 'The Triumph' is not well written but it shows something of that 'quest for beauty and compassion and friendship' which he considered to be the best part of himself in 1917:

When life was a cobweb of stars for Beauty who came
    In the whisper of leaves or a bird's lone cry in the glen,
On dawn-lit hills and horizons girdled with flame
    I sought for the triumph that troubles the faces of men.

With death in the terrible flickering gloom of the fight
    I was cruel and fierce with despair; I was naked and bound;
I was stricken: and Beauty returned through the shambles
        of night;
    In the faces of men she returned; and their triumph I found.

This lyric seems to have caught Owen's attention. The imagery in the first stanza may be compared with similar but sturdier phrasings in 'From my Diary, July 1914', probably composed at this time: 'Leaves / Murmuring by myriads', 'Birds / Cheerily chirping', 'Braiding / Of floating flames across the mountain-brow'. He had described the quest for beauty as 'almost my Gospel' in July, but now he could see that the quest could be fulfilled in war experience.[23]

Owen was Sassoon's follower in writing about the war, but in verse about beauty the two poets were more nearly equal. In fact, there is one small instance of Sassoon imitating Owen rather than the other way about. At the end of August Owen wrote the last of his sonnets on a subject agreed with Leslie Gunston and Olwen Joergens, reading it to Sassoon a fortnight later:

> S. has written two or three pieces 'around' chance things I have mentioned or related! Thus the enclosed scribble is a copy of what he wrote after I had read three sonnets on 'Beauty' (subject) by E. L. G., O. A. J. and me.[24]

The 'Sonnet to Beauty' (also entitled 'The Hour of Youth') was later revised to become 'My Shy Hand', in which 'Beauty' rather than the poet is the speaker:

> My shy hand shades a hermitage apart, –
>   O large enough for thee, and thy brief hours.
> Life there is sweeter held than in God's heart,
>   Stiller than in the heavens of hollow flowers.
>
> The wine is gladder there than in gold bowls.
>   And Time shall not drain thence, nor trouble spill.
> Sources between my fingers feed all souls,
>   Where thou mayest cool thy lips, and draw thy fill.
>
> Five cushions hath my hand, for reveries;
>   And one deep pillow for thy brow's fatigues;
> Languor of June all winterlong, and ease
>   For ever from the vain untravelled leagues.
>
> Thither your years may gather in from storm,
> And Love, that sleepeth there, will keep thee warm.

The poem which Sassoon wrote after hearing the first draft of this is clearly 'Vision' (first entitled 'Poets'), with its echo of Owen's 'heavens of hollow flowers':

Men with enchanted faces, who are these,
Following the birds and voices of the breeze?
Men who desire no longer to be wise,
And bear eternal forests in their eyes.

They are all singing of beauty; yet their dreams
Are mute amid that silence hung with green;
Silence of drifting clouds with towering beams
Dazzling the gloom, silence on earth serene.

No song beyond that archway of the hours,
But beauty breaking in a heaven of flowers,
And everywhere the whispering of trees . . .
Men with triumphant faces, who are these?[25]

In turn, Owen's 'Six O'clock in Princes Street' seems to have been
written 'around' this (although it has more obvious similarities to
Yeats's 'When you are old'):

In twos and threes, they have not far to roam,
    Crowds that thread eastward, gay of eyes;
Those seek no further than their quiet home,
    Wives, walking westward, slow and wise.

Neither should I go fooling over clouds,
    Following gleams unsafe, untrue,
And tiring after beauty through star-crowds,
    Dared I go side by side with you;

Or be you in the gutter where you stand,
    Pale rain-flawed phantom of the place,
With news of all the nations in your hand,
    And all their sorrows in your face.

These interconnected 'scribbles' show the dilemma which Owen
and Sassoon felt themselves to be in. On the one hand, poets had
to 'follow the gleam', like Tennyson's Merlin, the enchanted
enchanter;[26] the 'triumph' which troubled their faces was that of
the vision of beauty. On the other hand, the miserable face of
the Edinburgh newsboy, symbolic of the winter of the present
world, was a reminder that 'tiring after beauty' through a 'cobweb
of stars' might be no more than 'fooling'. The true poet's desire

should be for wisdom and the spreading of wisdom; the sorrows of the nations needed better interpreters than the newspapers would provide. It might have been all very well for Yeats to hide amid a crowd of stars but the soldier poet in 1917 had to be, as it were, 'in the gutter'. Sassoon and Owen knew that beauty and truth were to be found in the faces of the troops, as Owen had begun to say already in 'Has your soul sipped', but the difficulty was to say it in print without allowing civilians to suppose that the troops were contented (or the poets perverted). To write angry satires against civilians and priests was in a sense an evasion of the poet's duty to record his true 'vision' of 'triumph'. Somehow these two tasks had to be brought together. Much of Owen's poetic effort in the bare year that was left to him after Craiglockhart was to go into fusing his lyrical, Romantic bent with his indignation at the war.

Three of his most fertile and deeply felt Craiglockhart poems show him beginning to combine his perception of war with his earlier poetry. 'Anthem for Doomed Youth', 'Disabled' and 'Dulce et Decorum Est' have their roots in his pre-1917 life and writing. All three demonstrate his Romantic inheritance although he failed to put it to full use in 'Anthem'; all three develop the theme of lost youth which he had been exploring in his lyrics before his shellshock. The last stanza of Sassoon's 'Prelude: The Troops', with its lament for the 'unreturning army that was youth', would have shown him how the subjects of 'The Unreturning' and 'Happiness' could be applied impersonally to war and soldiers, enabling him to build on his regret for the lost joys of boyhood and to work towards regaining and strengthening the maturity which he knew Beaumont Hamel had brought him.

'Anthem for Doomed Youth' was completed by 25 September after advice from Sassoon, who recognised for the first time that Owen's talent was out of the ordinary. The differences between the first draft and the last (there were at least seven drafts) show how Owen began to bring his lyrical writing into step with his opinions about the war. The result was a sonorous but rather confused poem; there was still much to get straight, but he had made a start:

> What passing-bells for these who die as cattle?
> – Only the monstrous anger of the guns.

> Only the stuttering rifles' rapid rattle
> Can patter out their hasty orisons.
> No mockeries now for them; no prayer nor bells;
>   Nor any voice of mourning save the choirs, –
> The shrill, demented choirs of wailing shells;
>   And bugles calling for them from sad shires.
>
> What candles may be held to speed them all?
>   Not in the hands of boys but in their eyes
> Shall shine the holy glimmers of good-byes.
> The pallor of girls' brows shall be their pall;
>   Their flowers the tenderness of patient minds,
>   And each slow dusk a drawing-down of blinds.

This sonnet has come in for some sharp criticism as a relapse into Owen's youthful Romanticism and as an unintentional glorification of death in war.[27] The first point seems to me misleading. The poem's language is certainly Keatsian ('Then in a *wail*ful *choir* the small gnats *mourn*'[28]) but the allusions are meant to be noticed, revealing the battlefield as a demented parody of the Romantic landscape. The second point, that 'Anthem' has a sanctifying effect on its subject, is more accurate, except that the first draft (Plate 11) shows that originally this effect was by no means unintended. The draft begins by asking, 'What minute bells for these who die so fast?' Answer: 'Only the monstrous/ solemn anger of our guns'. The only possible response to the slaughter (of British troops) is that 'our' guns should hurl angry 'insults' at the enemy. The poet explains his function, which is to arouse grief and remembrance, by saying that 'I will light' the candles which will shine in boys' eyes (meaning that his poems will produce tears in the eyes of soldiers' sons or younger brothers). The funeral ceremonies will not be church services, since battle conditions prevent such things, but the majestic 'requiem' of British artillery in France and the sadness of bereaved families in England. Whether such rites are adequate or not, the poem does not say; they are the only possible ones, that is all.

Sassoon would have seen the first draft's shortcomings at once. Even if Owen did not mean it to be a statement in support of the British war effort, it could be used in that way. It was uncomfortably close to popular war poems such as Laurence Binyon's 'For the Fallen' or Beatrix Brice's 'To the Vanguard'.[29] So Sassoon cancelled 'solemn' in favour of 'monstrous' and changed 'our guns' and 'majestic insults' to 'the guns' and 'blind insolence'. Owen followed these pointers in subsequent drafts, removing the anti-German and sanctifying elements from the octave by making shells 'demented', introducing the 'patter' of rifles (the word derives from the meaningless repetition of paternosters) and describing doomed youth as 'cattle' for whom any rites would be 'mockeries'. These changes do not fully conceal the tone of the first draft, but they might have been enough if the sestet had been different. If Sassoon sensed that the last six lines were unsatisfactory, he would not have known Owen well enough to see what was wrong. The difficult transition from battlefields to home is admirably managed by means of bugles, which were familiar in both places (memorial services in 'sad shires' often ended with a bugle call). But the strongest objection to 'Anthem' is that its sestet betrays Owen's hard-won maturity by slipping back into the nostalgia that he had expressed in 'A New Heaven' before he had seen the trenches. The sestets of both sonnets propose that dead soldiers can find immortality in the memory and affection of their families. The first draft of 'Anthem' even says that the men's wreaths will be 'Women's wide-spread arms', but after Beaumont Hamel Owen had said in 'Happiness' that he had gone *beyond* 'the scope / Of mother-arms' (or of 'the wide arms of trees'). It is not difficult to sense the presence of Colin, Mary and Mrs Owen among the weeping boys, pale girls and patient minds in the final sestet of 'Anthem', but there could be no return even in death. The hardest thing of all for a soldier to accept was that even his own family would not understand or remember; Owen was to come to terms with this after Craiglockhart.

It is unjust to treat 'Anthem' as a 'late' poem on a level with 'Strange Meeting' or 'Insensibility', as some critics have done. It should be seen as Owen's first attempt to bring his own style into line with the views he was learning from Sassoon. Unlike his more obviously Sassoonish poems, 'Anthem' draws extensively on what he had heard and read in Bordeaux as a way of resisting his friend's overwhelming stylistic influence. The elegiac tone, elaborate sound-patterns and elegant metaphors are more Tailhadesque than Sassoonish. Phrases in the first draft – 'solemn', '[priest-words] requiem of their [burials]', 'choristers and holy music', 'voice of mourning', 'many [candles shine]' – seem to derive from his account of a French funeral service in 1914:

> The gloom, the incense, the draperies, the shine of many candles, the images and ornaments, were what may be got

anywhere in England; but the solemn voices of the priests was
what I had never heard before. The melancholy of a bass
voice, mourning, now alone, now in company with other voices
or with music, was altogether fine; as fine as the Nightingale –
(bird or poem).

A Craiglockhart fragment which contains material used for
'Anthem' experiments with imagery of 'deep' artillery and
'wailing . . . high' shells, like bass and alto voices. This fragment
also shows how Owen reached the metaphor of steady artillery
fire as funeral bells: the 'measured [smiting]' of the guns' 'iron
mouths' suggested the regular beat of minute or passing bells,
rung once a minute for every year of the dead man's life. The
bells also derive from the phrase 'Pacific lamentations of slow
bells' in another Craiglockhart fragment, from which they can
be traced back through 'A Palinode' (1915) to the 'lamentation
pacifique' of the angelus in *Madame Bovary*. Further connections
with Owen's stay in Bordeaux can be seen in the echo from his
1914 Decadent poem 'Long ages past ' ('on thy brow the pallor
of their death') and in the similarity between the bright-eyed
boys and candle-bearing acolytes such as he had seen at Mérignac
and read about in poems by Tailhade and others. These associations with the French Decadence hint at sinister meanings
behind the poem. Music, beauty and love have become deadly
mockeries of themselves. The girls may be *femmes fatales*, luring
youth to its doom. The domestic funeral rites may be as much a
parody of true religious observance as the rites on the battlefield.
These implications are not brought out clearly (most modern
readers will probably not recognise them at all) but they
are indications of how Owen would reconcile his Romantic
inheritance with his understanding of the war.[30]
    The possibilities of applying literary tradition to a war that
seemed to break with all tradition were taken much further in
Owen's next poem, 'Disabled'. Sassoon was beginning to be
impressed, making an unsuccessful effort to get 'Anthem' into the
Nation in October[31] and showing 'Disabled' to Graves, who came
up on a visit in that month. Graves was even more impressed,
writing to Owen soon afterwards with encouragement and some
typically Georgian advice. 'Disabled' was Owen's first thoroughly
original war poem, bringing his new skills and knowledge into
active partnership with his pre-Sassoon verse. As has been
suggested, the poem grew from several related 1916–17 sources

(see above, pp.79–82), including 'Lines to a Beauty seen in Limehouse'. The wounded man can be recognised as the East End 'Beauty', redrawn, disfigured, and perhaps containing elements of the poet himself. Whether or not Sassoon knew of the poem's origins, he would certainly have sympathised with its implications.

The admiring poet had gazed in vain at the Limehouse youth's handsome face and bare knees (an odd detail, but appropriate to a sculptured 'idol'), knowing that his tribute was ignored and that the 'half-god' would spend the night with someone else, presumably a woman. In 'Disabled', the young soldier and former footballer has had 'an artist silly for his face' and has been idolised by women; someone had told him 'he'd look a god in kilts', no doubt having seen him in football shorts. Now he is dressed in a grey parody of sports clothes, with short trousers and no sleeves because he is limbless. 'Voices of boys' are heard in the park – the next generation is perhaps playing football in its turn – until 'gathering sleep' 'mothers' them away from him. For him, sleep and the old happiness are unreturning. The greyness of his hospital suit and of the twilight are contrasted with the 'purple' of his spilt blood, Owen's favourite colour here carrying Romantic, Tailhadesque implications of youthful sexuality and sacrifice. The Limehouse 'Beauty' was unwise to turn to women. Instead of receiving the poet's blood sacrifice, he has become one himself; he has been drained of his blood and potency because he listened to 'the giddy jilts', the Fatal Women who persuaded him to volunteer. The 'hot *race* / And *leap* of purple' which 'spurted from his thigh' was for them an erotic climax (and for him a final athletic triumph); now they turn from him 'to the strong men that [are] whole'. The poem can be read both as a bitter comment on the role of women in wartime and as a study in late-Romantic themes.

'Disabled' contains several literary allusions. The ironic parallel between the soldier's past and present states is now generally agreed to be taken directly from Housman's 'To an Athlete Dying Young'. Another late-nineteenth-century influence is apparent in the Swinburnian juxtaposition of purple (sexual passion) and grey (impotence and repression). There is a less predictable allusion to the classical legend of Adonis, another beautiful youth who was wounded in the thigh – a sexual maiming – while absenting himself from his mistress in pursuit of sport. In Andrew Lang's translation of Bion's lament for Adonis, which Owen bought in December 1917 and might have glanced at before, 'the

dark blood *leapt* forth', leaving Adonis's chest *'purple* with blood'.[32]
It would be consistent with later poems if these associations were
meant to be observed. Readers of Owen's generation might also
have seen ironic relationships between the poem and recruiting-
posters such as those which showed sweethearts urging young
men to enlist or Mr Punch advising a footballer to become a
player in the 'Greater Game' in France. In making its comment
on war, the poem combines the immediacy of Sassoonish satire
with Romantic and Greek material of the kind Owen had used
in 'Perseus' and other work before 1917. At the end of the poem,
the dusk is the Decadent last evening, but the repeated question,
'Why don't they come?', sounds like an echo of a poster which
showed soldiers in need of reinforcements under the slogan, 'Will
they *never* come?' It was too late for anyone to come; the hour of
youth was unreturning.

The bitterness against women shown in 'Disabled' was
continued into 'Dulce et Decorum Est' in mid October. Again,
Owen drew on both literary tradition and contemporary propaganda. 'Dulce et Decorum Est' was originally drafted as a
'counter-attack' against the recruiting-verses of Miss Jessie Pope,
'a certain Poetess' whose doggerel was frequently published in
the right-wing press. More widely, the attack is against civilian
heroic notions in general. Many patriotic versifiers had quoted
the Latin tag or used it as a title; Owen himself had alluded to
it approvingly in 'The Ballad of Peace and War' some years
before. Now, after Beaumont Hamel, the proposition that death
for the fatherland was sweet and decorous seemed nauseating, so
he answered it with a correspondingly revolting 'trench life
sketch', perhaps thinking of Sassoon's most recent work in which
the horrors of war were being described more vividly than in *The
Old Huntsman*. There is a controlled and powerful anger in 'Dulce
et Decorum Est' which for some readers will be the poem's most
valuable quality. The control is partly achieved by a tight formal
discipline. The first half of the poem is a sonnet in all but its final
rhymes, but at the point at which one expects a final couplet
there is instead the first half of a quatrain, so that the sonnet does
not end but instead makes the reader pause in anticipation:

> Dim, through the misty panes and thick green light,
> As under a green sea, I saw him drowning.

The second half of the quatrain comes after a space as an isolated

pair of lines, repeating the image of drowning but placing it now
not on the battlefield but in dreams:

> In all my dreams, before my helpless sight,
> He plunges at me, guttering, choking, drowning.

These two lines begin another group of fourteen, but this second
half of the poem bears no resemblance to a sonnet and is held
together only by its rhymes. The organisation and clarity of the
first half is replaced by confused, choking syntax and a vocabulary
of sickness and disgust, matching the nightmare which is in
progress. Owen is directly facing the central experience of his war
dreams, the sight of a horrifying face which, Gorgon-like, renders
him a 'helpless', paralysed spectator. Some lines which were
eventually cancelled tell the reader to think, as the poet thinks,
of how the dying man's head was once 'like a bud, / Fresh as a
country rose, and keen, and young', an image which has sometimes
been regarded as a Romantic intrusion into the poem. The
Romanticism is far from irrelevant, however. The soldier had
been another beautiful youth but now his bud-like face is
compared to that of a devil sickened by 'sin'. Another draft line
mentions 'vile incurable sores on sin-kissed tongues'; 'sin-kissed'
became 'young corrupted' and eventually 'innocent', a word
which makes less sense than those which preceded it, removing
what was presumably a reference to venereal disease.[33] These
traces of sexual beauty and 'sin' in the poem are puzzling in
isolation and Owen was right to reduce them, but they show how
his nightmares worked. Once again, beauty and horror meet in
the damned and sated figure who is both Gorgon and victim,
destroyer and destroyed. The 'white eyes writhing' ('twinging' in
one draft) are a version of the sufferers' eyes in 'Purgatorial
passions' which 'twinged and twitched'; the simile of the bud
comes from 'The cultivated Rose' ('Contused with sap of life like
a bursting bud'); the submarine imagery ('flound'ring . . . green
sea . . . drowning') is familiar. The Dunsden 'phantasy' has
reshaped itself in a wartime setting so that the public statement
of the poem is given its force not simply by Owen's anger against
Miss Pope but by the pressing intensity of his private, ever-
recurring vision (the personal and traditional meanings of the
haunting face and its petrifying stare have been discussed in
previous chapters).[34] Owen's true courage is evident in his
exposing his imagination to war experience as material for poetry;

there is nothing comparable in Sassoon's work except perhaps 'Repression of War Experience', which deals only with waking horror, and 'Haunted', a poem clearly based on a shellshock nightmare but making no direct reference to war.

Rivers's steady work on his patient met with some success. On 24 October Graves wrote to Edmund Gosse that Sassoon had not modified his views about the war but was now applying to be sent back to France, to show that he was not afraid and to share the suffering of the troops.[35] This decision, reached after an intense mental struggle, put Sassoon in a position which many of his modern readers find puzzlingly hard to excuse. The alternatives which he had faced were either to maintain his protest, which was certain to become increasingly a matter of theory and public debate, or to recognise the intense loyalties which bound a serving officer to his men. Most poets who accepted the demands of comradeship felt that protesting was likely to do more harm than good. Sassoon attempted the compromise of returning to active service while remaining opposed to the war as a poet; in doing so, he pleased neither side. Graves said in his letter to Gosse that Sassoon

> thinks he is best employed by writing poems which will make people find the war so hateful that they'll stop it at once at whatever cost. I don't. I think that I'll do more good by keeping up my brother soldiers' morale as far as I can.

On the other side, Russell, Lady Ottoline and the *Cambridge Magazine* were deeply disappointed by Sassoon's decision. Owen tried to dissuade his friend but felt much the same conflict in himself; in the end he made the same decision as Sassoon, directly influenced by his example.

By the time he left Craiglockhart at the end of October, Owen had probably learned as much as he could from Sassoon. His last poem there, 'Soldier's Dream', was little more than a copy of ' "They" ' and other epigrams in *The Old Huntsman*. It was time to move on. But his debt as both soldier and poet to his friend was immense. Sassoon taught him to think for himself about the war and its politics. He taught him how to use experience in poetry and how to be true to it; indeed, like Harold Monro and Graves, he taught him how to be a Georgian. He introduced him to Hardy, Barbusse and other writers. As a later chapter suggests,

1. 'There was born . . . My poethood' : the woods in spring at Broxton, Cheshire.

2. 'As Perseus fearfully beheld the form / Of Gorgon, mirrored in the stilly well.'   Edward Burne-Jones, *The Baleful Head* (detail).

3. A 'most dervishy vertigo' : a self-impression of the lay assistant, Dunsden Vicarage (16 November 1912).

4. Owen, Mme Léger and Nénette (front row, second, third and fourth from the left) at one of Tailhade's lectures at the Casino, Bagnères-de-Bigorre (August 1914).

5. Laurent Tailhade introducing Owen to Flaubert in the villa garden at La Gailleste (September 1914).

6. 'By Hermes, I will fly . . .' : a statuette of Hermes bought by Owen, probably in Bordeaux in 1915, and later given to Leslie Gunston.

7. Second Lieutenant WES Owen, Manchester Regiment (1916):
one of a set of portraits taken by his uncle, John Gunston, to mark
his commission.

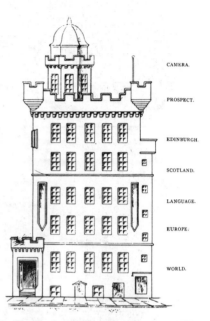

CAMERA.

PROSPECT.

EDINBURGH.

SCOTLAND.

LANGUAGE.

EUROPE.

WORLD.

8. Captain A J Brock, RAMC, probably in his room at Craiglockhart War Hospital, 1916–17.

9. Outlook Tower, Castle Hill, Edinburgh : a diagram taken from Patrick Geddes, *Cities in Evolution,* 1915.

10. 'Shell Shock!' : a cartoon from *The Hydra* (December 1917).

Anthem for Dead Youth.

What passing bells for those who die so fast?
— Only the (monstrous anger of the guns.
Let the blind insolence of their mouths
Be as the priest words of their burials.
Of choristers and holy music, none;
Nor any voice of mourning, save the wail
The long-drawn wail of high far-sailing shells.

What candles may we hold for those lost souls?
— Not in the hands of boys, but in their eyes
Shall many candles shine; and sunlight light them.
And Women's wide-spread arms shall be their wreaths,
And pallor of girls' cheeks shall be their palls.
Their flowers, the tenderness of mortal minds.
And every Dusk, a drawing-down of blinds.

First Draft
(With Sassoon's amendments.)

11. The first draft of 'Anthem for Doomed Youth' (September
1917).

ENLIST TO-DAY.

HE'S
HAPPY &
SATISFIED

ARE YOU ?

12. 'When I behold eyes blinded in my stead!' : 'Blinded for You!', a war-time postcard, sold in aid of the National Institute for the Blind.

13. 'Happy the man who...' : a recruiting poster.

Happy With His Vanity Bag Hat.

14. 'Pictures of these broad smiles appear in Sketches / And people say : They're happy now, poor wretches' : a photograph in the *Daily Sketch* (16 September 1918) showing 'Happy wounded Canadians at a casualty station'.

15. Taken at Hastings on the eve of embarkation (30 August 1918) :
'such a badly printed photograph...but looking at a little distance
it is *him*...'.

he is likely to have talked about homosexuality. Above all, he and Owen shared their poetic ideals as their poems show, discussing not only the monstrosity of the war but also the lyrical impulse, the poet's vision and what Sassoon had described to Nicholson as 'the quest for beauty and compassion and friendship'. They were followers of 'the gleam', a difficult role in wartime. In the end, it led them back to France, which may have been why Owen wrote to Sassoon in November 1917, parodying Tennyson:

> I am Owen; and I am dying.
> I am Wilfred; and I follow the Gleam.

Earlier in the same month he told Sassoon,

> you have *fixed* my Life – however short. You did not light me: I was always a mad comet; but you have fixed me. I spun round you a satellite for a month, but I shall swing out soon, a dark star in the orbit where you will blaze.

There are echoes here of 'following gleams' in 'Six O'clock in Princes Street' and of the image of the poet as meteor or errant star (no longer 'lawless') in 'O World of many worlds'. The references to dying and short life imply that Owen had already decided he would have to go back to war, but his youthful Romanticism had always required him to expect an early death. As in 'Storm', the flash of illumination may be fatal but it will make men 'cry aloud and start'. He said in September that some of his poems might 'light the darkness of the world', echoing the first draft of 'Anthem' in which he undertook to 'light' the many candles of grief. He compared his poetry more modestly to a single candle in 1918, saying when Sassoon was invalided home that he would throw his 'little candle' on Sassoon's torch and return to the front.[36]

Sassoon's 'torch' blazed with anger and protest but the anger sprang from his compassion for his men and from his vision of their beauty. In his own and Owen's opinion, his truest vein as a war poet was lyrical and elegiac. If the final stanza of 'Prelude: The Troops' was the piece which Owen described in September as 'the most exquisitely painful war poem of any language or time' (one observes the Romantic words 'exquisite' and 'pain'), it may be taken as an example of Sassoon's writing 'from the heart' as well as from experience ('I write straight from my experiences and heart', Owen said soon after meeting him, describing a letter home[37]). Its echo of Charles Sorley's 'pale battalions' of 'the mouthless dead' places its author with Sorley and Graves. In imagery and diction the stanza is less original than their best wartime work, but it lacks their shield of stoicism. Sassoon's emotion, somehow guaranteed even by the faults in the verse, stands unprotected, movingly exposed:

> O my brave brown companions, when your souls
> Flock silently away, and the eyeless dead
> Shame the wild beast of battle on the ridge,
> Death will stand grieving in that field of war
> Since your unvanquished hardihood is spent.
> And through some mooned Valhalla there will pass
> Battalions and battalions, scarred from hell;
> The unreturning army that was youth;
> The legions who have suffered and are dust.

7 New Influences: Georgians and Others

After Owen was discharged from Craiglockhart on 30 October 1917, he devoted all his spare energies to poetry. Brock had taught him to work, Sassoon had given him purpose and confidence. On 3 December, for example, he finished one poem, drafted three more and determined to get up early next day to 'do a dawn piece'. By the end of the month he was ready 'to revise now, rather than keep piling up "first drafts" ', wishing he could give his 'art' the six hours a day it needed. In February he remembered the Broxton bluebells which had 'fitted me for my job'. That job involved the risk of shellshock nightmares: 'I confess I *bring on* what few war dreams I now have, entirely by *willingly* considering war of an evening. I do so because I have my duty to perform towards War.'[1] In order to define his task he read writers whom Sassoon admired, looking for guidance in both style and ideas from, among others, the Georgians, Barbusse and, almost certainly, Hardy and Russell. Now that there was good reason to bring his 'art' into the immediate service of the age, it was more than time to take account of contemporary writers.

The first moderns to look at were obviously the Georgians, since he already knew three of them personally and had read Brooke. His new posting, on light duties at Scarborough, allowed time for visits to London and the Poetry Bookshop, where he was pleased to find himself remembered. He bought Monro's latest collection, *Strange Meetings* (1917), with its interesting title, and *Georgian Poetry 1916–1917*. This new volume of the anthology, published by the Bookshop in November, included work by Sassoon, Graves, Monro, Robert Nichols, John Masefield, W. W. Gibson, Walter de la Mare and John Drinkwater. Owen eventually possessed at least fifteen volumes by these Georgians and their original leader, Brooke; this was by far the largest representation of modern verse in his shelves, and most of it was

bought and read in November–December 1917. The lessons to
be learned from the Georgians were straightforward. They
all respected Brooke's pre-war language – 'musical, restrained,
refined, and not crabbed or conventionally antique, reading
almost like ordinary speech', as Graves had described it in 1916.[2]
Nineteenth-century diction, vagueness and insincerity were to
be rooted out, and satirised if necessary; realism, technical
accuracy, and fidelity to experience were the modern goals in
composition, and the lifestyle that went with them was expected
to be perhaps unconventional but always plain, sincere and
courageous. One can see the Georgian spirit at work in Owen's
little epigram 'Schoolmistress' (1918), in which a teacher who
has just read the Victorian epic 'Horatius' refuses to acknowledge
the greeting of a modern soldier named 'Orace. The gibe is
against civilian and Victorian double standards, yet by calling
his soldier Horace Owen identifies him as a modern hero.[3]
Similarly, Graves's 'The Legion', which Owen warmly admired
in November, compares Tommies to Romans driving the
barbarians out of Gaul yet makes the Romans sound like ordinary
modern soldiers rather than 'conventionally antique' heroes.

Owen was in the Bookshop in November when copies of
Graves's new book, *Fairies and Fusiliers*, arrived. He wrote 'Asleep'
on the same day, a poem which includes two unintentional echoes
from the book. Graves wrote to him encouragingly, hoping (in
vain as it turned out) to put him in touch with Robert Nichols:
'You must help S. S. and R. N. and R. G. to revolutionise English
Poetry – So outlive this War.' Edward Marsh was told that 'I
have a new poet for you, just discovered, one Wilfred Owen . . .
the real thing; when we've educated him a trifle more. R. N. and
S. S and myself are doing it.' Later, perhaps early in January,
Graves sent Marsh 'the few poems of Owen I can find: not his
best but they show his powers and deficiencies – Too Sassoonish
in places: Sassons is to him a god of the highest rank.' This letter
is marked 'Please return enclosures' without any indication of
what they were. Presumably Graves hoped that Marsh would
accept some of Owen's work for the next *Georgian Poetry*, but
Marsh must have shown his usual caution because Graves wrote
again, 'Owen, I told you, is fearfully uncertain: but he can see
and feel, and the rest will be added unto him in time.' Part of
the 'education' which Graves offered thus seems to have been
advice against being excessively 'Sassoonish'; from January
onward, that quality became less marked in Owen's verse.[4]

Graves also advised him in December to 'write more optimistically . . . a poet should have a spirit above wars', apparently in response to 'Wild with all Regrets', which Owen had written earlier in the month and dedicated to Sassoon. This may have prompted Owen to take up a poem provisionally entitled 'The Unsaid' and give it the new title of 'Apologia lectorem pro Poema Disconsolatia Mea' ('A defence to the reader of my disconsolate poem'), adding to the manuscript a half-remembered quotation from Barbusse's *Under Fire*: 'If there be a bright side to war, it is a crime to exhibit it.'[5] Whereas Graves wrote for soldiers, as he had told Gosse, Owen still had a civilian audience in mind; 'Apologia' explains that an 'optimistic' account of fighting might lead civilians to believe that soldiers were content in the trenches. That this could happen is evident from plenty of civilian verse, including, for example, a poem in John Oxenham's *The Vision Splendid* (1917), a book Owen had read at Craiglockhart, in which a civilian asks, 'What did you see out there, my lad?' and a Tommy answers,

> I have seen Christ doing Christly deeds . . .
> I have sped through hells of fiery hail
> With fell red-fury shod;
> I have heard the whisper of a voice;
> I have looked in the face of God.

To which the civilian thanks God for His grace and says the soldier has a right to his 'deep, high look'. 'I, too, saw God through mud', Owen admitted in 'Apologia' – but he would have included Oxenham among the complacent civilians who are rebuked at the end of the poem. He was not sure that 'the Unsaid' should be said, but he felt as well qualified as Graves to lay claim to a 'spirit above wars', remembering the 'exultation' he had felt in No Man's Land: 'I, too, have . . . sailed my spirit light and clear / Past the entanglement . . . And witnessed exultation'. To emphasise the point in his 'apology' to Graves, he borrowed imagery from 'Two Fusiliers', a poem in *Fairies and Fusiliers* which describes how Graves and Sassoon had found love at the front and 'Beauty in Death'. Sassoon had testified in his dreamier way to similar insights in 'The Triumph', while Nichols, the third of the Georgian trio, told Mrs Owen after the war that his own poem 'The Secret' was an 'exact parallel' to 'Apologia'.[6]

'Apologia' denies that beauty and love are as 'old song' had described them; they are now to be found in the experience and comradeship of battle. This seems to be a Georgian rejection of nineteenth-century poetry, but the critic has to be careful here. The poet does not reject Romantic values such as beauty and truth but gives them a new location. The manuscripts show Owen trying out some of the Romantic and Decadent phrasing he had used in erotic verse before meeting Sassoon: 'Sweet ran our sweat' (the pararhyme is from 'Has your soul sipped'); '[glorious lovely] ecstatic [seemed was] the [crimson] purple of our murder'; '[You shall not jest] with them whose wine is fate.'

> Unmeet for you is their mirth
>   By your safe hearths. Close lightning keeps them warm.
>   Only to gods and men grown sick of earth
> Hilarious sound the thunders of this storm.[7]

There are traces of 'Perseus' here, including the rape of Danae and its parallel in 'Storm' ('those hilarious thunders of my fall'). He had recognised in 'The cultivated Rose' that beauty could be found in hell. He did not use this pre-1917 material in drafts of 'Apologia' in order to make some ironic contrast with his earlier verse; its effect is rather to separate the soldiers' knowledge from the civilian's in much the same way that the Romantic poets often saw themselves as possessing a mystery and suffering to which ordinary people had no access. The 'men grown sick of earth' are now soldiers but in literary tradition they had been poets, 'the wise' whose secret knowledge of beauty gave them the power to laugh like gods, their spirits 'floating' far above worldly torment. Owen was to develop the parallel between soldiers and poets in some of his 1918 work.

His new Georgian allegiance led him to reinterpret rather than to reject Romanticism but it did prevent him from approving Leslie Gunston's first and only book. It was bad luck for Gunston that *The Nymph, and Other Poems* came out only a week or two after *Fairies and Fusiliers*. Owen thought the book premature, remarking that it contained too much fictitious kissing. 'I think every poem, and every figure of speech should be a *matter of experience*.' 'Nothing great was said of anything but a definite experience.'[8] He was echoing a new master and an old, not only Sassoon but also Keats, who resolved 'never to write for the sake of writing or making a poem, but from running over with any little knowledge or experience which many years of reflection may perhaps give me'.[9] Owen's 1918 claim that he was 'not concerned with Poetry' was an affirmation of Georgian, Keatsian and Geddesian values, and a dismissal of his own and his cousin's old habit of writing for the sake of writing. Although he never ceased to use his knowledge of Aestheticism, he had no more time for notions of art for art's sake, especially when they resulted in the sort of 'Poetry' that was still appearing in great quantities in the *Poetry Review* and other publications from the house of 'Erskine Macdonald'. There would be no more sonnets on agreed subjects. 'They believe in me, these Georgians', he told his cousin, and referred proudly to 'We Georgians' in January. Gunston was warned not to become a lagoon left by 'the ebbing tide of the Victorian Age' but was doubtful about *Fairies and Fusiliers* and Owen's curious new rhymes. 'Graves's technique is perfect,' Owen replied, and 'I suppose I am doing in poetry what the advanced composers are doing in music.'[10] This determination to be modern and 'advanced' without abandoning accuracy of technique was typical of his Georgian ambitions in the winter of 1917–18. In contradicting 'old song' in 'Apologia', Owen was referring in particular to Gunston's poem 'L'Amour', which had said that 'Love is the binding . . . of lips, the binding of eyes'.[11] 'Apologia' denies this: 'love is not the binding of fair lips / With the soft silk of eyes' as is told in 'old song' but the comradeship of soldiers, 'wound with war's hard wire'. The image from 'L'Amour' is set aside in favour of one taken from Graves's 'Two Fusiliers'. This was not only a literary gesture but also a sign that Owen still thought Gunston ought to be in uniform. But he continued to feel his old affection for his cousin, whose friendship and encouragement had been valuable to his poethood.

Owen's determination not to become a Victorian 'lagoon' himself is apparent in 'Wild with all Regrets' (5 December, later redrafted as 'A Terre'), an unmistakably Georgian poem, its title an ironic reference to Tennyson[12] and its form and style deriving from Masefield by way of Sassoon. The model for this and other Georgian monologues was Masefield's *The Everlasting Mercy* (1911), which had been famous before the war as an example of contemporary realism; Owen may never have read the original but he owned a copy of Sassoon's part-parody, part-imitation of it, *The Daffodil Murderer*.[13] He had marked several pleasing passages in Sassoon's poem, such as:

> I thought, 'When me and Bill are deaders,
> 'There'll still be buttercups in medders . . . .

He used his kind of verse in a number of Sassoonish poems, including 'The Chances':

> 'Ah well,' says Jimmy, and he's seen some scrappin',
> 'There ain't no more than five things as can happen. . . .

But Sassoon had smoothed Masefield's style, going on to tame it still further in the title poem of *The Old Huntsman*, another monologue by a man facing death:

> What a grand thing 'twould be if I could go
> Back to the kennels now and take my hounds
> For summer exercise; be riding out
> With forty couple when the quiet skies
> Are streaked with sunrise, and the silly birds
> Grown hoarse with singing; cobwebs on the furze
> Up on the hill, and all the country strange,
> With no one stirring; and the horses fresh,
> Sniffing the air I'll never breathe again.
>   . . . . .
> You've brought the lamp, then, Martha? I've no mind
> For newspaper tonight, nor bread and cheese.
> Give me the candle, and I'll get to bed.

'Wild with all Regrets' is similar in structure and in some of its subject matter but Sassoon's even tone has been roughened again:

> We said we'd hate to grow dead-old. But now,
> Not to live old seems awful: not to renew
> My boyhood with my boys, and teach 'em hitting,
> Shooting and hunting, – all the arts of hurting.
> – Well, that's what I learnt, – that, and making money.
> Your fifty years in store seem none too many,
> But I've five minutes. God! For just two years
> To help myself to this good air of yours!
> One Spring! Is one too hard to spare? Too long?
> Spring air would find its own way to my lung,
> And grow me legs as quick as lilac-shoots.
>   *   *   *
> Yes, there's the orderly. He'll change the sheets
> When I'm lugged out. Oh, couldn't I do that?

'This "Wild with all Regrets" was begun & ended two days ago, at one gasp', Owen told Sassoon. 'If simplicity, if imaginativeness, if sympathy, if resonance of vowels, make poetry I have not succeeded. But if you say "Here is poetry", it will be so for me. What do you think of my Vowel-rime stunt . . . ?"[14] Pararhyme was now not a device for Decadent lyrics but a progressive experiment, a 'stunt' akin to 'what the advanced composers are doing in music'; the harshness and lack of conventional 'beauty' in the poem are deliberate attempts at a modernity which was in fact more extreme than most Georgians would have liked. The once-celebrated ruggedness of *The Everlasting Mercy* seems a tame affair by comparison.

By the end of 1917 Owen felt himself to be a Georgian and a fully fledged poet:

> I go out of this year a Poet, my dear Mother, as which I did not enter it. I am held peer by the Georgians; I am a poet's poet.
> I am started. The tugs have left me; I feel the great swelling of the open sea taking my galleon.[15]

It was a good Masefield-like metaphor and an exciting moment. In the same letter he said he had felt 'sympathy for the oppressed always', which was probably true although it had not often been evident in his verse before Craiglockhart. He had said in a school essay ten years earlier that the Romantics had brought 'a new sympathy with man especially the poor' into literature, and he had found the same quality in Sassoon, Monro and other Georgians. It was not enough to be 'a poet's poet'; like all true poets he was needed as a spokesman. One of the last of the galleon's 'tugs' was Brooke's friend W. W. Gibson, whose *Battle* (1915) Owen read in December. The little poems in *Battle* have rarely had credit as the first things of their kind but Gibson was writing 'trench life sketches' sympathetic to the common soldier long before Sassoon, even though he never actually saw the trenches.[16] Owen may have looked at some of his other books. The introductory poem in *Fires* (1912) symbolises Gibson's sudden change a few years before the war from being a conventional poet of dream and fancy to writing about the sufferings of the poor:

> Snug in my easy chair,
> I stirred the fire to flame.
> Fantastically fair,
> The flickering fancies came, . . .

Amber woodland streaming;
Topaz islands dreaming; . . .
Summers, unreturning; . . .
Till, dazzled by the drowsy glare,
I shut my eyes to heat and light;
And saw, in sudden night,
Crouched in the dripping dark,
With steaming shoulders stark,
The man who hews the coal to feed my fire.

When the newspapers reported a disastrous colliery explosion at Halmerend on 12 January 1918, the poem from *Fires* may have provided the framework for Owen's 'Miners', in which another Georgian sits before his fire and is made aware of the fate of miners rather than of the pleasant dreams which he had been expecting. 'Miners' was the first product of the fully launched poet in 1918, but his consciously Georgian phase was already nearing its end.

Owen's reading in the winter of 1917–18 was not confined to the Georgians. There seem to be traces in his later work of the war poems in Hardy's *Moments of Vision*, which came out at the end of November ('The Pity of It', for example, may be compared with Owen's repeated 1918 phrase, 'the pity of war').[17] Sassoon's admiration for Hardy had probably persuaded Owen to read *The Dynasts*, too, that great epic–drama about war and the pity of war. In tracing the movements of the Napoleonic armies, Hardy describes them from above in stage directions: 'The view is from a vague altitude'. Owen seems to take this up in the first line of 'The Show': 'My soul looked down from a vague height, with Death'. His ensuing description of a battlefield where troops 'writhed' and 'crept' like 'caterpillars' is similar to many of Hardy's descriptions, in which armies move like 'caterpillars' ('a dun-piled caterpillar, / Shuffling its length in painful heaves along'; 'The caterpillar shape still creeps laboriously nearer . . . '). The various Spirits which comment on the action of *The Dynasts* watch from the Overworld 'the surface of the perturbed countries, where the peoples . . . are seen writhing, crawling, heaving, and vibrating' in the beautiful but meaningless patterns which seem in themselves the single 'listless aim' of the controlling 'Immanent

Will'.[18] The Will is solely interested in art for art's sake, like the 'everliving' gods in the quotation from Yeats which Owen chose as an epigraph for 'The Show'. The weary gods breathe dreams on the mirror of the world, the Will weaves its patterns – both are exclusively 'concerned with Poetry'. But in *The Dynasts*, as elsewhere in Hardy, the patterns are observed by the Spirit Ironic and the Spirit of the Pities, and it is the latter which has the truest response. There is no lack of irony in Owen's 1918 poems, but Hardy would have confirmed what experience had already suggested, that 'the pity of war' was 'the one thing war distilled'.

In the first draft of 'The Show' the caterpillars are brown, blue, grey and green, the colours of British, French, German and American uniforms, for the detached view with which the poem begins is based on political awareness as well as on an attempt to see events from the position of 'the Elders and the gods, high witnesses of the general slaughter'. The gods on Olympus used to look down and 'smile in secret, looking over wasted lands'[19] according to literary tradition, and the older generation seemed to be doing much the same in 1917–18, but the poet, too, had to be able to separate himself from the conflict and see it as a whole. Owen met H. G. Wells in November, one of the leading writers about the war and its politics, an advocate of internationalism, efficiency, the defeat of militarism by military means, and the need for personal and communal dedication to a new world order. Owen read at least two of his books in December; if he also knew *The World Set Free* he would have found a description of the last war of the world that used Hardyesque imagery: 'If some curious god had chosen to watch ... he would have noticed ... the long bustling caterpillars of cavalry and infantry, the maggot-like waggons ... crawling ....'[20]

The distance between the poet and his subject in 'The Show' soon closes, however, because this is no Wellsian fantasy but a 'true resumption of experienced things', the high viewpoint having been earned by participation in the very battle which is being described. In this and other ways the poem is much closer to Henri Barbusse's *Le Feu* than to anything by Wells, and its preliminary title, 'Vision', was perhaps a deliberate acknowledgment of Owen's debt to Barbusse's first chapter, 'The Vision', an aerial vision of war-torn Europe watched from above by spirits of dead soldiers ('I can see crawling things down there'). At the end of the book, the battle is seen from the ground through the experience of the squad of French soldiers around whom the story

has centred, the author having revealed himself as one of them.
Similarly the poet in 'The Show' begins by looking down from a
height on crawling armies and ends by falling to earth, where he
finds himself at the head of his own platoon. Nothing before *Le
Feu* had given such an appallingly vivid description of trench
warfare or combined it with such passionate political conviction.
The English translation, *Under Fire*, appeared in June 1917 and
Sassoon was reading it by mid August; he lent it to Owen, who
seems to have read it at Craiglockhart and again in December.[21]
Barbusse was fiercely Socialist, presenting his soldiers as victims of
profiteers and war-makers, expounding the 'Difference' between
those who fought and those who grew rich, and stressing the
impossibility of communication from soldier to civilian. The
squad is cut off from its exploiters but forms an intense
comradeship within itself, thereby becoming a Marxist paradigm
of the working class as a whole. The book may well have been
the first Socialist polemic Owen had read; its message would have
been supported to some extent by Sassoon's reluctant (and in the
event temporary) commitment towards the left. However, both
Englishmen would have noticed one thing lacking from *Under
Fire*. Whether Barbusse was portraying reality or his political
convictions, the book gives no evidence of a close relationship
between French officers and other ranks. The mutual affection
and respect between British subalterns and their men described
so often (by officers – not so often by privates) in Great War
literature seems to have been generally regarded as peculiar to
the British Army and a mainstay of its morale; Owen and Sassoon
would have felt that strengthening that bond was one way in
which an officer really could ease his men's ordeal. That was
another reason why complete detachment was neither possible
nor desirable. An officer poet, feeling 'sympathy for the oppressed',
had a double duty to lead and to plead, as Owen accepted in
1918. By contrast, the callousness of civilians seemed beyond
pardon.

'The Show' and similar poems were designed to shock civilians
out of callousness into recognition of actuality, not to arouse pity
but to convey the knowledge which could be pity's foundation.
Owen was to clarify his thoughts about this process in the spring
but in the winter his understanding of 'pity' was only beginning
to take shape. The first stage was to shatter civilian complacency,
so that priority went to writing poems that would be, like
Sassoon's, 'perfectly truthfully descriptive of war'. When he came

to revise 'Wild with all Regrets' into 'A Terre' in April, he referred to it as a 'photographic representation'. This may shed light on the legend that he used to carry photographs of mutilated soldiers with which to shock civilians, a story based on the evidence of a single witness, Frank Nicholson, who met him in Edinburgh. Nicholson recalled that Owen put his hand to his pocket to produce the pictures 'but suddenly thought better of it and refrained'; they talked of pararhyme instead. It is difficult to imagine how a subaltern could have obtained such photographs without falling foul not only of the military censor but also of his fellow soldiers.[22] Nicholson may have been under a misapprehension, since he never actually saw what Owen had in his pocket. The evidence which the poet intended to produce may in fact have been one of his new poems in pararhyme, a 'photographic representation' of war's horrors, possibly even an early draft of 'The Show'. It was through the realism of poems such as these that Owen intended to assault the civilian conscience.

If using shock tactics to make civilians see was a short-term aim, the poet's long-term duty was to bring healing and peace to the world, as Shelley had taught Owen years before. It was inconsistent with Shelleyan ideals that poets should participate in war, but the immediate task of protesting had to be done by soldiers because their experience entitled them to tell the truth with unique authority. However, there was a case for Owen and Sassoon to stay out of all further fighting now that they had earned their right to speak, because violence remained inexcusable and a dead poet was of no use to anybody. Graves told Owen to 'outlive this War'. It seemed likely that a shellshock case would not be sent out again anyway ('I *think* I am marked Permanent Home Service', Owen said in November).[23] But Sassoon had decided in October to apply for active service again. Owen tried to dissuade him, though without much hope; writing to him in November, he commented on Graves's 'Letter to S. S. from Mametz Wood', asking, 'If these tetrameters aren't enough to bring you to your senses, Mad Jack, what can *my* drivel effect to keep you from France?'[24] The 'drivel' referred to here seems to be a poem which stated the argument for poets staying out of the fighting in order to work for peace. There is only one surviving poem which meets this description, the untitled piece beginning 'Earth's wheels run oiled with blood'.

'Earth's wheels' has often been regarded as a part-draft of 'Strange Meeting' but it seems to have been originally composed as a poem in its own right.[25] Owen used the same type of paper and the same strict pararhymed couplets for it as he did for 'Wild with all Regrets' early in December, so that 'Earth's wheels' seems to belong to that period (Sassoon left Craiglockhart on 26 November and reported for duty on 11 December). As in Graves's 'Letter to S. S.', one officer seems to be addressing another: 'We two', 'Let us forgo our rank and seniority', 'Let us break ranks'. The military metaphors are maintained, first coming from trench warfare ('dig') and then taking on biblical characteristics. One of the five drafts is as follows:

> Earth's wheels run oiled with blood. Forget we that.
> Let us lie down and dig ourselves in thought.
>
> Beauty is yours and you have mastery.
> Wisdom is mine, and I have mystery.[26]
>
> Let us forgo men's minds that are brutes' natures.
> Let us not sup the blood which some say nurtures.
>
> Be we not swift with swiftness of the tigress;
> Let us break ranks and we will trek from progress.
>
> Miss we the march of this retreating world
> Into old citadels that are not walled.
>
> Then, [when] their blood hath clogged their chariot wheels
> We will go up and wash them from deep wells
>
> What though we sink from men as pitchers falling
> Many shall raise us up to be their filling
>
> Even from wells we sunk too deep for war
> Even as One who bled where no wounds were.

This very literary piece, full of allusions appropriate to an exhortation from one Romantic poet to another, needs some explanation. The two officers are poets, masters of beauty and truth. The speaker urges that they should take no more part in war, which has become, in Bergsonian terms, a machine; the 'war-blood' (as another draft describes it) now oiling the machine will eventually bring it to a halt. Men have become brutish.

What is supposed to be 'progress' is actually a retreat into outdated, indefensible positions. War is an Old Testament activity, but in the New Testament Christ ('One') is said to bring water from a living well, fulfilling the prophecy that his people would 'draw water out of the wells of salvation' in the day of peace.[27] Christ 'bled where no wounds were' in his 'agony and bloody sweat' before the Crucifixion; the blood which redeems is not blood shed in war, so the popular notion of the soldier-as-redeemer is false. Owen's imagination was still based on the biblical texts and images which he had memorised daily in his younger years, but 'Earth's wheels' also shows the influence of Shelley, the self-declared 'atheist'. Shelley said that *The Revolt of Islam* had grown 'as it were from "the agony and bloody sweat" of intellectual travail', seeing himself as a sort of poet Christ, so it may be that Owen's 'One' is to be understood as a similar figure.[28] The 'wells we sunk too deep for war' are both biblical and Shelleyan:

> Those deepest wells of passion or of thought
> Wrought by wise poets in the waste of years.[29]

In the *Revolt*, Cythna promises Laon to pour

> For the despairing, from the crystal wells
> Of thy deep spirit, reason's mighty lore,
> And power shall then abound, and hope arise once more.[30]

This, then, is the programme which Owen envisaged for himself and Sassoon at the end of 1917. They should be on the side of spirit, not machinery, refusing to take part in violence even if that meant the end of their military careers. They might be able to achieve little until the war machine exhausted itself but in the end their store of beauty and wisdom would be a source of purification and new life for society. Meanwhile they should isolate themselves like Laon and Cythna, using the time for thought and preparation. Owen read Wells's *What is Coming?* in December or January and would have noticed its author's hope that 'we may presently find . . . devoted men and women ready to give their whole lives, with a quasi-religious enthusiasm, to this great task of peace establishment'.[31]

'Earth's wheels' was to be adapted in the spring for inclusion in 'Strange Meeting', where the dead poet describes the pro-

gramme he 'would have' followed if he had lived. The political
prophecy in both poems may owe something to Wells but its
main source seems to be Bertrand Russell, whose work had
strongly influenced Sassoon. Sassoon may well have lent some of
Russell's books to Owen; he certainly wanted to put him in touch
with Russell in November.[32] There is an essay on 'The Danger
to Civilization' in *Justice in War-Time* (1916) which makes a
political forecast very like the one in 'Strange Meeting'. Russell
expects that 'universal exhaustion' will set in if the war continues
('when much blood had clogged their chariot-wheels'), making
true progress impossible ('this retreating world'). Peoples will
have become used to passive obedience or violence, and
governments will have grown accustomed to autocratic power,
so that, when the ruling classes cut back on education after the
war in an attempt to keep the populace ignorant and forcibly
disciplined, either 'apathy or civil war' will result ('men will go
content with what we spoiled, / Or, discontent, boil bloody, and
be spilled'). Russell says past progress has largely been maintained
by young teachers, most of whom are now dead or worn out ('the
undone years, / The hopelessness . . . I would have poured my
spirit without stint'). Russell's other 1916 book, *Principles of Social
Reconstruction*, gives advice to the few who are prepared to 'lie out
and hold the open truth' (to use a phrase from an 'Earth's wheels'
draft). He admits that little can be achieved in the short term.
The aim must be long-term, to prepare for a saner society which
future generations can enjoy. The wise man follows a 'consistent
creative purpose' and is not diverted by external pressures, but
in wartime his reward is bound to be loneliness.

> To one who stands outside the cycle of beliefs and passions
> which make the war seem necessary, an isolation, an almost
> unbearable separation from the general activity, becomes
> unavoidable. . . . The helpless longing to save men from the
> ruin towards which they are hastening makes it necessary to
> oppose the stream, to incur hostility, to be thought unfeeling,
> to lose for the moment the power of winning belief.[33]

Russell held this ground resolutely throughout the war, despite
a prison sentence in 1918 and bouts of near-despair.

But Russell had never led men in battle. Sassoon was persuaded
by Rivers and the call of the troops to abandon his protest as
'futile' and to return to the line, whence he assured his anxious

correspondents that he had made the right decision. Graves, now on home service after a breakdown, wrote a long poem expressing sympathy for what he supposed to be Sassoon's lonely misery, but in reply Sassoon insisted that he was much happier than he had been in 1917. He told Lady Ottoline in May that he was 'driven by an intense desire to do all I can to train the 150 men I am in charge of – because I *know* that it is the only means of mitigating their wretchedness when they get into the Inferno'. Owen understood this position although he was less single-minded himself, advocating 'breaking ranks' in 'Earth's wheels' and allowing friends in 1918 to try to find him a home posting. In February he quoted a Russell-like statement that 'few have the courage, or the consistency, to go their own way, to their own ends', but that might have described Sassoon, who had freely chosen to be true to himself by following a course which served neither pacifism nor militarism. Owen had said at Craiglockhart that 'I hate washy pacifists as temperamentally as I hate whiskied prussianists. Therefore I feel that I must first get some reputation of gallantry before I could successfully and usefully declare my principles.' He was right about his usefulness: a shellshocked officer suspected of cowardice would not have carried any weight with the public without further trench service. He was never really in much doubt. At the end of 1917 he remembered the expression on the faces of the troops ('a blindfold look . . . like a dead rabbit's') and said that 'to describe it, I think I must go back and be with them'. If it was true that 'every poem, and every figure of speech should be a *matter of experience*', the poet had to be with his men at the front.[34]

Owen said in November that poetry for Sassoon had become 'a mere vehicle of propaganda', implying some doubt about using poetry for that purpose, but in fact the propaganda element had already begun to fade from his friend's verse. The dominant force in Sassoon's war poems from that November onwards was not political protest but sympathy with the troops; he told E. M. Forster in the following year that soldiers were 'the only thing in the war that moves me deeply. When I see them in large masses they seem like the whole tradition of suffering humanity'. He wanted to write poems about this in 1918; thinking of *The Dynasts*, Whitman, and the paintings being done by official war artists, he wondered if he might become an official war poet and attempt 'larger canvases, sort of Whitmanesque effects of masses of soldiers' as a detached observer. He said later that this was unconsciously a move towards the kind of work that Owen was doing by then. Owen would not have accepted this, any more than he accepted that poetry should be 'a mere vehicle of propaganda'; if he was less sure of himself as a soldier than Sassoon, he was more certain and determined as a poet.[35]

'The Show' may stand for his achievement in his last winter. It is a large 'canvas', portraying 'masses of soldiers' from the 'vague altitude' of *The Dynasts*, but the poet does not remain an observer. The language is as elaborately patterned as in 'From my Diary' but no longer experimental or decorative, its discordant pararhymes and sound sequences being similar to the sort of work that the 'advanced composers' were producing in music. Technique is not present for the sake of 'Poetry' but as a means of controlling and organising experience, for the poem is not only a Romantic vision[36] but also one of those shellshock dreams that Owen '*willingly*' brought on in the course of following what he believed to be his 'duty . . . towards War'. He was still haunted by the ground at Savy Wood 'all crawling and wormy with wounded bodies',[37] and in the poem this memory combines with an earlier horror, the tormented face, now filling the whole landscape as the maggot-infested face of war:

> Across its beard, that horror of harsh wire,
> There moved thin caterpillars, slowly uncoiled.
> It seemed they pushed themselves to be as plugs
> Of ditches, where they writhed and shrivelled, killed.
> . . .
> I saw their bitten backs curve, loop, and straighten.
> I watched those agonies curl, lift, and flatten.

It is testimony to the help given him by Brock and Sassoon, and to his own dedication as a poet, that Owen was able to write about this twofold nightmare without wavering. 'The Show' is as vivid as anything in *Under Fire* or *Counter-Attack* and more concentrated. As a Georgian, he writes of 'real experience', sending his mind back even to the scene of his shellshock. Paralysed by the sight of the rotting face, the poet falls towards it until he finds himself as another version of it, a severed head, when Death

> picking a manner of worm, which half had hid
> Its bruises in the earth, but crawled no further,

Showed me its feet, the feet of many men,
And the fresh-severed head of it, my head.

The image is of a platoon commander lying dead or helpless while his leaderless men find what shelter they can. Sooner or later the memory of that failure on the railway embankment would have to be exorcised by a return to France. Meanwhile, although as an officer he had no real choice, as a poet he still had work to do at home. The newly launched poet, now 'come to the true measure of man'[38] and believed in by leading Georgians, went into 1918 firmly committed to his 'duty . . . towards War'.

8 The Pity of War

In the early months of 1918 Owen decided on the subject of his poetry, stating it in his Preface in the spring as 'War, and the pity of war'. In settling on the subject of 'pity' he was returning to the beliefs of the Romantics, although he was no doubt also influenced by Hardy and the new direction in Sassoon's work. His 1918 poetry has a resonance of feeling, language and ideas which comes from his knowledge of the Romantic poets; whereas his pleasure in being 'held peer by the Georgians' at the end of 1917 soon ceased to represent an ambition to write in an exclusively Georgian way, his allegiance to his first masters had never faltered. His thoughts about the subject and function of his poetry in 1918 are reflected in his plans for publishing a book; the fragmentary Preface and two lists of contents which he drafted in the spring can be seen as a commentary on the poems he was writing at the time.

There was little fighting on the Western Front between mid December and mid March, so that he began to feel more optimistic about his future, even starting to buy furniture for use after the war. His months in Scarborough were busy and sociable. But in February things looked 'stupefyingly catastrophic on the Eastern Front' as Russia collapsed and Germany began to assemble her forces for a final attack in the west. On 12 March Owen was posted to a training-camp at Ripon, one tiny movement in the Army's efforts to gather reinforcements. 'An awful Camp – huts – dirty blankets – in fact WAR once more. Farewell Books, Sonnets, Letters, friends, fires, oysters, antique-shops.' Nine days later the German offensive opened, throwing the British line into desperate retreat. As in 1914, national opinion rallied behind the war effort, leaving only the most dedicated pacifists still demanding peace negotiations. By the end of March Owen was 'trying to get fit', 'Permanent Home Service' having ceased to be either possible or desirable. His writing-career might have ended here, had camp routine been less generous in its allowance of free time, but on

136

the 23rd he found a room in a cottage in Borrage Lane which he was able to rent for use during his long free evenings. Almost all his finest poems (except 'Spring Offensive') and his plans for a book seem to have been composed or revised in this secret 'workshop'. He reported on the 31st:

> Outside my cottage-window children play soldiers so piercingly that I've moved up into the attic, with only a skylight. It is a jolly Retreat. There I have tea and contemplate the inwardness of war, and behave in an owlish manner generally.

Fine spring weather, secrecy and an attic room being the ideal conditions for his poetry to grow, 1918 brought the last and by far the most fruitful of the creative springs of his poethood.[1]

Despite his lightness of tone, that 'owlish' process of contemplating 'the inwardness of war' was a strict, intensely serious discipline, learned, one may guess, from Wordsworth's account of composing from 'emotion recollected in tranquillity'. Owen was 'haunted by the vision of the lands about St Quentin crawling with wounded', the very ground he had advanced over in 1917. 'They are dying again at Beaumont Hamel, which already in 1916 was cobbled with skulls.' Meanwhile the children played soldiers and 'all the Lesser Celandines opened out together' in the lane '(my Lane)'. It was now that Owen earned Murry's later description of him as 'not a poet who seized upon the opportunity of war, but one whose being was saturated by a strange experience, who bowed himself to the horror of war until his soul was penetrated by it, and there was no mean or personal element remaining unsubdued in him'. In order to achieve this imaginative state, Owen trained himself in the impersonal, poetic insensibility which he describes in the poem of that title; he had been working towards this throughout the winter ('Insensibility' may have been in draft before he went to Ripon) but the process was completed at Borrage Lane. Wordsworth had said,

> poetry is the spontaneous overflow of powerful feelings; it takes its origin from emotion recollected in tranquillity; the emotion is contemplated till by a species of reaction the tranquillity gradually disappears, and an emotion, kindred to that which was before the subject of contemplation, is gradually produced and does itself actually exist in the mind.

Owen had been working on similar lines in February when he had willingly brought on war dreams in order to perform his 'duty . . . towards War'. The task was both painful and dangerous for a poet of strong imaginative sensations such as he was, but the first draft of 'Insensibility' shows him working out a solution. Poets cannot shirk their 'duty', but since a mere 'hint', 'word' or 'thought' can smother their souls in blood they must acquire the vision of the common soldier, his senses dulled by the 'scorching cautery of battle'. Owen commented in March that the 'enormity of the present Battle numbs me' and said later in the year after returning to the front that his senses were 'charred'. This dullness of sensation became a necessary preliminary to writing poetry, a means of keeping control over bloodiness and more than shadowy crimes. That 'blindfold look . . . like a dead rabbit's' of the troops was not only a petrifying memory but also a clue to the way in which a poet could still function. However 'great' or otherwise Owen's Borrage Lane poems may be judged to be, his method of working deserves to be recognised as an extraordinary undertaking – a young subaltern training by day for the fighting that would almost certainly kill him, and in the warm spring evenings walking down a country lane to shut himself away in a windowless room and open his 'inward eye' to the intensity of those feelings and experiences that had brought him close to madness a year before.[2]

That his guide in all this was Wordsworth is confirmed by the deliberate literary references in 'Insensibility'. Like Wordsworth's 'Intimations of Immortality' the poem is a 'Pindaric' ode, a form developed in the eighteenth century, and is concerned with the loss of poetic imagination.[3] Several of the key words ('imagination', 'feeling', 'simplicity') belong to late eighteenth-century critical debate, while the opening phrase, 'Happy are men who', is the classical *Beatus ille* construction used, for example, by Pope in 'Happy the man whose wish and care / A few paternal acres bound'.[4] The sanity and order of eighteenth-century literature is used as ironic contrast, showing up the inverted values of war. As the simple Augustan swain was 'happy' in his little world, so the modern soldier is 'happy' when he is back at home 'with not a notion / How somewhere, every dawn, some men attack'. Wordsworth had lamented his own blunted vision but for Owen such a limitation was essential in 1918, when the problem for the 'wise' was an excess of imagination rather than a lack of it.[5] Nevertheless, the aim of poetry was still as Wordsworth had stated it, to reach and ennoble the human heart. Wordsworth ends his ode by claiming that the 'meanest flower that blows' could still give him 'Thoughts that do often lie too deep for tears'. Owen echoed the language but broadened the statement, making it more Shelleyan than Wordsworthian; thoughts about men who 'fade' and 'wither' in war should arouse 'The eternal reciprocity of tears'.

'Insensibility' seems to refer to two great Romantic manifestos, not only Wordsworth's Preface to *Lyrical Ballads* but also Shelley's *Defence of Poetry*. The last stanza contains Owen's first use in a war poem of the word 'pity'. Shelley argues that poetry should arouse man's imagination, making him understand other people and sympathise with them, putting himself 'in the place of another and of many others'. Poetry promotes love, which develops from sympathy and is the key to all moral goodness. Furthermore the power of poetry is such that it 'turns all things to loveliness; it exalts the beauty of that which is most beautiful, and it adds beauty to that which is most deformed; it marries exultation and horror, grief and pleasure, eternity and change; it subdues to union under its light yoke all irreconcilable things'. In the intensity of poetry even the most terrible subjects can be beautiful. One can see how this can be applied to Owen's work; in 'Disabled', for example, the mutilated soldier has lost all his physical beauty, yet the poem 'adds beauty to that which is most deformed' until the reader becomes aware of the man as a fellow human being still fit to be loved. The word 'exultation' occurs in 'Apologia', where it is 'married' with horror. This imaginative process is not to be confused with what Owen dismisses in 'Insensibility' as 'poets' tearful fooling', the sentimental versifying of writers who are more interested in being poetic than in their subject ('Above all I am not concerned with Poetry'). Yet even the true poets could not reach all hearts. The last stanza of his ode condemns civilian 'dullards' (originally 'these old') who, like the 'wise' but for very different reasons, make themselves insensible by choice. This Romantic distinction between the 'wise' and dullards follows that made by Shelley in the Preface to *Alastor* between poets, who suffer from too acute a consciousness of humanity, and men, 'who, deluded by no generous error, . . . loving nothing on this earth, . . . yet keep aloof from sympathies with their kind, rejoicing neither in human joy nor mourning with human grief; these, and such as they, have their apportioned curse'. Owen delivers the curse in one of his most elaborately composed stanzas, using pararhyme and other sound-patterns now with practised ease:

> But cursed are dullards whom no cannon stuns,
> That they should be as stones.
> Wretched are they, and mean
> With paucity that never was simplicity.
> By choice they made themselves immune
> To pity and whatever moans in man
> Before the last sea and the hapless stars;
> Whatever mourns when many leave these shores;
> Whatever shares
> The eternal reciprocity of tears.

In my judgment, this stanza is not quite as successful as some modern critics have made it out to be. Its technical complexity is remarkable – only Owen could have written it – but the imagery lacks substance. The 'last sea' and the 'hapless stars' are no more than clichés, as in Arthur Symons's 'Beyond the last land where the last sea roars' (and Owen's own 'Timeless, beyond the last stars of desire', a line which he had tried out as an ending for 'My Shy Hand').[6] The rest of 'Insensibility' is much more solid, the images coming from hard experience – 'Their spirit drags no pack', 'alleys cobbled with their brothers' (a memory of bones frozen into the streets of Beaumont Hamel[7]) – but in attempting to articulate a fullness of emotion which any twentieth-century poet would have found difficult to handle the last stanza comes too close to Tennysonian cadences and nineteenth-century vagueness. He was to be more successful, and no less original in his technique, in later 1918 poems.

He had probably begun to think about the poetry of 'pity' before the end of 1917, since it is clear that he set himself to read several famous elegies during the winter and following spring. He may have had no very scholarly idea of what an elegy was, remembering only from reference books that it originally meant a song of grief over a dead man and from Tailhade's example that *élégiaque* meant something other than *aristophanesque*, but when he described his war poems as 'These elegies' in his Preface he did not use the word at random. In December he read Lang's translation of the elegies by Bion and Moschus that had been Shelley's model for *Adonais*. In the spring he considered calling his book 'English Elegies' or 'With Lightning and with Music' (a phrase from *Adonais*). Dr Bäckman has pointed out some convincing parallels between the rhetorical structure of 'Asleep' (late 1917) and that of Milton's 'Lycidas', and between the tramp in 'The Send-Off' (Ripon) and the swain in Gray's 'Elegy Written in a Country Churchyard'. 'Futility' (probably Ripon) seems to reflect a famous passage in *In Memoriam*, while Tennyson's description of Hallam as 'strange friend' may be the source of that paradoxical phrase in 'Strange Meeting'.[8] 'Hospital Barge' (December) was written after a 'revel in "the Passing of Arthur"' and is a less serious effort in Tennyson's elegiac mode. Some of Owen's other poems have affinities with the elegies in *A Shropshire Lad* and *The Dynasts*. As early as November 1917 he was aware of a difference between his own verse and Sassoon's, saying that Sassoon wrote 'so acid' while he wrote 'so big'.[9] He described 'Miners' in January as 'sour' but the poem also has a 'bigness' which Sassoon had not attempted, a largeness of expression which is elegiac rather than satirical, universal rather than immediate, even though the poem was inspired by an actual event. His readings in elegy are reflected particularly clearly in the oddly undistinguished 'Elegy in April and September',[10] composed in those months in 1918. On the back of the April draft he made a note about Matthew Arnold, including the titles of two elegies, 'Thyrsis' and 'The Scholar-Gipsy', and '1. lofty, restrained, dignified. / 2. wistful agnostic'. His own 'Elegy', with its search for a lost poet–friend among woods and fields, is an imitation of Arnold's two poems; indeed, the five adjectives which he attached to Arnold could be applied to much of his own 1918 work, which conforms to 'a solemn dignity in the treatment' like the sonnets he had thought of collecting under the title 'With Lightning and with Music' some three years earlier. These traces of Shelley, Tennyson, Arnold and Gray in his late work are a reminder that these were the four poets he had reverently described in a verse letter from Dunsden in 1911. His reading in the winter of 1917-18 was not only in new writers such as the Georgians but also in old ones studied long before.

Critics have argued that elegy is a false response to war because it offers consolation. Although Owen's poems are not wholly devoid of consolation, he said firmly in his Preface that 'These elegies are to this generation in no sense consolatory'. One value of elegy was that it could provoke pity, appealing to hearts which might have been unmoved by satirical attacks. Having read Sassoon's book of press-cuttings, he knew that some contemporary reviewers had dismissed Sassoon's acidity as the product of immaturity and nervous strain. Edmund Gosse had remarked disapprovingly in October 1917 that such verse would tend to weaken the war effort, quite failing to see that that was precisely what it was intended to do.[11] It was possible to miss the target entirely with poems in Sassoon's style, devastating though they seemed to be. In any case, the elegiac convention allowed for protest; if Milton had attacked priests in 'Lycidas' and Shelley had savaged critics in *Adonais*, it was legitimate to attack civilians in war 'elegies'. To accept the elegiac element in Owen's 1918 poems is not to deny that they contain social criticism, but it is to recognise that their subject is indeed 'pity' and that in defining it they make extensive use of literary tradition.[12]

At Borrage Lane he returned to 'Wild with all Regrets', expanding it into 'A Terre'. He had originally imagined the dying officer in the poem as bookish ('But books were what I liked. Dad called me moony')[13] but in the finished December 1917 draft the man is given a thoughtless, sporting past. In the 1918 version the bookishness is restored, the officer quoting from *Adonais* to a friend who seems himself to be the author of a 'poetry book'. The reference to Shelley suggests that the officer's viewpoint is Shelleyan throughout in its condemnation of class distinction, blood sports and war. His somewhat obscure remark that his soul is 'a little grief' lodging for a short time in his friend's chest may be compared with the lament in *Adonais* (xxi):

> Alas! that all we loved of him should be,
> But for our grief, as if it had not been,
> And grief itself be mortal!

The hope of becoming one with nature after death, the state which Shelley claimed for Keats, is a 'poor comfort' in the cruel, impoverished world of war but it is nevertheless the 'philosophy of many soldiers'. Although Shelley did not mean it in the sense in which the 'dullest Tommy' holds it, as a democrat he might have been pleased to find it so widely valued. There is a scepticism about poetry in 'A Terre', Owen's as well as Shelley's (a typically Owenish metaphor is scornfully relegated to the friend's 'poetry book'), but this is not so much a condemnation of Romanticism as a Romantic unease like the doubts about the poet's usefulness which Keats had expressed in *The Fall of Hyperion*. Perhaps poetry *was* mere 'tearful fooling'; the poet could only assert his faith by continuing to write, not for the sake of 'Poetry' but for the sake of humanity.

Some of the poems which Owen wrote at Borrage Lane are less concerned than his earlier war pieces had been with conveying realistic detail to civilians. Instead he takes a subject which a civilian could see or imagine and reveals its significance. Thus, for example, 'The Send-Off' describes a draft of soldiers setting out for France and suggests that they are victims of a 'hushed-up' conspiracy. The situations in these poems are representative rather than specific, sometimes entirely without topical reference so that they become applicable to any war, although they remain in origin 'matters of experience'. 'Arms and the Boy', for instance, may have been suggested by the irony of the children's playing soldiers outside the cottage. Since a boy is not armed by nature, society must provide him with man-made weapons:

> Lend him to stroke these blind, blunt bullet-leads,
> Which long to nuzzle in the hearts of lads,
> Or give him cartridges whose fine zinc teeth
> Are sharp with sharpness of grief and death.
>
> For his teeth seem for laughing round an apple.
> There lurk no claws behind his fingers supple;
> And God will grow no talons at his heels,
> Nor antlers through the thickness of his curls.

This is Owen's 1918 voice, no longer that of Georgian realism though still ironic towards the older generation. War's greatest cruelty is seen to be its destruction of youth and beauty. Its relationship to the young soldier is presented in sexual terms, consumingly urgent on one side ('long to nuzzle', 'famishing for flesh') and innocently exploring on the other ('try', 'stroke'). The imagery is strongly physical, with particular emphasis on parts of the body; the title literary; the language rich but dissonant in the manner of 'advanced composers' of contemporary music. Yet the universality of the statement does not weaken its bitterness, for the boy and the 'blind, blunt' bullets are not mere generalities. In the Easter Sunday letter which mentions children playing soldiers, Owen refers to two boys whom he often remembered:

> I wonder how many a *frau, fräulein, knabe und mädchen* Colin will kill in his time?

Johnny de la Touche leaves school this term, I hear, and goes to prepare for the Indian Army.

He must be a creature of killable age by now.

God so hated the world that He gave several millions of English-begotten sons, that whosoever believeth in them should not perish, but have a comfortable life.

The feeling in 'Arms and the Boy' comes from that kind of reflection, but it is rendered into impersonal terms as a result of deliberate contemplation of the 'inwardness of war'.

Like earlier poems, Borrage Lane work demonstrates how strongly Owen was moved by the waste of young life. One of his first attempts there, 'As bronze may be much beautified', which was begun on Good Friday but never finished, seems to have been first conceived as a lament for a particular youth:

> As women's pearls needs [be refreshed in deep sea,]
>
> There he found brightness for his tiring eyes
> And the old beauty of his young strength returned –
>
> Dropped back for ever down the abysmal war.

Perhaps a young soldier, Antaeus-like, was to have been pictured as regaining contact with earth and sea, only to be killed by the war machine and lost underground. In 'Futility' the loss of potential new life is related to all natural renewal, just as Antaeus was the embodiment of all life in nature:

> Move him into the sun –
> Gently its touch awoke him once,
> At home, whispering of fields half-sown.
> Always it woke him, even in France,
> Until this morning and this snow.
> If anything might rouse him now
> The kind old sun will know.
>
> Think how it wakes the seeds –
> Woke once the clays of a cold star.
> Are limbs, so dear achieved, are sides
> Full-nerved, still warm, too hard to stir?

Was it for this the clay grew tall?
– O what made fatuous sunbeams toil
To break earth's sleep at all?

F. W. Bateson (1979) condemned 'Futility' as an elegant technical exercise in which 'prosodic gadgets' count for more than truth. The 'Hell of trench warfare was already becoming . . . an abstraction' to Owen, according to Bateson, whereas in 'the great war poems' of 1917 grief is an 'authentic' 'personal experience'. It is true that the 1918 poems are less immediately 'realistic' and colloquial than those of 1917, becoming boldly original in technique and firmly grafted into literary tradition, but there is no fault here unless one is determined that all war poems must be of the Georgian kind. Bateson's comments rest on careless reading and a very hazy notion of the dates of Owen's poems. He understands 'Move him into the sun' as a 'curiously inhumane' order to carry a dead or dying man out of a hospital bed into the snow; were that correct, Owen would indeed be shown to have lost touch with true feeling and experience. The setting of the poem seems clear enough: the man has just died from exposure after a night in the open (like 'Exposure', 'Futility' may draw on an actual memory) and his body is to be moved out of the shadow of a parapet or some other object into the light and warmth of the rising sun. Bateson also complains that the second stanza is scientifically nonsense and that it puts the blame for war on the sun rather than on man, but Owen is making use of myth not science (according to ancient legend the sun's rays brought living creatures out of mud). The sun symbolises the source of life; if Owen is blaming it, he is blaming God, which was at least a defensible position to hold in 1918. Much literature has to be dismissed if poets are not to be permitted to question the Creator or to conclude that life is futile. The questions in the poem are not answered, however, so the blame may still be humanity's. Man was made for life and for sowing seed which the sun can ripen; when he turns to killing, nature has to reject him, his potency is lost and he dies. This is consistent with the mythic pattern which Owen had been working on for years. In the 'Perseus' sequence the young lover, 'full-nerved', commits some strange 'Wrong' which results in his being cut off from fertilising sunlight and paralysed (or, as it were, frozen) in darkness. The powerful cadence of the last lines of 'Futility' is passionately felt and far from mere elegance. As Owen felt his young strength returning in the sunlight of the Yorkshire spring while men died again at Beaumont Hamel, his memories of the 'Hell of trench warfare' were very far from becoming an 'abstraction'.

The way in which 'the pity of war' could be evoked, and the relationship between it and anger and disgust, are suggested in the organisation which Owen planned for the book of war poems which was his immediate objective at Ripon. All that remains of this plan is a fragmentary Preface and two rough lists of contents, together with a note of some possible titles ('Disabled and Other Poems' being his preferred choice).[14] The earlier list, written on the same type of paper as the final drafts of 'Insensibility' and 'Strange Meeting' and the first draft of 'Exposure', probably dates from April or early May.[15] It gives an idea of the stage his work had reached after perhaps a month's labour in the cottage. 'Strange Meeting' is included among nineteen or so 'Finished' poems; 'The Send-Off' and 'Mental Cases' await their final titles, as does the first version of 'Exposure'; 'The Sentry' is still 'Only Fifty Yards', the phrase with which the 1917 draft ends. Except for 'Spring Offensive', 'Training', 'The Kind Ghosts' and 'Smile, Smile, Smile', which were written later, almost all his war poems were by now either complete or in draft. It follows that much of his effort in May must have been devoted to polishing existing poems, producing the numerous fair copies overlaid with massive revisions that have been such a minefield for his editors. It was probably in late May or early June that he drew up his second, more detailed list and roughed out a Preface.

The Preface has become so famous that its fragmentary nature is often forgotten. The following version of it attempts to show some of Owen's first thoughts and revisions.[16]

Preface

This book is not about heroes. English Poetry is not yet fit to speak of them.
Nor is it about [battles, and glory of battles or lands, or] deeds or lands nor anything about glory or honour any might, majesty, dominion or power [whatever] except War.
[Its This book] is Above all I am not concerned with Poetry.

[Its The] My subject is War, and the pity of [it] War. The Poetry is in the pity.

[I have no hesitation in making public
                                publishing such]
[My] Yet These elegies are [not for the consolation] to this generation in no sense consolatory to this [a bereaved generation]. They may be to the next. [If I thought the letter of this book would last, I woul might have used proper names;] All a poet can do today is [to] warn [children] That is why the true [War] Poets must be truthful.

If I thought the letter of this book would last, I [wo] might have used proper names: but if the spirit of it survives – survives Prussia – [I] my ambition and those names will [be content; for they] have achieved [themselves ourselves] fresher fields than Flanders,
                    far be, not of war
                                        would be
            sing

A following sheet bears the words 'in those days remembers' (perhaps a continuation of the Preface) and the second table of contents. Owen's declaration of poetic intent is in the tradition of the Preface to *Lyrical Ballads* and the *Defence of Poetry*. (It also contains an echo of Keats's poem about art, the 'Ode on a Grecian Urn': 'These elegies are to this generation in no sense consolatory . . . They may be to the next' / 'When old age shall this generation waste, / Thou shalt remain, . . . a friend to man'.) As Wordsworth, Shelley and Keats required, the poet rejects 'Poetry' for its own sake and dedicates himself to the betterment of humanity. There were requirements peculiar to 1918, including the need to refrain from consoling the older generation. It was perhaps because most existing poetic language was consolatory in its effect that Owen considered English poetry to be not 'yet' fit to describe the heroism of soldiers, but the word 'yet' implies that he would have liked to see a kind of poetry fit to speak of heroes and in 'Spring Offensive' he was to move towards a poetry of that kind. The later part of the Preface shows him reaching out to 'children', the future generations who were in the event to be his audience.

The list of contents which accompanies the Preface is arranged in a careful sequence, each poem being given a 'Motive' and each group of poems a further label. The first group, which seems in the manuscript to consist of thirteen titles, is 'Protest', beginning with 'Miners' and including 'Dulce et Decorum Est', 'S. I. W.' and 'The Dead-Beat', as well as some newer work such as 'Aliens' (a title for 'Mental Cases' which was suggested by Owen's friend Charles Scott Moncrieff at the end of May[17]). The fourteenth poem, 'The Show', is at the halfway point, its Motive of 'Horrible beastliness of war' sharply contrasting with the next two, 'Cheerfulness' ('The Next War') and 'Cheerfulness & Description & Reflection' ('Apologia'). Two more 'Description' poems follow, then five of 'Grief' and four which all seem to be included under 'Philosophy'. This arrangement was devised to take the reader through a developing, coherent experience. He would begin with 'Miners', a poem based on a civil disaster but leading into the subject of war, then move on to 'Arms and the Boy', a statement of the 'unnaturalness of weapons' and the exploitation of the younger generation. Then comes a series of angry protests at the madness, lies and callousness of war. 'The Show' brings the first half of the book to a climax of horror. Then the mood changes. Having been taught to protest, the reader may now see in 'The Next War' and 'Apologia' that war has a cheerfulness understood only by those who have faced its true nature. 'Exposure' and 'The Sentry' provide more description but check any inclination to make light of the troops' suffering. Then the poems of 'Grief', which include 'Anthem' and 'Futility', establish the mood of elegy, introducing the final 'Philosophy' section ('Strange Meeting', 'Asleep', 'A Terre', 'The Women and the Slain'). The reader was thus to proceed from protest through grief to meditation, a pattern Owen himself had gone through since meeting Sassoon. No stage was invalid but none was complete in itself. Protest was unproductive unless it led to grief, and grief in turn had to bear fruit in new attitudes. Owen seems to have thought of the entire experience as 'pity' – not grief alone but grief arising from re-educated knowledge and feeling, producing a positive, active frame of mind.

'Disabled and Other Poems' was never put together and Owen's notes for it should not be taken as his final thoughts about his poetry. He was never entirely convinced that publishing in wartime would be useful, and his cousin's precipitate rush into print had persuaded him that poetry could not be hurried. He said at the end of May that he could 'now write so much better than a year ago that for every poem I add to my list I subtract one from the beginning of it. You see I take myself solemnly now, and that is why, let me tell you, once and for all, I refrain from indecent haste in publishing.'[18] His departure from Ripon on 5 June perhaps prevented his taking his plans for a book any further but what mattered was to write rather than to publish. His audience was posterity. Poems such as 'Exposure' and 'Strange Meeting' show that he lost hope for the war generation, concluding that civilians could never understand and soldiers could never explain. All that a true war poet could do was to warn children, who might find consolation later in the knowledge that a true voice had managed to speak. His poems might prove that there was something indestructible in the human spirit, but that would be consolation only if future generations acted on his warning and loved their fellow men. On the whole we have not so acted, which may be why we find his 1917 poems of protest more immediately forceful than his later work; prevented by conscience from discovering any kind of consolation in his poems, we are ill at ease when he introduces them as 'elegies'. Nevertheless he meant us to see that protest is an essential stage, but not a final one, in the process of perceiving that 'pity of war' which he described in a draft of 'Strange Meeting' as 'the one thing war distilled'. The task of poetry was the Shelleyan one of arousing imagination, enabling people to share in the experience of others, to sympathise with them and love them, for war was above all a failure of love.

9 To Suffer without Sign

The caution of Owen's commentators about his sexual nature is understandable but no longer necessary or particularly helpful. His interest in young male beauty was one of the sources of his poetry. Having grown up in a world in which homosexuality was unmentionable, if not unthinkable, he may have repressed his preferences for some time, concealing them from himself as well as everyone else (hence, perhaps, the recurrent references to guilt and secrecy in his poems, and the bizarre sexual confusions in works such as 'The time was aeon' and 'Perseus'). His stay in France in 1914–15 seems to have brought him some sexual experience but he is unlikely to have come across any intellectual defence of homosexuality (except as one of the 'strange' sins of the Decadence) until he met Monro or, more probably, Sassoon. Sassoon eventually introduced him into one of the very few literary circles in which 'Uranianism' was accepted and easily discussed. An awareness of these matters illuminates Owen's 1918 poems and his thoughts about religion and war.[1]

After Wilde's disgrace, homosexuals had been obliged to be strictly respectable, so that little advice or leadership was available – except from Edward Carpenter, the celebrated apostle of Socialism and free living, whose essays in defence of Uranian love drew many grateful young men towards him. Sassoon visited him in 1911; Monro spent hours talking to him in 1910–11 when the *Before Dawn* poems were being written; Graves wrote to him in 1914; others who came within his influence included E. M. Forster and D. H. Lawrence.[2] Carpenter preached a gospel of 'beauty and compassion and friendship' (to use the terms in Sassoon's letter to Nicholson), maintaining that Uranian men tended to be musical and artistic, with a highly developed capacity for aesthetic and emotional sensation; they were quick to feel affection for children and pity for the unfortunate, and they could be 'overcome with emotion and sympathy at the least sad occurrence'.[3] Believing (like Geddes) that humanity was out of touch with its environment, Carpenter looked forward to a new order of freedom made possible by the hard work and endurance of the few who were already in a right relationship with nature. Sassoon took a saying of his as a watchword: 'Strength to perform, and pride to suffer without sign.' It is most unlikely that Carpenter was not mentioned at Craiglockhart.[4] There is no certain evidence, but I guess that Sassoon lent Owen some of the prophet's works and that Owen discovered, as Sassoon had done in 1911, that the artistic temperament which he had sought for and found in himself could be seen as essentially homosexual, or, to reverse the equation, that his sexual tendencies were to the benefit of art and humanity. This insight would have been of immense help to his confidence as both man and poet.

When Owen left Craiglockhart, Sassoon gave him an introduction to Robert Ross in London. Sassoon would not have done that lightly. The spirit of Oscar Wilde was still alive in Ross's elegant flat in Mayfair, for Ross had been Wilde's most devoted friend and was still his loyal defender. Other loyal disciples included More Adey, whom Owen met, and Robert Sherard, author of three books about Wilde. Owen read at least one, probably two, of Sherard's books during the winter, as well as *De Profundis* and some or all of Wilde's verse, following his usual habit of reading up an author with whom he found he had personal links. Ross was wholly out of sympathy with middle-class orthodoxy but Wilde's fate had taught him the necessity for extreme discretion. Loathing the war, he had been the first person to encourage Sassoon to write satires but had nevertheless collaborated with Graves in covering up Sassoon's protest in 1917. His nervousness was understandable, because he had suffered repeated persecution from Lord Alfred Douglas in revenge for his publishing *De Profundis*, an abridged version of Wilde's letter to Douglas from prison. Douglas, who had turned violently against his dead lover, maintained that Ross's editing had made a monster sound like a martyr. He had forced Ross to sue him in 1914 by calling him 'the High Priest of all the sodomites in London'; then he triumphantly produced fourteen witnesses. No police prosecution followed, however, as it had done for Wilde, partly because Ross had powerful friends and partly because there had been an advance in liberal opinion since 1895. Witnesses who spoke for Ross included Wells, whose reformist views on sexual matters had been expressed in several novels (Owen read at least two of these in 1917–18), and Arnold Bennett

(Owen read his scandalous *The Pretty Lady* in April 1918). But
Ross's troubles were not over. In January–March 1918 his friend
and secretary, Christopher Millard, Wilde's bibliographer, was
tried and imprisoned for a homosexual offence. At the same time
a Member of Parliament named Pemberton Billing astonished
London with a story that the German Secret Service had listed
in a 'Black Book' no fewer than 47,000 prominent British people
whose private lives were suspect; as an example of national
corruption, he pointed to a current production of Wilde's *Salomé*.
The leading actress sued him for libel and lost, to the loud delight
of press and public, the trial in late May and early June becoming
a grotesque display of hatred for homosexuality and 'the Oscar
Wilde cult'. There was extensive press coverage. Douglas gave
evidence, taking the opportunity to renew his attacks on Ross.
No doubt Pemberton Billing believed that a victory had been
achieved for public morality. Several lives were ruined, including
Ross's, whose health suffered mortal damage (he died in October).
Sassoon and others were distressed for their friend and furious
that attention could so easily shift from the war to a silly scandal.
Owen would have been well aware of all this, although he could
not mention it in letters home; the hysterical public response to
the trial would have strengthened his scorn for civilian opinion
and confirmed his sense of belonging to a secret caste. One side
effect of the affair may have been the loss of headway that seems
to have occurred at the time in plans for 'Disabled and Other
Poems', since Ross had been encouraging Owen as he had Sassoon
and had promised to arrange for publication.[5]

The importance of Owen's friendship with Ross has not been
generally understood. It was not only that 'Owen, the poet' was
introduced to a number of talented literary people, including
Wells and Bennett, but also that in getting to know Ross he came
as near as was possible to knowing Wilde himself. Ross was a
man of culture, charm and excellent conversation, keeping open
house for his many devoted admirers. He lacked Wilde's egotism
and ostentation but stood for Wilde's values in art and life. His
flat must have seemed like a fortress; within its walls of outward
respectability (a fragile protection when court cases were in the
news), Owen was among friends with whom he could talk openly
and yet in secret. Thanks to Sassoon and Ross, the confusions of
earlier years were resolved at last, giving way to a self-knowledge
and self-confidence that are reflected in the strength and sureness
of his 1918 poems. Many, though not all, of Ross's regular visitors

were homosexual. Some believed in the war, some did not, but
there was general contempt for 'screaming scarlet Majors' (as
Ross called them),[6] 'old men' and unsympathetic civilians. As
might be expected among Wilde's followers, there was much
lively conversation about art and literature, and an assumption
that the outside world was ignorant and wilfully deaf to truth.

In mid May Ross introduced Owen to Osbert Sitwell, a young
officer then discovering himself as a poet and Uranian. Sitwell's
early verse had been modelled on *Salomé*[7] but in 1917 he had
started publishing anti-war poems influenced by Sassoon, whom
he knew and admired. Owen had probably seen his work in the
Nation and may have imitated it. 'The Parable of the Old Man
and the Young' is similar to Sitwell's 'The Modern Abraham'
(*Nation*, 2 February 1918), while the last stanzas of 'Apologia'
and 'Insensibility', and Owen's statement that true poets must
be truthful, make the same sort of point as 'Rhapsode' (*Nation*,
27 October 1917):

> Why should we sing to you of little things –
> You who lack all imagination?
> . . .
> We shall sing to you
> Of the men who have been trampled
> To death in the circus of Flanders;
> . . . .
> You hope that we shall tell you that they found their
>   happiness in fighting,
> Or that they died with a song on their lips,
> Or that we shall use the old familiar phrases
> With which your paid servants please you in the Press:
> But we are poets,
> And shall tell the truth.

Sitwell's anger was concentrated against what the *Cambridge
Magazine* called 'old men' or, as he himself put it, 'grand old
men . . . Who sacrifice each other's sons each day'.[8] His anger
was rooted in his detestation for his own father, a feeling which
Owen may have understood. Mrs Owen had encouraged her son
to keep at a distance from his father; while there is little evidence
of hostility in what was certainly an uncomfortable relationship,
Owen was always prone to regarding other men as substitute
fathers as though his own had proved inadequate. It may be that

an unconscious personal resentment is expressed in the harsh treatment of fathers, including God the Father, in such poems as 'Parable', 'S. I. W.', 'Soldier's Dream' and 'Apologia'.[9] Owen had other things in common with Sitwell, including sexual preferences and a taste for Aestheticism and the Decadence. He did not try to imitate the clumsy Modernist 'free verse' of 'Rhapsode' but he does seem to have been impressed by the more talented work of Sacheverell and Edith Sitwell, whose flamboyantly 'advanced' poetry was in much the same French-inspired style as T. S. Eliot's 1917 volume, *Prufrock*. Edith's anthology *Wheels*, the first wholly British collection of Modernist poetry, was deliberately anti-Georgian (Owen does not call himself a Georgian in later 1918 letters). The fragment 'The roads also', which Owen wrote after Ripon, is clearly a Modernist street scene in the style of numerous Sitwell poems. It is almost the only clue on which one might base an answer to the frequent question as to what he would have written had he lived. Unlike Sassoon, he would have admired Eliot and made use of his style, as he had made use of Monro's, Tailhade's, Sassoon's and others', without remaining a mere imitator; he might have helped to bridge the post-war gap between Modernism and the native English tradition. But his interest in *Wheels* and Sacheverell Sitwell's work did not begin until July and August, too late for it to bear much fruit.[10]

There were other new friends in 1918 besides Sitwell. In January Owen went to Graves's wedding, where he was introduced as 'Mr Owen, Poet' or 'Owen, the poet', and people had heard of him because 'Miners' was in that week's *Nation*. He met Edward Marsh, Graves perhaps hoping that the anthologist would accept the new-found poet as a Georgian, and Charles Scott Moncrieff. The evening was spent at Ross's with Roderick Meiklejohn, one of Sassoon's regular wartime correspondents, and 'two Critics', one of whom was Scott Moncrieff. And it was then or not long afterwards that Scott Moncrieff fell in love with Owen and, being the eccentric but brilliant literary imitator he was, conceived the odd idea of expressing his 'passion' in a series of sonnets modelled on Shakespeare's to 'Mr W. H.'. The relationship did not work out as he wished, for Owen seems to have been flattered by the sonnets (he kept one of them, dating it 19 May) but reluctant to respond to their appeal. Scott Moncrieff was a promiscuous, quirky and difficult man; Sitwell and Sassoon detested him, and his view of the war as a chivalric enterprise was quite unlike theirs. He had deplored Sassoon's war poems in a 1917 review.

On the other hand, he warmly admired Owen's poetry for its adventurous use of pararhyme, assonance and other devices, becoming the first critic to mention him in a review (on 10 May). When he began work in the summer on the *Song of Roland*, the first of his major translations, he drafted a dedication to 'Mr W. O.' (who was thus recognised as the translation's 'onlie begetter', as 'Mr W. H.' had been of Shakespeare's sonnets), 'my master in assonance', adding that 'lessons are to be found in the Song of Roland that all of us may profitably learn – To pursue chivalry, to avoid and punish treachery, to rely upon our own resources, and to fight uncomplaining when support is withheld from us.' Similar heroic values are implicit in some of Owen's war poems; he had enough in common with Scott Moncrieff to find his admiration and encouragement welcome if somewhat excessive.[11]

Owen wrote several lyrics in 1918 which may have been more intelligible to Ross and his friends than they are now. Scott Moncrieff was given a manuscript of 'I am the ghost of Shadwell Stair', a little poem apparently based on Wilde's 'Impression du Matin'; he sent a copy of it after the war to Marsh with the following note:

> The above is a copy of the manuscript. More Adey insisted on reading 'fade' and 'laid' in the last verse, for grammatical reasons. During the influenza epidemic in 1918 I tried to turn it into French prose, rhymed. I give the first verse, on account of the last word, which Owen welcomed rather as tho' it put the key in the lock of the whole . . . .
>
> 'Je suis le petit revenant du Bassin; le long de quai, par l'abreuvoir, et dans l'immonde abbatoir j'y piétine, ombre fantassin.'[12]

This disproves the suggestion made in the *Complete Poems* and elsewhere that the ghost is a female prostitute. The 'petit . . . fantassin' (little infantryman) is 'little' Owen himself, who must have returned since his enlistment to that dock area which he had walked through in 1915 and where, then or later, he had seen the Limehouse boy. The poem is not a mere copy of Wilde, as it would have been if Scott Moncrieff had written it, but 'a *matter of experience*' as the 'key' reveals. Nevertheless, despite a characteristic emphasis on 'physical sensation' and familiar traces of the Little Mermaid and her statue (trembling, purple light, water, shadows, unsatisfied waiting), the poem remains enigmatic. Other post-Craiglockhart pieces which seem to be similarly in code include 'Reunion' and a group of three curious ballads, 'Page Eglantine', 'The Rime of the Youthful Mariner' and 'Who is the god of Canongate'.[13] As in earlier lyrics there is a recurrent suggestion of secret knowledge that cannot be fully communicated, a theme which reappears in the war poems ('the truth untold', 'Wisdom is mine and I have mystery', 'secret men who know their secret safe', 'The Unsaid', 'Why speak not they of comrades . . . ?'). 'Who is the god of Canongate' seems to be about the secret world of 'rent boys', well known to Wilde, Ross and Scott Moncrieff, its subject being a 'little god' who walks the pavements barefoot (a street boy or Eros) and is visited in his room by barefoot men (pilgrims to the shrine who need to be secret). Mentions in the manuscript of Covent Garden, 'Bow St. cases' (Wilde was charged at Bow Street) and Canongate associate the 'god' of Canongate with other strange figures seen in London and Edinburgh streets — the 'phantom' in Princes Street, the 'ghost' of Shadwell Stair, the 'half-god' in Limehouse, and the handsome youth in the lost or projected poem which provided material for 'Disabled'.[14] The god of love, who had been hopelessly out of reach in 'To Eros', 'Perseus' and the Limehouse poem, is no longer an inaccessible deity.

The tradition left by Wilde to his followers was largely borrowed from French Decadent literature, so Owen was on familiar ground. Art was an autonomous, mysterious world closed to all but the privileged few, a world where bourgeois morality had no authority and where 'strange' sins might be permissible in the interests of beauty. But it was also a world permeated with religious values and imagery. When Owen read *De Profundis* and 'The Ballad of Reading Gaol' in the winter of 1917–18, he found that the older Wilde had much to say about Christ. Since another of the questions commonly asked about Owen is whether he was still a Christian in 1917–18, this may be an appropriate point at which to attempt an answer.

I have said in earlier chapters that Owen was predisposed by his own nature and his knowledge of Decadent authors towards imagery of martyrdom and passive suffering. He compared dying soldiers to Christs as early as December 1914 and in the successive versions of 'The Ballad of Peace and War'. In May 1917, after he had been in the trenches, he modified this to correspond with a general attitude among soldiers that they were not dying for civilians but for their comrades; Christ was in No Man's Land, Owen now said, and there 'men often hear His voice: Greater love hath no man than this, that a man lay down his life – for a friend'. He recognised, however, that soldiers were killers as well as victims, so that they had to ignore Christ's essential command, 'Passivity at any price! . . . Be bullied, be outraged, be killed; but do not kill.' So he felt he was 'a conscientious objector with a very seared conscience'. These phrases are often quoted. What tends to be ignored are his later comments. He read John Oxenham's verse in June, remarking that Oxenham 'evidently holds the Moslem doctrine – preached by Horatio Bottomley, but not by the Nazarene – of salvation *by death in war*'. (Bottomley was a jingo journalist who supported the war with extreme claims for its religious and moral value.) In August he discussed this further in an 'important' letter, saying that while he wore his star as a Second Lieutenant he could not obey Christ's command or share in the hypocrisy of churchmen who were urging on the war effort; 'thinking of the eyes I have seen made sightless, and the bleeding lads' cheeks I have wiped, I say: Vengeance is mine, I, Owen, will repay'. 'I fear I've written like a converted Horatio Bottomley', he went on, but he was now convinced that the notion that 'men are laying down their lives for a friend' was 'a distorted view to hold in a general way'. So he resolved to suppress 'The Ballad of Peace and War'. This August letter is a remarkable one, its defiant contradiction of a biblical text ('Vengeance is mine; I will repay, saith the Lord') being similar to the post-Dunsden declarations in 'O World of many worlds' and 'The time was aeon'. The manuscript shows signs of strong feeling; Owen's handwriting slopes across the page and there are marks round the phrase 'lads' cheeks' that are either blots or tears. Those blinded eyes, and the bleeding cheeks once imagined in the 'Ballad of Peace and War' but now vividly remembered from experience, belong to the same wounded 'lads' who are described in 'Has your soul sipped', 'Greater Love', 'The Sentry' and other poems. Owen's allegiance to poetry, 'physical sensation' and 'the Flesh' made it impossible in 1917–18, as in 1912–13, to accept Christian doctrine or to respect clerics who refused to recognise that impossibility; he was a poet and soldier, and could be false to neither role. There were 'no more Christians at the present moment than there were at the end of the first century' – and he could not be one of them.[15]

He found this comment about primitive Christianity echoed in
De Profundis and quoted Wilde's version of it in a February
1918 letter (but carefully avoided mention of Wilde by name).[16]
Rejecting Pauline doctrine and the established churches did not
prevent respect for Christ as a person. *De Profundis* describes
Christ as the supreme example of pity and individualism, the
perfect artist who apprehended beauty through sorrow and acted
as the 'external mouthpiece' of 'the entire world of the inarticulate,
the voiceless world of pain'. Christ had spoken for those 'who are
dumb under oppression' and had 'made of himself the image of
the Man of Sorrows'. In this interpretation of Jesus the man,
Wilde was defining the role he saw for himself as artist; Owen
would have seen how it parallelled the role of Sassoon and himself
as soldier poets. Some of his comments on his poetic task after
Craiglockhart are similar to Wilde's: he had felt 'sympathy for
the oppressed always', he was 'the poet of sorrows' and he wanted
to act as 'pleader' for 'men, that have no skill / To speak of their
distress, no, nor the will'. No doubt he saw that Wilde's letter
from prison was grossly self-regarding, but there were things to
learn from it.[17]

It may have been reading Wilde that persuaded Owen to
complete 'Greater Love' in 1918, a poem that had perhaps been
in draft for some time. He was always uneasy about it but its
subject had fascinated him throughout the war. The various
drafts reveal the poem's Decadent qualities. In a form and style
taken from Swinburne and Wilde, it compares dying soldiers on
the one hand with Christ and on the other to 'anyone beautiful'
or 'any beautiful Woman'. It began as a kind of Decadent love
lyric, like 'Has your soul sipped', without religious allusions. The
'beautiful woman' has appropriately late-Romantic features (pale
hand, drooped eyelids, red lips), as have the soldiers, whose
bleeding kisses and trembling cramps of death are more 'beautiful'
(later 'exquisite') than her beauty. There are echoes here of the
'Perseus' fragments; the Cultivated Rose had a mouth like blood
and a 'Dangerously exquisite' trembling beauty, and the speaker
in P1 suffers an agonising cramp like 'the first / Terrific minute
of the hour of death'. The old treachery of Eros is answered in
'Greater Love' when the poet himself rejects 'Love' and uses
erotic language to describe violent death. As in the face of
Medusa, the greatest beauty lies in the greatest horror; as the
first draft puts it, 'Your drooped lids lose their lure / When I
recall the blind gaze of men dead'. Several critics have explained

the Romantic language of the poem as an ironic dismissal of 'old song', but in my opinion Owen means what he says: young men dying in battle are more beautiful than any civilian lover. There is no irony in the comparison. He had said much the same in 'Has your soul sipped', as had other soldier poets elsewhere.[18]

Amendments to the poem brought it even closer to the stock poetic notions of 1917–18 by adding the comparison with Christ. Owen chose the title from what was perhaps the war's most frequently quoted biblical text.[19] He changed the original tenth line ('Where love seems not to care'), substituting 'God' for 'love' and thereby allowing later readers to detect an allusion to Christ's 'My God, my God, why hast thou forsaken me?' The original last line was 'O Love, love them, kiss them, and touch me not', where 'Love' is presumably the beautiful woman, but it was rewritten as 'Weep, woman, weep; you may weep but touch them not' and then 'Weep, you may weep, for you may touch them not'. Rosemary Freeman (1963) compared this final version to the risen Christ's words to Mary Magdalene, 'Woman, why weepest thou? . . . Touch me not' (but this may be an over-ingenious reading; the men cannot be touched because they are dead and in France, not necessarily because they are risen). Owen also replaced 'Rifles' with 'Your cross' in the last stanza, reinforcing the theme of sacrifice by alluding to his own very real memory of 'eyes blinded in my stead'. The later drafts of the poem thus imply that soldiers are Christs carrying crosses, being crucified for the redemption of others and, perhaps, rising again. But Owen knew perfectly well that too much of this had been said by sanctimonious people at home, including Bottomley. There was, for example, a popular postcard (Plate 12) showing an injured soldier and the caption 'Blinded for You!', one of many pictures which gave a sacrificial, Christian significance to death in battle. Numerous sentimental poems and newspaper articles had used the 'greater love' text for the purposes of consolation and propaganda. It is not surprising that 'Greater Love' appears only under 'Doubtful' in Owen's second list of contents at Ripon, or that he was never satisfied with the wording of the poem (the third stanza, for example, remains pointlessly repetitious, despite numerous attempts at its last line).

Nevertheless 'Greater Love' is remarkable for the intensity with which it makes its double contrast, recording that element in 'the truth untold' which was hardest to get across to civilians. As an officer away from the front, the poet thinks with guilt and pity of 'eyes blinded in my stead'; as one in sympathy with Wilde and Sassoon, he expresses his conviction that the suffering men are beautiful like lovers and he rejects 'any beautiful woman'. The theme of martyrdom was at the heart of Wilde's aesthetic creed, with Christ the suffering man as its central symbol. If there was a God, though, he was as deaf to human suffering as Hardy's Immanent Will. Owen described this deity in a flippant Wildean ballad, 'A Tear Song' (1918), in which God ignores human prayers but takes a choirboy's tear as a pearl for his jewel box. This God the Father is a pitiless maker and collector of the beauty born of suffering, an 'old man' immune by choice to 'whatever moans in man'. One similarity between Christ and soldiers was that they were sons sacrificed by fathers (or by fatherly officers), a point readily seen by Osbert Sitwell; the famous passage in Owen's letter to Sitwell about teaching Christ to lift his cross by numbers, which has often been quoted as proof of Owen's religious faith in 1918, is a typical example of Wildean imagery, written to impress its recipient.[20] Sitwell, who was at best an agnostic, used Christian themes in his poems to satirise civilians and churchmen. If Owen can be claimed as a Christian in 1918, it is only in some very loose sense of the word that would not have been acceptable to, say, the Vicar of Dunsden. There is no evidence that he acknowledged the divinity of Christ, for example, except occasionally in metaphor. The religious imagery in his 1918 work is not a sign of faith; its function is to reveal the beauty and bitterness of man's life on earth.

Owen's use of the Decadent tradition is especially clear in 'The Kind Ghosts', the only manuscript of which is dated '30/7/18 Scarboro':

> She sleeps on soft, last breaths; but no ghost looms
> Out of the stillness of her palace wall,
> Her wall of boys on boys and dooms on dooms.
>
> She dreams of golden gardens and sweet glooms,
> Not marvelling why her roses never fall
> Nor what red mouths were torn to make their blooms.
>
> The shades keep down which well might roam her hall.
> Quiet their blood lies in her crimson rooms
> And she is not afraid of their footfall.

They move not from her tapestries, their pall,
Nor pace her terraces, their hecatombs,
Lest aught she be disturbed or grieved at all.

Presumably the sleeping figure is Britannia, representing the
Nation at Home, especially women, as unaware as ever of doomed
youth. The hecatombs, vast public sacrifices, have been of men
slaughtered as cattle. It has been suggested that Owen meant
'catacombs' but perhaps he thought 'hecatombs' meant *places* of
sacrifice; he probably had in mind pagan sacrifices such as those
in *Salammbô*, where parents sacrifice their children to Moloch.[21]
The sleeping figure is a languid Fatal Woman much like
Salammbô herself, unaware of the death she brings to her lovers.
As in Venus's palace in 'Laus Veneris' (Owen bought a copy of
Poems and Ballads in August), her chambers 'drip with flower-like
red'. Her roses are the bleeding mouths often described by
Swinburne and the Decadents (Wilde wrote of beautiful men
'whose wounds are like red roses'[22]). The sado-masochism, erotic
reverie and elaborate sound-effects of Decadent poetry find what
is perhaps their final expression in 'The Kind Ghosts' in an
extraordinary but strangely appropriate context. The nation's
love for her young men is here the unseeing but consuming love
of a *femme fatale*. The 'ghosts' which 'well might' haunt her palace
do not do so, because the truth will remain untold. Soothed by
the conventional music of the poem, the reader lingers in the
garden of England where the roses bloom for ever – and then
sees what they are.

By the time Owen came to write 'The Kind Ghosts' he knew
he would soon have to return to France. There had seemed to
be a chance that Scott Moncrieff would be able to get him a
home posting, a plan that was talked of as late as mid June, but
on 4 June he was graded fit and sent to his battalion at
Scarborough for full duties. The pressures of camp life and the
loss of his Borrage Lane retreat made serious reading and writing
almost impossible.[23] Then in late July came the decisive news
that Sassoon was home with a severe head wound. Owen's letters
undergo a striking change from this moment; usually long and
conversational, they now became brief but charged with repressed
emotion. Sassoon wrote gloomily to him from hospital,

Overtures are already beginning to make me exchange pride
and clean soul for safety and the rest of it.
But I am too feeble to be able to think it out at all. I only

feel angry with everyone except those who are being tortured at the war.[24]

In the same spirit, now that Sassoon was out of action, Owen decided that it was his duty to 'throw my little candle on his torch, and go out again'. It was his duty because he saw himself as a spokesman, a role he had claimed repeatedly in his poetry from 'On my Songs' and 'The Poet in Pain' to 'Insensibility' and 'The Calls'. Many of the writers he admired had been spokesmen,[25] including not only Dickens and Shelley, both of whom he read again when he went back to France, but also Tailhade, Wilde and other Aesthetes who had made *épater le bourgeois* an obligation on the artist. Wilde had spoken out for the oppressed and inarticulate in 'The Ballad of Reading Gaol' and in his forceful newspaper articles about prison conditions. The satirical war epigrams of Sassoon and Sitwell were directly in the Wilde tradition in their onslaught on bourgeois complacency. But now Sassoon had fallen silent, as he admitted in 'Testament', a little poem which he sent to Owen in the summer: 'O my heart, / Be still; you have cried your cry; you have played your part'. In response, Owen took his place, saying he was glad to be returning to the front because he would be 'better able to cry my outcry, playing my part'.[26]

Presumably Owen applied to be drafted. The ensuing order was issued on 9 August but he was taken off the draft two days later because a medical inspection revealed a cardiac irregularity.[27] The old scares about his heart were apparently true, but a weakness which might have kept him out of uniform three years earlier was discounted now that the Army was desperate for men. Scott Moncrieff pressed the case for a home posting, but later recorded that the authorities refused it on the grounds that special privileges should not be available for anyone who had been sent home with loss of nerve. Owen may have known nothing of this ill-judged attempt at rescue until afterwards. Within a day or two he was on embarkation leave, finding time to see Sassoon and Sitwell in London. On the day before he sailed, he accompanied his mother on a brief visit to his youngest brother at Hastings. The three Owens were photographed together and Mrs Owen later had copies made from this last, crude portrait of him (Plate 15): 'The eyes are touched up badly but looking at a little distance it is *him* he had that look of high *purpose* in his dear gentle eyes. Oh! how he hated war and all its horrors – but he felt he *must* go out

again to share it with his boys....'[28] Now that he was free he
could assure her she was 'absolute' in his affections, writing
devoted letters that were meant only for her. He travelled on
alone to Victoria, where he was met by Scott Moncrieff.
Scott Moncrieff's account of their final evening together seems
to conceal a painful memory:

> I was sickened by the failure to keep him in England, and
> savage with my own unhealed wounds.... If I was harsh with
> him, may I be forgiven, as we tramped wearily round the
> overflowing hotels. In the end a bed was found.... After a
> few intense hours of books and talk in my lodging, I escorted
> him to his. As we reached it, he discovered that he had left his
> stick behind, but insisted that it was too late now to return.

Scott Moncrieff's attempt to keep his friend off the draft had
succeeded only in uncovering the 1917 accusation of cowardice,
perhaps to Owen's distress and anger. The contrast between Scott
Moncrieff and Sassoon was now obvious: the former ill-tempered
with his friends (he was notoriously quarrelsome), sexually
demanding, preaching knightly virtue yet urging a fit friend to
stay at home; the latter angry with the war and all its cant,
decent in all his behaviour, proud 'to suffer without sign', a
selflessly courageous soldier devoted to his men. It seems that
Owen refused to spend the night at Scott Moncrieff's lodging.
Perhaps it was the bitterness of that parting and jealousy of
Sassoon that drove Scott Moncrieff to write clumsily about Owen
afterwards, making the cowardice story public instead of spreading
the poet's fame as he had once hoped to do. By revealing the
War Office's ruling and omitting any mention of Sassoon, he
gave the impression that Owen would have accepted a job at
home even at the last moment, but Owen's August letters show
that this was not the case.

Owen embarked next day, pausing for a bathe at Folkestone,
where there emerged from the sea a 'Harrow boy', beautiful,
sympathetic, hating the war, a vision of all that was worth fighting
for, all that the war was destroying. 'And now I go among cattle
to be a cattle driver...', 'a Shepherd of sheep that do not know
my voice'.[29] Men in France had developed a strong sense of
separateness from people at home; the sense of being isolated and
under attack which all Ross's friends must have felt that summer
could easily transform itself into a shared companionship with

the Nation Overseas. A similar transformation had developed since August 1917 as Owen had learned to draw on his erotic verse for his war poems, the beautiful tortured youths in 'The cultivated Rose' and other pieces turning into soldiers, blood-red mouths into Britannia's roses, and the guilt of a love that was 'Against the anger of the sun' ('Reunion') into the guilt of war which the sun had to destroy ('Spring Offensive').

When he went back to France he took his recently purchased copy of Swinburne with him. In October his colleagues teased him for talking too much with some local girls; the 'dramatic irony was too killing', he told Gunston, 'considering certain other things, not possible to tell in a letter./ Until last night though I have been reading Swinburne, I had begun to forget what a kiss was.'[30] A much more serious incident a few weeks before had provided Swinburnian imagery which would have become a poem if there had been time. He had been sent out with a servant, 'little Jones', a talkative, sympathetic and devoted companion. At the beginning of October the battalion went into action, Jones receiving a head wound in the first hour. Three letters give a notion of the poem that was never written.[31] To 'My dear Scott Moncrieff':

> I'm frightfully busy . . . and many glorious Cries of the blood still lying on my clothes will have to be stifled.
> . . . I find I never wrote a letter with so much difficulty as this. Perhaps I am tired after writing to so many relations of casualties. Or perhaps from other causes.

To 'Very dear Siegfried':

> the boy by my side, shot through the head, lay on top of me, soaking my shoulder, for half an hour.
> Catalogue? Photograph? Can you photograph the crimson-hot iron as it cools from the smelting? That is what Jones's blood looked like, and felt like. My senses are charred.
> I shall feel again as soon as I dare . . . .

To 'My darling Mother':

> Of whose blood lies yet crimson on my shoulder where his head was – and where so lately yours was – I must not now write.

The three relationships are distinct. 'Cries' for Scott Moncrieff, because at this formal level Owen was the pleader crying his cry on behalf of 'many' soldiers. It was difficult to know what to say to Scott Moncrieff after their last meeting. Metaphor for Sassoon, and an implied rebuke in answer to some suggestion that catalogues of horrors were material for poetry. More intimately still, Mrs Owen could be told of the parallel between Jones and herself, though she would not have fully understood it. The extreme of 'physical sensation' could only be dealt with under the protection of 'insensibility', the poet's senses 'ironed' in a 'scorching cautery of battle'. The method, the material and the developing imagery are clear; the poem remains a phantom.

10 'Strange Meeting'

The 'Perseus' manuscripts, with their associated poems and fragments, show that Owen's imagination had often been voluntarily or involuntarily occupied with images of bodily entry into hell, a theme which begins to emerge as early as 1911–12 in his 'Supposed Confessions'. The means of entry was a paralysing descent, sometimes into smothering water ('Down-dragged like corpse in sucking, slimy fen'), sometimes into a dark cave. Once in the underworld – an obsessive but not very scholarly compound of the classical Hades, Dante's Inferno, Tannhäuser's Venusburg, and the Hell familiar to an ex-Evangelical – the living visitor from the upper world met a staring face or faces, either in darkness or under a sun 'for ever smouldering blood'. At least three of Owen's trench experiences gave substance to these horrors: his fifty hours in the flooded dug-out where the sentry was blinded, his fall into the ruined cellar, and his sheltering in the 'hole just big enough to lie in' on the railway embankment. He was trapped and helpless in all three places. Ensuing shellshock manifested itself in terrifying dreams, probably often of approaching or pursuing faces, dark tunnels and falling from heights, although the subject matter eventually became less warlike, sometimes reverting to his pre-war 'phantasies'. Soon after his return to barracks after Ripon, war dreams began again, perhaps caused, he thought, by 'the hideous faces of the Advancing Revolver Targets' (presumably mobile dummies which moved towards the marksman as he fired).[1] He had brought on such dreams in the winter by carrying out his 'duty' as a poet and now they were revived by his military duties; the constant threat of nightmare, pressing upon his determination both to give words to the pity of war and to prove himself as a soldier, was accepted for duty's sake.

The smothering hole could be generalised as a shell crater, the individual experience translated into the experience of an anonymous squad:

166

Cramped in that funnelled hole, they watched the dawn
Open a jagged rim around; a yawn
Of death's jaws, which had all but swallowed them
Stuck in the bottom of his throat of phlegm.

They were in one of the many mouths of Hell
Not seen of seers in visions; only felt
As teeth of traps; when bones and the dead are smelt
Under the mud where long ago they fell . . . .

Several critics have pointed out that the jaw image comes from
Under Fire ('the story of a squad') and 'The Charge of the Light
Brigade', noting that Owen was reading Barbusse and Tennyson
in December 1917. As I have suggested elsewhere, 'Cramped in
that funnelled hole' can be recognised as work for the 'dawn
piece' which he planned to write in that month. The dawn piece
was later to take shape as 'Exposure', early drafts of which are
in a setting of 'soft mud'.[2] The word 'Cramped' can be associated
not only with the soldiers' frozen helplessness in 'Exposure' and
Owen's possible collapse in his embankment shell hole but also
with P1, where the speaker is 'wrenched with inexplicable
weakness' as he enters hell, and with the paralysis described in
the 'Supposed Confessions'. In Romanticism poetic vision is often
accompanied by sudden bodily weakness, as in one of the poems
most dear to Owen in his early days, 'The Ancient Mariner':

> Forthwith this frame of mine was wrenched
> With a woeful agony,
> Which forced me to begin my tale . . . .

As in *The Fall of Hyperion*, the helplessness, the 'cramp', leads to
torment but also to re-creating experience in poetry, although
Owen, being 'not concerned with Poetry', stresses that the story
he is telling is not a vision but a concrete experience which has
actually been suffered with all its 'physical sensation' ('not a
dream, / But true resumption of experienced things', as he had
insisted in that visionary poem, 'The time was aeon').

The funnelled hell mouth, part of the hideous face of war
described in 'The Show', has 'all but' swallowed the soldiers. In
'Exposure' Owen changed the metaphor from a mouth to a house.
He had first thought of writing his 'dawn piece' during a night
in November 1917 when he had tramped the streets of York,

searching vainly for a hotel bed: 'the . . . hotels would not open
to my knocking'. In 'Exposure' the soldiers return home in
imagination but find that 'on us the doors are closed'. The
popular song said, 'Keep the home fires burning . . . . Though
your lads are far away they dream of home', but when these lads
dream of home they find that the fires have been left to die out
by families who have locked up and gone to bed oblivious. Owen
had once liked to imagine that soldiers could 'die back to those
hearths we died for' and even in the trenches had felt protected
from the 'keen spiritual Cold' by the love that reached him from
home.[3] Now, with the maturity that had come with the frost at
Beaumont Hamel, the troops see that they have already joined
the Unreturning. The first full draft of the poem (almost certainly
Ripon work) ends with an outline for an extra stanza:

> Blasts of the shells, blasts of the wind, these are our house.
> Whether we feel no   or yet are   creatures
> Our [hope is waiting] in deepness of dark craters
> We wait till [at our feet] [suddenly] the torn ground gulfs for us
> And our door opens[4]

After the repeated refrain of 'nothing happens' the poem was to
have ended with an expected event, the opening of the hell door
at the bottom of a crater. This ninth stanza was not completed
but Owen attempted to introduce further house imagery into the
eighth, comparing frozen bodies to bricks and plaster, rubble
awaiting the 'picks and shovels' of the burying-party. Scrapping
most of this, he was left with the burying-party and the corpses,
'their wide eyes ice' (later revised to 'All their eyes are ice'),
thereby returning to the pervasive image of his dreams, the
freezing or petrifying stare. Six years earlier he had described the
Little Mermaid trying to 'thaw' the 'cold face' of her drowned
statue when 'still the *wide eyes* stared, and nothing saw'. Descent,
stare and paralysis were all part of the same complex of imaginative experience.

His table of contents gives 'Description' as the poem's 'Motive'.
As description 'Exposure' is magnificent, its richly elaborate
technique, Romantic in sound and phrasing, working by a kind
of negative principle to evoke the unnatural nothingness of the
scene, the deprivation which the soldiers suffer in mind and body.
The opening echo of Keats ('Our brains ache' / 'My heart aches')
has often been noticed; there are several other literary parallels,

including one between the men's returning to their home fires as 'ghosts' in vain and Tennyson's lotos-eaters, who say that they might as well not return home for 'surely now our household hearths are cold' and 'we should come like ghosts to trouble joy'. These deliberate allusions set up an ironic relationship between the dazed weariness of the soldiers and the languor of the 'Nightingale' ode or 'The Lotos-Eaters'. Keats and the lotos-eaters had relaxed in beautiful landscapes which nevertheless held implicit lessons of morality and mortality. Now the threat is explicit; nature, the moral guardian of man and all life, is compelled to make war on the war-makers, seeming another German army as her clouds advance from the east like grey-uniformed storm troops.

In 'Mental Cases', written at Borrage Lane, the poet has passed the 'door' and sees some of hell's inmates. The title, which took some time to select, is one of several examples of Owen's skill in choosing impersonal military and medical terms – 'Disabled', 'S. I. W.', 'Conscious', 'Exposure', 'Spring Offensive' – for ironic effect. The method is modern irony and understatement but the feeling behind it is learned from earlier writers, especially Dickens, as is evident from the style of a 1912 comment on the word 'Cases', referring to a sick pauper child at Dunsden: 'This, I suppose, is only a typical *case*; one of many *Cases*! O hard word! How it savours of rigid, frigid professionalism! How it suggests smooth and polished, formal, labelled, mechanical callousness!'[5] The subjects of 'Mental Cases' are shellshocked soldiers, as in Sassoon's 'Survivors' and like some of the patients both poets saw at Craiglockhart. Yet Owen's figures have a further dimension, unlike Sassoon's. Words such as 'twilight', 'purgatorial',[6] 'Ever', 'Always' and the opening of what was to have been a fourth stanza, 'Time will not make', all prevent the hell of the poem from being read as a mere simile. The poet is being shown round eternal hell by a guide whom he is questioning: 'Who are these?' The question and its answer are a hellish parallel to the biblical 'What are these which are arrayed in white robes? . . . These are they which came out of great tribulation', and to Dante's questioning of his guide and brother poet, Vergil, in the *Inferno* ('Instructor! who / Are these, by the black air so scourg'd?').[7] Owen's instructor addresses him as 'brother', perhaps as both poet and officer, like Sassoon at Craiglockhart explaining how officers and the educated classes had dealt war and madness to the unfortunate; but readers tend to assume that 'brother' refers to them rather than the poet, so that 'Mental Cases' works as both a traditional poetic vision and a modern public reproach.

Much of the first stanza comes almost word for word from 'Purgatorial passions' (1916) and is thus descended from the various portraits of damned lovers which seem to have been associated with 'Perseus'. The twilit, shuddering figures in 'Mental Cases' are successors to the trembling victims of lust in the 1916 fragment as well as to the lovesick Danae, the Cultivated Rose with his shaking fingers and sweating scalp, the Flesh in 'The time was aeon', and the Mermaid's statue. Only the foxglove simile in 'Purgatorial passions' is missing, but that is replaced by the Decadent image of the bleeding sun in a comparison between dawn and a reopened wound which the young Rupert Brooke would have envied. Decadent authors had often used exquisitely subtle language to describe loathsome subject matter; Tailhade would have appreciated the meticulous care that lies behind the jagged vocabulary and elaborate assonance and alliteration of 'Mental Cases'. But these literary devices are used as means of defining experience, not as ornaments. The third stanza describes tormented eyes (which in draft work 'shrink and smother', 'smother' being one of Owen's nightmare words as in 'Dulce et Decorum Est') and the 'hilarious, hideous' falseness of fixed, corpse-like smiles, symptoms also recorded in Dr Brown's accounts of shellshock patients. These faces of war, like the pock-marked face of the trench landscape in 'The Show' or the 'dead rabbit' look of the troops, were among those 'phantoms of the mind' which Brock saw in Owen's poetry. The impassioned power of 'Mental Cases' lies in the authority and forceful language with which the 'pleader' makes his appeal, an authority which is based not only on the poem's relationship to traditions of Dantean vision and Romantic horror and protest but also on the poet's own knowledge of neurasthenic dreams.

Having entered hell and seen its occupants, the poet might be expected to hear one of them speak. Such an episode seems to be described in a strange manuscript probably written no earlier than late 1917. Conceivably work for 'Perseus', this fragment is only a very rough outline that was never revised:

> With those that are become
> Before the Future and later the Past
> For whom the present is an Absence.
> and there was one

Whose fingers pinched upon my arm
(As some old hag hissing a tale
[For an unwilling and disgusted listener]
Grips an unwilling youth with vicious finger-bones)
That he might stare into me, madmanlike.
And this was the tale
'You are not he but I must find that man.
He was my Master [whom I worshipped]
Him I wounded unto slow death. God!
Where is he that I ask if he knew me
I loved [him] the more near than brotherly
"For each man slays the one he loves ⎫
The coward ⎬
    The brave ⎭
But I must find the hand that pushed mine
[I will tear it] Emperor's or King's[8]

This has several things in common with 'Mental Cases' and
'Strange Meeting'. 'Those' ('Who are these?') are spirits in a
timeless hell, cut off from the past ('the undone years') and the
future (the 'hopelessness'). The 'madmanlike' figure, who is a
Romantic tale-teller as obsessive as the Ancient Mariner ('He
holds him with his skinny hand ... He holds him with his
glittering eye'), stares at his listener as the dead man stares with
'fixed eyes' in 'Strange Meeting'. Having fatally wounded his
master ('Wisdom was mine, and I had *mastery*'), he seeks to be
recognised by him ('I *knew* you in this dark'). The words
'wounded' and 'hand' reappear in a draft version of a famous
line in 'Strange Meeting': 'But I was wounded by your hand, my
friend'. (Owen had often associated hands and hand contact with
guilt.) The murderer's hand has been 'pushed' by an Emperor's
or King's (meaning, presumably, that he has been a soldier in a
war attributable to the Kaiser or King George – as in 'Strange
Meeting', nationality is unimportant). He has killed someone he
'loved', and he quotes from Wilde's 'The Ballad of Reading Gaol':

> Yet each man kills the thing he loves,
>   By each let this be heard,
> Some do it with a bitter look,
>   Some with a flattering word.
> The coward does it with a kiss,
>   The brave man with a sword!

Wilde's lines confirm the ambiguity of 'so you frowned . . . through me as you jabbed and killed' in 'Strange Meeting',[9] where the poet is told he has killed a friend with both a bayonet (a sword, the brave man's weapon) and a frown ('a bitter look'). A further gloss on the relationship between the two 'strange friends' is provided by the fragment's reference to more than brotherly love. Wilde's stanza helped to make some sense out of those nights in 1911–12 which Despondency had steeped in 'bloodiness and stains of shadowy crimes'; as the Limehouse poem and 'Has your soul sipped' had already implied, Eros might only be satisfied by the blood sacrifice of the poet or his idol or perhaps both.

The setting of 'Strange Meeting' (probably another Borrage Lane poem[10]) is, as Blunden noted, 'only a stage further on than the actuality of the tunnelled dug-outs'.[11] Indeed, Owen at one stage opened the poem with

> It seemed from that dull dug-out, I escaped
> Down some profounder tunnel, older scooped
> Through granites which the nether flames had groined.

'That' dug-out may have been a remembered version of the one described in 'The Sentry', where he had been tempted to let himself drown. The final draft is less specific in its first line and it omits reference to hell fire:

> It seemed that out of battle I escaped
> Down some profound dull tunnel, long since scooped
> Through granites which titanic wars had groined.
>
> Yet also there encumbered sleepers groaned,
> Too fast in thought or death to be bestirred.
> Then, as I probed them, one sprang up, and stared
> With piteous recognition in fixed eyes,
> Lifting distressful hands, as if to bless.
> And by his smile, I knew that sullen hall, –
> By his dead smile I knew we stood in Hell.
>
> With a thousand pains that vision's face was grained;
> Yet no blood reached there from the upper ground,
> And no gun thumped, or down the flues made moan.

'It seemed': again this is a dream poem, drawing on the Romantic tradition of visionary poetry and on Owen's personal knowledge of nightmare. He had memories of being trapped in holes at the Front and in a London tunnel not unlike the one in the poem. He knew his mother's fear of enclosed spaces, and, as Dr Backman has pointed out, the whole family seems to have had a memory of meeting a ghostly stranger under tunnel-like trees.[12] The poem may have been influenced by Barbusse's appalling description of an underground dressing-station; more certainly it can be related to Sassoon's 'The Rear-Guard', which Owen published in the *Hydra* at Craiglockhart. Sassoon describes an experience of a fellow officer in the Hindenburg Tunnel who had asked directions from a recumbent figure, had found that he was looking at the agonised, staring face of a corpse, and had fled in horror. Owen would also have known Graves's poem, 'Escape', an early draft of which is addressed to Sassoon;[13] Graves mythologises his own narrow escape from death as a descent into Hades, where Proserpine sends him back to sunlight up 'the corridor'. As in other poems, Owen outdoes the two Fusiliers, since his protagonist is answered by the corpse and stands his ground; his escape is into the tunnel, not out of it.

The cavern in 'Strange Meeting' derives from classical and Romantic myth as well as from the Western Front. It has been cut in ancient times by 'titanic wars' (or 'plutonic flames' in one draft), presumably in that war of the Titans which Keats had intended to describe in *Hyperion* (the first of all wars, according to myth). Like many legendary caves it contains sleepers. Owen may have been rereading Spenser's account of the Cave of Morpheus, which is 'farre from enemyes' and full of murmuring sound (but 'No other noyse, nor people's troublous cryes, / As still are wont t'annoy the walled towne').[14] The messenger sent to awaken Morpheus has to push him to get a reply. No noise of war reaches Owen's tunnel, although, in a cancelled line, 'slumber droned all down that sullen hall'. The approach of the protagonist who 'probes' the sleepers, and perhaps the 'citadels that are not walled', suggest a link between the Spenser passage and 'Strange Meeting', and the triangle is completed by Keats's description of the Cave of Quietude in *Endymion*, a cave that was certainly based on Spenser's. Endymion falls from bliss into despair and then into the psychological condition represented by the Cave of Quietude, a 'deep' 'hell' where 'Woe-hurricanes beat ever at the gate' but no sound penetrates. Calm sleep is possible in this state of spiritual deadness but no one can enter the refuge who strives for it.[15] Owen's

cavern symbolises a similar state, though less calm; it is a
Romantic metaphor as well as a classical setting. The innumerable
caves in Romantic literature are often images of the mind,
especially in Shelley ('the inmost cave / Of man's deep spirit').
When Owen returned to the front in 1918 he referred to it as
'Caverns & Abysmals' such as Shelley 'never reserved for his
worst daemons' (he was reading Shelley again).[16] The word
'strange' in his title is a late-Romantic adjective, however, and
fin de siècle writers provide many descriptions of cloistral silence
(Dowson), subterranean descents (Wells) and infernal, hopeless
landscapes (Thomson). Owen had always been interested in such
subjects; in drafts of 'The Unreturning', for instance, the poet
calls for the dead but no 'sleeper out of Hades woke' (a draft of
'Strange Meeting' revises this to 'all was sleep. And no voice
called for men'). Imagery of silence, sleep and descents into
darkness reflect the *ennui* and despair of the late-nineteenth-
century sensibility; as Brock later suggested, Europe had been
neurasthenic long before 1914.[17]

It is 'hopelessness' (originally 'lethargy') which the Other, the
dead man, gives as the principal reason for his mourning,
associating it with 'the undone years', presumably the years of
youth which are wasted ('the soils of souls untilled', as the first
draft defines 'the pity of war', echoing 'Futility') and those of
maturity which are not now to come. 'The old happiness is
unreturning': this had been a constant theme since Beaumont
Hamel.

> 'Strange friend,' I said, 'here is no cause to mourn.'
> 'None,' said that other, 'save the undone years,
> The hopelessness. Whatever hope is yours,
> Was my life also; I went hunting wild
> After the wildest beauty in the world,
> Which lies not calm in eyes, or braided hair,
> But mocks the steady running of the hour,
> And if it grieves, grieves richlier than here.
> For by my glee might many men have laughed . . .

The early 'hope' which the Other has lived for and lost has to
do with the näively Keatsian ambitions expressed in some of
Owen's lyrics. Poems such as 'My Shy Hand' and 'The Fates'
had described a timeless beauty, in pursuit of which the poet had
hoped to 'miss the march of lifetime', 'the vain untravelled

leagues'. Owen had rejected such hopes as signs of immaturity in 1917; he remained true to Keats, but to the Keats of *The Fall of Hyperion*.

> And of my weeping something had been left,
> Which must die now. I mean the truth untold,
> The pity of war, the pity war distilled.
> Now men will go content with what we spoiled,
> Or, discontent, boil bloody, and be spilled.
> They will be swift with swiftness of the tigress.
> None will break ranks, though nations trek from progress.
> Courage was mine, and I had mystery,
> Wisdom was mine, and I had mastery:
> To miss the march of this retreating world
> Into vain citadels that are not walled.
> Then, when much blood had clogged their chariot-wheels,
> I would go up and wash them from sweet wells,
> Even with truths that lie too deep for taint.
> I would have poured my spirit without stint
> But not through wounds; not on the cess of war.
> Foreheads of men have bled where no wounds were.

This draws heavily on 'Earth's wheels', the exhortation which, as I have suggested, Owen wrote for Sassoon in December. In that poem he had set out a programme of breaking ranks from militarism in order to defend truth until post-war society was once again ready for the peaceful message of the poets. The influence of Bertrand Russell on both 'Earth's wheels' and 'Strange Meeting' has already been discussed (see above pp.132–3). The most conspicuous change from the first poem to the second is the tense ('will' to 'would have'). The Other's pessimistic outline of future events still echoes Russell's prophecies, just as Sassoon still sympathised with Russell's position;[18] but like all soldiers Owen and Sassoon were not free, in conscience or in practice, to stay out of the front line. The hopes outlined in 'Earth's wheels' are now abandoned. One thing that is not abandoned, however, is the sense of apartness described in the earlier poem and by Russell (and the Romantic poets). The two characters in 'Strange Meeting' are cut off by being out of sympathy with the war, by being in hell, by being poets, by their status as soldiers and by the love which seems to be the fatal bond between them.

Apart from Russell, the principal influence on this part of the poem is Shelley, to whom Owen deliberately alludes. Just as the aspirations in 'Earth's wheels' had reflected the ideals of Laon and Cythna in *The Revolt of Islam*, so 'Strange Meeting' is influenced by a passage in *The Revolt* (v.i–xiii) in which Laon is recognised by a friend in a camp full of sleeping men, and then successfully stops a battle by stepping defenceless in front of the first raised spear, receiving its point in his 'arm that was uplifted / In swift expostulation'. As a result of his intervention, friends and enemies are reconciled like brothers 'whom now strange meeting did befall / In a strange land'. Owen, who had known *The Revolt* since Dunsden, must have intended the source of his title and the irony in his allusion to be recognised. The Other has tried to parry a bayonet ('Lifting distressful hands') but has been killed, dying as a fighting soldier not as an unarmed pacifist. Owen's 'strange meeting' takes place after death, unlike Shelley's. The poet in uniform could not hope to emulate Laon.

The last lines of the poem bring a change in style, from the ornate, semi-biblical language of the rest of the speech to a slow pacing of monosyllables that is movingly dramatic:

> I am the enemy you killed, my friend.
> I knew you in this dark: for so you frowned
> Yesterday through me as you jabbed and killed.
> I parried; but my hands were loath and cold.
> Let us sleep now . . .

Despite their simplicity, these lines complete the ambiguity and 'strangeness' of the poem. The syntax allows two meanings: the Other has been killed not only by a bayonet but also by a frown, the murderous concentration on the poet's face which represents the inhumanity of war, the pitilessness which kills. As Wilde had concluded, 'each man kills the thing he loves', some by 'a bitter look' and some by 'a sword', a look being crueller and more cowardly than a sword. The poet's frown is the means by which the Other recognises him, just as he learns from the Other's 'dead smile' that they are in hell. Each remembers his opponent's expression at the moment the bayonet struck; by looking at one another face to face in hell, they discover the truth. The ambiguous placing of 'through me' draws attention to the double significance of many details in the poem. This meeting in the underworld is a replica of yesterday's action. The poet may have entered hell during an infantry attack, like the men in 'Spring Offensive'; he would still be carrying his rifle and perhaps uses it to 'probe' the sleepers, frowning as he peers into the dark. The Other lifts his hands as if to parry the blow, his face contorted as at the moment of death. He speaks paradoxically: 'I am the *enemy* you killed, my *friend*'. The 'tunnel' (Hindenburg Tunnel or hell) contains sleepers, deep in 'thought' if they are alive or 'death' if they are not, who are 'encumbered' (by packs or war memories). The Other seems both to parry and to bless; he smiles, in agony or in welcome, suffers 'pains' which may be physical or spiritual, and can see despite 'this dark'.

The ambiguities of the poem centre on the identity of the Other. It has been common to regard him as Owen's double, an *alter ego* whose poetic creed and career are those of his author. However, until a late stage of revision, Owen thought of him as 'a German conscript', an enemy counterpart rather than a *Doppelgänger* – and not quite a counterpart, either, since Owen himself was a volunteer and had already turned away from some of the ideals which the Other adumbrates. Tailhade had warned young recruits that they would be expected to kill men whose lives had been similar to their own. Nevertheless, Owen would have been aware that encounters between a man and his other self are common in Romantic literature (they occur in Shelley and Dickens, for example). He may well have read an article by W. C. Rivers in the *Cambridge Magazine* (January 1918) which discussed Yeats's recent use of the double, relating it to literary tradition and Freud. Rivers observed that the double is sometimes represented as having the power to cast its original into hell and that in several stories, including *The Picture of Dorian Gray* (which Owen must have known), the original stabs his other self, thereby causing his own death; traditionally, meeting one's double is likely to be fatal. The event in Owen's poem cannot be reduced to a meeting between a man and his double – he had no intention of presenting war as a merely internal, psychological conflict – but neither is it concerned with the immediate divisions suggested by 'German' and 'conscript' or 'British' and 'volunteer'. The poem is larger and stranger than that. The two men are not identified, except that at first one is alive in hell and the Other is dead ('Whatever hope *is* yours, / *Was* my life also'). The meeting does seem to be fatal, however, since at the end the Other invites the poet to join him in sleep. This sleep is itself ambiguous, being death and rest yet also consciousness and torment. If the idea of the double is present at all, it may be in the mysteriously sexual element in this encounter between two men who meet, discover each other and sleep. There is a trace here of the narcissism evident in Owen's descriptions of those other sufferers in twilight, the Cultivated Rose and the casualty in 'Disabled'. The poet sees himself in the Other but the Other is an independent being.[19]

If the many doubles which have been cited with reference to 'Strange Meeting' are not all strictly relevant, there are other literary parallels which seem convincing, including Dante's pitying recognitions of the agonised faces of spirits who have had to 'abandon hope' in hell.[20] The tortured face and 'fixed eyes' of Owen's 'vision' have no lack of antecedents in Gothic fiction and Romantic poetry. In Landor's *Gebir* (III. 135), for instance, the hero descends into a cavernous underworld and is told how the dead meeting the dead have 'with fixt eyes beheld / Fixt eyes'. Tortured, hypnotic eyes are stock Romantic properties; the Ancient Mariner, the last chapter of *Salammbô* and, above all, Keats's vision in the second *Hyperion*, provide examples which Owen knew well. In Keats's 'Lamia' the philosopher's relentless stare reveals the truth and kills delight. In *The Fall of Hyperion* the goddess of memory unveils her dying yet immortal face and unseeing eyes, thereby allowing the poet to share in her knowledge of the titanic wars of long ago and of the fallen Titans lying 'roof'd in by black rocks . . in pain / And darkness, for no hope':

> deathwards progressing
> To no death was that visage; it had past
> The lilly and the snow; and beyond these
> I must not think now, though I saw that face . . . .[21]

The first draft of 'Strange Meeting' mentions the whiteness of the Other's face: 'With a thousand fears his [strange, white] face was grained'. The *Hyperion* passage is also echoed in 'The Sentry', another description of seeing fixed eyes in a dug-out: 'I try not to remember these things now'. Murry said in 1919 that *The Fall of Hyperion* was undoubtedly Owen's source: the 'sombre imagination, the sombre rhythm [of 'Strange Meeting'] is that of the dying Keats . . . this poem by a boy with the certainty of death in his heart, like his great forerunner, is the most magnificent expression of the emotional significance of the war that has yet been achieved by English poetry'.[22] Owen could have wished for no greater compliment.

It may be that 'Strange Meeting' does not fully tell 'the truth untold'. The *Cambridge Magazine* said on 24 February 1917 that 'anybody who has intimate friends at the front must know . . . that a great deal remains untold . . . often the most important part of psychological experience'. Critics who are interested in such matters may consider, for example, the relationship between this poem, 'Earth's wheels', 'With those that are become' and Owen's feelings for his friend and master, Sassoon. Certainly, 'Strange Meeting' is one of his most intensely personal poems, despite the grand, impersonal language of its central sections, yet it is also one of his most political and wide-ranging statements. Like the manuscripts of his other 1918 hell pieces ('As bronze', 'With those that are become', 'Mental Cases', possibly 'Exposure', and 'Spring Offensive'), the two drafts of 'Strange Meeting' show signs of his intending to continue the poem. 'Let us sleep now . . . .' is scribbled in as an afterthought in an unusually shaky hand. The classical tradition represented death as a tranquil, silent sleep but the late-Romantic use of the image made the state less desirable; in Swinburne's *Atalanta in Calydon*, for instance, the dying Meleager describes himself as 'gone down to the empty weary house / Where no flesh is nor beauty nor swift eyes'. (Murry might have adduced Swinburne as another source of Owen's sombre imagination and rhythm.) The 'sullen hall' of 'Strange Meeting' is not the Cave of Quietude and the pain there is not only that of loss; the sleepers are 'encumbered' and groaning, carrying with them into an eternity of damnation the wounds and dreams of war.

11 'Spring Offensive'

In his letters from France in September and October 1918 Owen
used the word 'serene' to describe himself, privately reassuring
his mother and Sassoon that his nerves were now in 'perfect
order'. 'You would not know me for the poet of sorrows.' He had
no intention of reassuring anybody else, since civilians were not
to imagine that any soldier overseas was contented, so he marked
a particularly cheerful letter 'Not to be hawked about'. It no
longer seemed right to draw attention to his own experiences;
whereas in 1917 he had asked for parts of his letters to be
circulated, now even his letters about fighting were marked
'Strictly private' and 'Not for circulation as a whole'. While he
committed himself to serene activity, his anxious mother had
taken to her bed. The last paragraph he ever wrote (31 October)
sums up their respective roles:

> I hope you are as warm as I am; as serene in your room as
> I am here; and that you think of me never in bed as resignedly
> as I think of you always in bed. Of this I am certain you could
> not be visited by a band of friends half so fine as surround me
> here.

'The shades keep down which well might roam her hall'; Mrs
Owen's passivity was not to be disturbed.[1]

He was clear about the political nature of the war's closing
stages, telling Sassoon that he might find himself in front of a
firing-squad if he wrote poems in the dug-outs or talked in his
sleep[2] but soon finding that his opinions were widely shared.
There had been much talk in Britain of 'the Nation at Home'
and 'the Nation Overseas', but the latter now seemed the only
true nation, still worth loyalty long after honour had gone from
the home front. He was glad to be 'back here with *the Nation*',
away from civilians who were still supporting the war without
having to fight it.[3] He still disapproved of the Gunston brothers,

telling Leslie that 'I must say that I feel sorry that you are neither in the flesh with Us nor in the spirit against war.'[4] In mid October an officer returned to the battalion 'from his first visit to London utterly disgusted with England's indifference to the real meaning of the war as we understand it'. Quoting Sassoon, Owen defined 'we' as 'every officer & man left, of the legions who have suffered and are dust'.[5] Delighted to find that the troops were turning against *John Bull* and the *Daily Mail*, which were clamouring for total victory, he felt a strong sense of solidarity with 'Us', soldiers who knew that they and the dead were all that was left of England:

(This is the thing they know and never speak,
That England one by one had fled to France,
Not many elsewhere now, save under France.)[6]

He was convinced that the war was once again being prolonged by prussianism at home and abroad, urged on by a vindictive newspaper campaign and atrocity stories. 'I have found in all these villages *no evidence of German atrocities*', he told Gunston. 'Do you still shake your befoozled head over the *Daily Mail* & the *Times*?' The damage he saw was caused by British guns, which were killing French civilians because the Allies were refusing to let Germany retreat in peace. The German offensive had made further war inevitable in the spring but circumstances were now very different. That he was by no means alone in his views is suggested by an official order that 'Peace Talk must cease in the Fourth Army'.[7] It was a little like Ross's circle on a much wider scale: a group of friends, who were alienated from press and public but held together by comradeship, a shared secret and common adversity. The bond seemed political as well as comradely; the description in Owen's last letter of his men, his 'band of friends', is unusually detailed and similar to some of Barbusse's accounts of the squad in that political book *Under Fire*.

'FRENCH SENATE THRILLED Clemenceau's Great Speech' (*Daily Mail*, 19 September). *The Times* provided a literal translation of the speech, in which the French Premier had announced France's refusal of an Austrian offer of peace talks on the grounds that peace now would be a betrayal of the troops who were still fighting. Owen incorporated parts of the translation into 'Smile, Smile, Smile', which he finished four days after the speech was reported.[8] The wounded soldiers in the poem smile 'curiously' over such newspaper items and keep the secret of the Nation Overseas.

> Pictures of these broad smiles appear in Sketches
> And people say: They're happy now, poor wretches.

This draft version of the ending is a clue to another newspaper source. On 16 September the *Daily Sketch* had published a picture (Plate 14) of three wounded men, each with a ghastly smile, the caption twice describing them as 'happy'.[9] The three smiles in the photograph may have suggested the poem's title; it was men such as these whom the music-hall song urged to 'Smile, smile, smile'. If the poem is read in conjunction with Owen's letter of 22 September to Sassoon, it can be seen to be a deliberately Sassoonish piece, a 'cry' in the style that Sassoon seemed by then to have abandoned. It is Owen's last poem in his friend's manner, but by this stage in his poethood he could not write anything that lacked his own stamp; the 'secret' is akin to the 'truth untold' in 'Strange Meeting', and the 'sunk-eyed wounded' with their limp heads leaning towards each other are described in the same terms as the damned in 'Purgatorial passions', whose eyes were sunk in 'chasms' and whose necks were 'bowed like moping foxgloves all'.

And indeed he was one of the damned himself, though with no visible wounds. Some of his companions had been with him in 1917 and remembered the railway embankment and the flooded dug-out. He revised his poem about the dug-out in September, sending it to Sassoon with the provisional title 'The Blind' (presumably Sassoon chose the final title, 'The Sentry'). That memory of an injured head was soon replaced in intensity by another when little Jones was shot in fierce fighting in October. The railway bank had been the site of Owen's alleged loss of nerve and he had not forgotten that he needed to 'get some reputation of gallantry' before he could speak out publicly against the war. The October battle gave him his chance. He was awarded the Military Cross, having 'behaved most gallantly' according to the citation, and was glad of it 'for the confidence it may give me at home'. He described his twofold task in a famous statement: 'I came out in order to help these boys – directly by leading them as well as an officer can; indirectly, by watching their sufferings that I may speak of them as well as a pleader can. I have done the first.'[10] He told his mother that he had shot one man with his revolver, captured a machine gun, and taken scores of prisoners, but he did not say what he had done with the machine gun. When the MC citation was published in the *Collected Letters*, one sentence in it was misquoted: 'He personally captured an enemy Machine Gun in an isolated position and took a number of prisoners.' The original citation, of which a typescript on War Office paper is preserved in Tom Owen's scrapbook, gives a different wording: 'He personally manipulated a captured enemy M. G. from an isolated position and inflicted considerable losses on the enemy.'[11] The poet who had hoped to wash the blood off war's chariot wheels had, in the words of his last poem, out-fiended the fiends and flames of hell with 'superhuman inhumanities,/ Long-famous glories, immemorial shames'.

He told his mother,

> I can find no word to qualify my experiences except the word SHEER. (Curiously enough I find the papers talk about sheer fighting!) It passed the limits of my Abhorrence. I lost all my earthly faculties, and fought like an angel.[12]

The language of 'Apologia pro Poemate Meo' and 'Insensibility' was being tested out afresh ('some scorching cautery of battle', 'power was on us . . . Not to feel sickness or remorse of murder', 'Seraphic for an hour'). The word 'sheer' was added to his poetic vocabulary for use in 'Spring Offensive'; he had probably seen it in two typical headlines in the *Mail*: 'ADVANCE BY SHEER FIGHTING The Better Men Win' (19 September) and 'SHEER FIGHTING Both Sides Pay the Price Huns Wait for the Bayonet' (3 October).[13] The poem which he would have written about Jones's wound would not have been in Sassoon's style ('crimson-hot iron . . . That is what Jones's blood looked like, and felt like. My senses are charred').

The only verse which Owen is likely to have written after the October fighting is the later part of 'Spring Offensive'. He had said in his Preface in the spring that English poetry was 'not yet fit to speak' of heroes and his own attempts to write about modern soldiers as, for example, Arthurs ('Hospital Barge') or Horatius ('Schoolmistress') had not been very successful, but in 'Spring

Offensive' he began to fashion a kind of poetry that would be fit
to speak of heroes while denying heroic qualities to war itself. His
last poem seems both a prologue to new writing and an epilogue
to all he had written before. It took him some time to write; he
seems to have begun it in the summer, since '[Attac] Spring
Offensive' appears in a list of titles probably drawn up when he
was choosing work to send to *Wheels* in August, but as late as 22
September he sent Sassoon a version of the first seventeen lines
only, asking 'Is this worth going on with? I don't want to write
anything to which a soldier would say No Compris!'[14] Despite
this awareness that the poem was more 'difficult' than some of
his work, he decided that it was indeed worth finishing, although
the ending which he wrote may be different from, and shorter
than, the one he originally had in mind; he added the last stanza
in a hurried, unrevised pencil and would certainly have rewritten
it had time allowed. The earlier part of the only complete draft
is extensively amended, with illegible, cancelled and alternative
wordings which will always puzzle his editors. There are signs
that he began to break up the stanzas into irregular paragraphs,
as he had done in several other poems, but this job was
not carried through. The version which follows is Professor
Stallworthy's text:

> Halted against the shade of a last hill
> They fed, and eased of pack-loads, were at ease;
> And leaning on the nearest chest or knees
> Carelessly slept.
>   But many there stood still
> To face the stark blank sky beyond the ridge,
> Knowing their feet had come to the end of the world.
> Marvelling they stood, and watched the long grass swirled
> By the May breeze, murmurous with wasp and midge;
> And though the summer oozed into their veins
> Like an injected drug for their bodies' pains,
> Sharp on their souls hung the imminent ridge of grass,
> Fearfully flashed the sky's mysterious glass.
>
> Hour after hour they ponder the warm field
> And the far valley behind, where buttercups
> Had blessed with gold their slow boots coming up;
> When even the little brambles would not yield
> But clutched and clung to them like sorrowing arms.
> They breathe like trees unstirred.

Till like a cold gust thrills the little word
At which each body and its soul begird
And tighten them for battle. No alarms
Of bugles, no high flags, no clamorous haste, –
Only a lift and flare of eyes that faced
The sun, like a friend with whom their love is done.
O larger shone that smile against the sun, –
Mightier than his whose bounty these have spurned.

So, soon they topped the hill, and raced together
Over an open stretch of herb and heather
Exposed. And instantly the whole sky burned
With fury against them; earth set sudden cups
In thousands for their blood; and the green slope
Chasmed and deepened sheer to infinite space.

Of them who running on that last high place
Breasted the surf of bullets, or went up
On the hot blast and fury of hell's upsurge,
Or plunged and fell away past this world's verge,
Some say God caught them even before they fell.

But what say such as from existence' brink
Ventured but drave too swift to sink,
The few who rushed in the body to enter hell,
And there out-fiending all its fiends and flames
With superhuman inhumanities,
Long-famous glories, immemorial shames –
And crawling slowly back, have by degrees
Regained cool peaceful air in wonder –
Why speak not they of comrades that went under?

The poem is loosely based on the assault at Savy Wood in April 1917, the action which preceded Owen's shellshock. Some details, such as the lack of bugles, correspond to those in his 1917 letters but there are differences.[15] For example, the weather in that April had been poor, the land around Savy had already been fought over and the advance was at walking speed, but the troops in the poem 'race' over an undamaged landscape in May sunshine. The reader is evidently meant to recognise the significance of Maytime and the double meaning of the title. May is the traditional setting for poetic experience, in medieval dream poems, for example, or the 'Ode to a Nightingale'.[16] The term 'spring offensive' was a standard one, referring to the 'Push' that could be expected when winter weather ended, but here the spring is the object as well as the time of the attack ('Halted *against* the shade of a last hill', '*against* the sun'). The opposition between attackers and season is crucial to the poem.

The setting is familiar. The unidentified soldiers (as he said in his Preface, Owen avoided using names and nationalities in his 1918 work) have walked eastwards up a valley, the dew making buttercup petals or pollen adhere to their slow (reluctant?) boots. Ahead, an abrupt, grassy hill is silhouetted against the flat glare of the rising sun, whose rays are already warming the landscape through which the troops have passed. Like 'Futility' and 'Exposure', this is another 'dawn piece'.[17] Some men take the opportunity of catching up on the sleep they have lost during the night march; others (the 'wise', perhaps, for many men in the Nation now shared the wisdom of poets) stand still as trees, taking the summer into their veins as a tree draws sap from its roots. The comfort is only bodily, a 'physical sensation', for the souls that are alert are acutely aware of the menace of the ridge and of the unsheltered plateau of 'herb and heather' beyond it. This landscape is recognisably Salopian, the blessing offered by the buttercups being very similar to that given by the 'croziers' of young bracken shoots in a metaphor which Owen had made years before on Caer Caradoc. Harold Owen even claimed that the buttercup image was coined near Shrewsbury, a claim which unfortunately merits some scepticism but which was no doubt based on an actual memory of the family's yellowed boots in the Monkmoor fields. The ridge in 'Spring Offensive' resembles Shropshire hills such as Haughmond, the Wrekin and the Caradoc, the first two of which Owen mentioned in October. He said he had been as agile in the fighting as his Welsh mountain forefathers. He must have been letting his mind dwell on his native border country where he had spent the happiest times of his adolescence, no doubt also remembering the steep, heather-topped hill at Broxton.[18]

The soldiers in the poem have ignored the buttercups and the 'sorrowing arms' of the brambles. In personifying nature Owen often imagined hands and holding: 'the grip and stringency of winter', 'Pale flakes . . . fingering', 'the wide arms of trees'. In the last of these examples, from 'Happiness', the tree branches had originally been 'mother-arms'; they offer an unfilled, unreachable embrace, matching the possessive love which Owen knew his mother still felt for him even though he was no longer 'a Mother's boy'. In 'Exposure' nature's embrace has the sinister caress of snow, for nature and love could become fatal when their more kindly aspects were set aside in the inevitable transition from innocence to adult knowledge. Spring was the moment of growth, the season for 'putting forth' poetry (as leaves to a tree) and for walking in the woods with young companions – with Rampton in 1912, Henriette in 1914, the Mérignac boy amid the surging foliage of 'his' woods in 1915. After another such ramble in 1916, Owen reported that 'we ate the Vernal Eucharist of Hawthorn leaf-buds. / These are the days when men's hearts (some men's) become tender as the new green.'[19] Spring's power was both redemptive and sexual, its new beauty innocent and erotic, parallelling the 'crucial change . . . from boy to man' in the human body that Owen had watched in Rampton. Many of his lyrics had used images of season and landscape to describe the beauty of youth. The speaker in P1 had imagined that there would be 'rest for ever' in the 'valleys' between his lover's shoulders. The valley in 'Spring Offensive' may have offered 'ease / For ever from the vain untravelled leagues' but it, like youth, has had to be left behind. In 1912 Owen had preferred 'the placid plains of *normal ease*' to the higher 'dangerous air where actual Bliss doth thrill' but in the 1918 poem the soldiers lie at ease only as a preliminary to climbing into hilltop air full of the 'even rapture of bullets'.[20]

The spring offensive runs counter to Geddesian and Romantic principles. Brock's allegory of neurasthenia was of Antaeus defeated by Hercules, 'the war machine', who tore the organism from its environment or, in Owen's poem on the subject, 'rooted up' Antaeus like a tree. The troops in 'Spring Offensive' see the blessing offered by nature, standing 'like trees unstirred' until the order to attack comes like a chill wind; then they are stirred, the contact is broken and they move into action, tree-like no more.[21] War has uprooted them but this time not against their will; they are not felled by a storm wind but choose to move, aware of the consequences. Brock's Ergotherapy was at odds with its own ideals, curing soldiers in order to send them back into action where, even if their nerves were in perfect order, their 'work' was to attack their environment once again. But the environment in the poem is not only the nature which Owen had studied with Brock in the Pentlands and with Gunston in Oxfordshire and

Shropshire but also the nature which the Romantics had taught him to revere: 'murmurous with wasp and midge', 'the summer oozed', 'drug' – these phrases are deliberate echoes of Keats's odes ('murmurous haunt of flies on summer eves', 'oozings', 'some dull opiate'). There may also be more distant echoes of a passage in *The Revolt of Islam* (vi. ix–x), where Laon and others gain 'the shelter of a grassy hill' and hold off a murderous enemy with 'stern looks beneath the shade / Of gathered eyebrows', standing 'firm as giant pine'.[22] The soldiers in 'Spring Offensive' reach the 'shade' of a grass-grown hill and direct challenging looks at their enemy, the sun. If the men who have pondered the warm field with its buttercups and brambles may be understood to be as wise as poets, since poetic wisdom was now the 'philosophy of many soldiers', and as observant as naturalists ('Do Plants Think?'), they perceive that they have chosen to attack the landscape of poetry and social health or, less figuratively, the ideals which had been preached by the Romantics and the Outlook Tower. They are unlikely, after all, to become 'one with nature, herb, and stone'.

The attack necessitates a spurning of nature, almost as though it were 'against' the hill and the sun. The only suggestion of a human enemy is the word 'bullets' in a line which Owen revised again and again without completing it to his satisfaction; had there been time he might have seen that the word itself was the cause of the trouble. He also tried many wordings for the challenge which the men make by looking at the sun, describing the reflection in their eyes as 'light', 'glory', 'radiance', 'mighty kindle', 'lift and sparkle', 'blaze' and 'flare', before hitting on the pun of 'lift and flare'. As a battle flag is lifted to 'flare' in the wind, so the soldiers, lacking flags, raise their eyes and let them flare in the sunlight. The 'kind old sun' is rejected like a lover. For a moment his smile is outshone by the 'mighty' smile of the advancing infantry. There is something here of the 'exultation' which Owen felt at Savy Wood. 'So, soon they topped the hill, and raced together' was originally less vigorous: 'Turning, they topped the hill, and walked together'. They had already turned to face the sun, so 'Turning' had to go. Owen tried 'Splendid', 'Proudly', 'Glorious', 'Lightly', 'Bright-faced', before settling on 'So, soon' and replacing 'walked' with 'raced', giving the line the speed and exhilaration which he wanted to convey.

'Bright-faced' is interesting. Christ's face shone 'as the sun' at the Transfiguration and God's face is said in Revelation to shine 'as the sun shineth in his strength'.[23] In *God the Invisible King* (1917), a book Sassoon and Owen talked about when they first met, H. G. Wells had said that the God of the new age that was beginning should be represented as 'a beautiful youth' already brave and wise', standing 'lightly on his feet in the morning time', his eyes 'as bright as swords', his lips parted with eagerness, his sword and armour 'reflecting the rising sun'; Christ was no meek victim but a militant hero with a countenance 'as the sun'.[24] Owen seems to be using a version of the soldier–Christ image quite unlike that of the crucified martyrs in 'Greater Love', for he was as clear as ever that he could not follow the Christ who preached 'Passivity at any price'. Wells's book expressed a religious optimism about the war which its author later regretted and which Owen would have found absurd, but the epithet 'Bright-faced' does suggest that Owen imagined the soldiers as being like new gods in their brief splendour. There is no orthodox Christianity in 'Spring Offensive' but there is a commitment to activity. The men's racing may be compared with 'Training' (late June 1918), where the poet prefers the 'clean beauty of speed and pride of style' of cross-country running to love and languor. Like the 'cold gust' in the later poem, 'Cold winds . . . Shall thrill my heated bareness'; but 'None else may meet me till I wear my crown'. (The 'crown' may have been that of martyrdom or an athlete's prize or, perhaps, the crown-shaped badge of a major, the reward for military success.) 'Bright-faced' also suggests sexual and poetic achievement, as in the 'Perseus' fragments, 'Storm', 'A Palinode' (where the 'blessed with gold' metaphor first appears) and other poems. In order to win beauty, poets and lovers had to risk 'the anger of the sun' ('Reunion'); the soldier must take a similar risk to prove his love for his comrades ('they raced *together*').

As the blood of the casualty in 'Disabled' had flowed with a 'race / And leap', so the men in 'Spring Offensive' 'race' over the exposed ground and 'leap' to unseen bullets. In both poems the words mark the climax, though in the second it is not passive. Owen wrote 'Leapt to unseen bullets' and 'Breasted the [surf of] even rapture of bullets' in his draft, implying excitement and voluntary pushing forward as in swimming. In response, the flashing sky[25] suddenly burns with fury against the soldiers and they fall into 'Caverns & Abysmals' such as Shelley 'never reserved for his worst daemons', 'abysmal war' opening 'sheer' at their feet like 'the end of the world',[26] the *fin du globe* image

becoming literal. The 'cups' are shell craters, some of the 'many
mouths of Hell', infernal chalices for the sacrificial blood. Such
is 'the anger of the sun' against the band of friends who have
spurned his love but been true to their own.

The 'last hill' has become the 'last high place'. With his
background of regular Bible study, Owen would have connected
'high places', mentioned over a hundred times in the Old
Testament, with sacrifice. Some pagan high places were associated
with Moloch, the deity to whom fathers sacrificed their children
by fire.[27] In *Salammbô*, the novel Owen read before enlisting, the
Carthaginians feed their sons into a red-hot statue of Moloch,
the god of sun, fire, blood and war, in the hope of bringing a
disastrous war to an end. Moloch also appears in Osbert Sitwell's
war poems as the god to whom 'old men' were sacrificing their
sons. Owen's reading of the war's last stages was that fathers were
making their sons pass through the fire ('hot blast and fury',
'flames') while women slept like Salammbô the *femme fatale*,
unaware of the 'hecatombs' around them. The work of leading
boys to the sacrifice was left to officers, priests of the modern
Moloch.[28] In the light of this, Joseph Cohen's interpretation of
'God caught them even before they fell' as a 'gesture . . . of
compelling blood-lust'[29] is a little less absurd than it seems and
certainly nearer the mark than the usual sentimental explanation
of the line as pious reassurance. Owen seems originally to have
planned to end the poem at this point with 'Of them *we* say God
caught them as they fell', as he had said of corpses in 1917 that
in 'poetry we call them . . . glorious'.[30] The first person was
misleading in 1918 but his underlining of '*we*' shows how the final
version of the line should be read: '*Some*' people at home may say
that God caught them, but what say those who fought in and
survived the offensive? '*Some* say': the myth is brushed aside,
rather as Milton dismissed the legend of Mulciber's fall ('thus
they relate, / Erring').[31] Pagan slaughter leaves no room for
Christian consolation.

   The soldiers have faced 'the stark blank sky', as Keats faced
the 'blank splendour' of Moneta's eyes. Their smile answers the
sun's; as in 'Strange Meeting' and 'Smile', Smile, Smile', there is
recognition and shared secrecy. Their door opens 'in deepness of
dark craters'. They enter hell in body as well as in spirit, following
the poets and heroes of legend who entered the underworld as
living men. Like the speaker in P1, they are cut off from the sun.
Their fall from light to darkness is 'sheer' like the abrupt descents

from delight into loss which characterise Keats's poetry. The sheer fall can be related to Owen's comparison of the sensation of going over the top at Savy Wood to 'those dreams of falling over a precipice, when you see the rocks at the bottom surging up to you'.[32] The nightmare is also sheer in the sense of being complete, for this is his final dream vision, sternly controlled and made impersonal. Hell's fire and darkness are outdone by 'bloodiness and stains of shadowy crimes', called by the world 'glorious' and rewarded with Military Crosses. The upper air purges itself of heat and fury, the sun can smile again, and the survivors regain the May landscape in wonder beneath 'an inoffensive sky',[33] carrying with them the knowledge which Owen had gained from Beaumont Hamel and more recently from his use of the machine gun.

'Why speak not *they* of comrades that went under?' Politicians and journalists were all too ready to speak but the anonymous soldiers stay silent, with 'strength to perform, and pride to suffer without sign'. 'Spring Offensive' takes English poetry a little nearer being fit to speak of heroes. In speaking for his men, the pleader does not after all restrict himself to writing about their sufferings or about passivity, although there is nothing in the poem that could be taken to imply a favourable view of war. The medium is narrative fiction in the epic tradition, as several critics have pointed out; Owen would probably not have described this poem as an 'elegy'.

He does not refer to himself in 'Spring Offensive' yet the poem came – as naturally as leaves to a tree – out of his inner life, rounding off his poetic career and at the same time giving promise of further achievement. Its setting corresponds to the landscapes where his imagination had first been touched by beauty, while its language and imagery show how much he had learned from the great nineteenth-century writers whom he had admired since his schooldays. His use of a short lifetime's experience and reading is characteristic; he brings poetic ways of seeing and evaluating to bear upon contemporary events, with the aim of speaking for the common man as the Romantics had done before him. His own courage is reflected in the men's lack of hesitation in the poem, just as his own self-awareness becomes theirs when they stand like trees; he no longer speaks about them from a distance as he had been obliged to do since Savy Wood, but shares in their action, seeing where it leads. Something of the 'Perseus' pattern survives in the stare, the sudden activity, the fall into hell and the unanswered final question. The question itself is a reminder not only of the ominous divisions that have split twentieth-century society but also of the strangeness and secrecy of his genius.

He was set apart all his life. First, he was his mother's favourite, isolated from the rest of the family. Then his poethood began in secret darkness, born out of a tradition which had made the poet both the prophetic voice of the people and a solitary, damned figure, a dreamer cast out from sunlight. Sexually, he belonged to a group which had to be separate and unseen, though it scorned the morality of the crowd. And he was a soldier in the Nation, that band of friends in France who seemed in the end to be like lovers, poets and heroes. At the moment of their greatest achievement they were superhuman and glorious, yet their deeds were shameful inhumanities and their reward was damnation whether they lived or died. He recorded their story, which was also his, in his poems: 'These elegies are to this generation in no sense consolatory. They may be to the next. All a poet can do today is warn. That is why the true Poets must be truthful.' He was true to his destined task of warning and pleading. For himself, after many doubts and troubles, he was content to follow the gleam into darkness.

Appendix A: Biographical Notes

For abbreviations used in the Appendixes and Notes, see pp.207–8.

1 CLYDE BLACK (1872–1948)

Arthur Clyde Henderson Black apparently became a lay assistant at Dunsden in 1912. Twenty years older than WO and perhaps a recent convert (he feared relapsing 'into his old ways'), he was 'solemn' but forceful, making the household retire to bed at ten. Perhaps he led the Revival; he certainly 'cornered' Willie Montague and elicited confessions of faith from him and at least two other parishioners within two days. (*CL*, 172, 166, 168, 170). London College of Divinity, 1915. Ordained, 1917. Parishes in East London and Sussex. Unidentified in *CL* and not mentioned in JS, but a significant figure in the story of WO's religious crisis.

2 THE LÉGERS

Albine (Nénette) Léger married Jean Loisy, poet, but died young. He recorded his grief in *De la mort à l'espérance* (Paris: Beauchesne, 1952) and quoted a chapter from his wife's unfinished novel (*Tout un monde*). The chapter is a thinly disguised account of Albine and her parents starting a Bagnères holiday *c.* 1909. The relationships match many details in *CL*, although Mme Léger is portrayed as entirely virtuous. M Loisy told me some years ago that Mme Loisy had remembered her mother's English tutor with sorrow and affection. She published three complete novels, as well as translations of *Middlemarch* and *The Rainbow*; even when WO knew her she had written 'astonishing' dramas (*CL*, 271). Her father, Charles Léger, was well esteemed in dramatic circles but had lost his parents' money in experimental theatre. Mme Léger restored the family fortunes by going into partnership with her father-in-law. Like her daughter, she died young. The house at La Gailleste still stands, sadly altered, and still belongs to the Cazalas family.

3   LAURENT TAILHADE (1854–1919)

Born at Tarbes, not far from Bagnères. The latter was a place he loved. 'C'est là . . . que, pour la première fois, j'ai communié de la beauté des choses' (Mme Laurent-Tailhade, 155–6). Accounts of his life are inconsistent. Son of a drunken father and pious, devoted mother, he was correctly pious himself at first, then joined Péladan's mystic order, and eventually became an atheist. 'Le Christianisme n'étant pour lui qu'une pollution de la raison humaine' (Kolney, 47). Settled in Paris, where he became a leading figure in the Decadence (although he later dismissed the word as meaningless and described members of the group with sceptical amusement). A principal contributor to *Le Décadent*, especially of satires against bourgeois philistinism. Friendly with many famous authors; he, Verlaine and Moréas are said to have been an inseparable trio. But also an extravagant dandy, welcome in the best *salons* for his brilliant conversation. His anarchist sympathies appear in his famous comment on the Vaillant bomb in 1893 (see Ch.3, above). A year later another bomb deprived him of his right eye. Imprisoned 1901–2 for criticising France's new ally, the Tsar. Fought many duels, some of them with people offended by his satires. His politics moderated in later life. Married at least twice but responsive to male beauty. By 1914 he was fat, greedy, short of breath and weakened by absinthe and opium. Nevertheless by November he was 'shouldering a rifle' (*CL*, 295) in company with Anatole France, aged seventy, who had joined the ranks to demonstrate publicly that he had abandoned his pacifist principles in the face of invasion; presumably Tailhade was making the same point.

OEF has two letters to WO, 1 Apr 1915 (published in *Yggdrasil*, Paris, July–Aug 1939) and 1 May, both urging him to stay in Paris. On 1 April, Tailhade says he has seen M Léger but had feared to seem indiscreet 'en demandant vos nouvelles'; invites WO to translate 'Les fleurs d'Ophélie' because WO had admired it; and says he has not forgotten 'cet aprés-midi, ni le chemin de La Gailleste, ni Baudéan, ni le Casino de Bagnères'. On 1 May, he asks WO for the name of his Paris hotel and hopes they will dine together. This Paris visit, not mentioned in JS, certainly occurred; WO stayed in a hotel for at least one night and met Tailhade, who gave him *Poèmes élégiaques* on 4 May and introduced him to a composer (CL, 352, 336; JS, 321).

Mme Laurent-Tailhade, *Laurent Tailhade intime* (1924). F. Kolney, *Laurent Tailhade: son oeuvre* (1922). Noël Richard, *Le Mouvement Décadent* (1968) 157–70; *À l'aube du symbolisme* (1961) 148–53. Articles by Ezra Pound and Richard Aldington, reprinted in C. N. Pondrom, *The Road from Paris* (1974).

Tailhade's many books include *Poèmes élégiaques* (1907) and *Poèmes aristophanesques* (1904), his collected poems; *Pour la paix/ Lettre aux conscrits* (1909), the former a ?1908 lecture, the latter an essay dated 1903; *Plâtres et marbres* (1913); *La Douleur* (1914); *Quelques fantômes de jadis* (1920); and *Petits mémoires de la vie* (1922).

4 ARTHUR JOHN BROCK (1879–1947) AND THE OUTLOOK TOWER

Son of a Scottish gentleman farmer. Edinburgh University (MB, ChB, 1901; MD, 1905). Wanted to be an artist (his mother, Florence Walker, had published verse) but his father forbade it. By 1901 he was under the spell of Patrick Geddes, writing to him about books and ideas. Most of his theories were to be based on Geddes's 'synthesising' principles. Active in Outlook Tower affairs for many years, lecturing on Bergson in 1914, leading natural-history expeditions into the Pentlands, organising the Open Spaces Committee, etc. During the war, served on a hospital ship to India, then at Aldershot. Became (as Temporary Captain, Royal Army Medical Corps) one of the three medical officers at Craiglockhart in 1917 (the others being Rivers and a Dr Ruggles, under the command of an apparently much-liked Major Bryce). After the war, ran his house in North Queensferry as a home for mental patients. He and his Swedish wife, one of the first women physiotherapists, are remembered as impractical, incessantly talkative, keen travellers, full of good works in the local community. AJB was a voracious reader with wide interests, and a prolific contributor to newspapers and professional journals. His articles show that the 'Ergotherapeutic' methods he used in 1917 were developed from his pre-war thought and practice.

Health and Conduct, with a foreword by Professor Patrick Geddes

(Williams & Norgate, 1923). This, Brock's major work, was published with a grant from Victor Branford, but so few copies were sold that the surplus was offered at a cheap rate by the Sociological Society (with its Le Play House label pasted over the original imprint). Translations: *Galen on the Natural Faculties* (Loeb, 1916); *Greek Medicine* (Dent, 1929). Articles (a few of many): 'Ergotherapy in Neurasthenia', *Edinburgh Medical Journal*, May 1911, 430–4; 'The "Moral Factor" in Physical Disease', *Practitioner*, 88 (1912) 315–21; 'The War Neurasthenic: A Note on Methods of Reintegrating him with his Environment', *Lancet*, 23 Mar 1918, 436; 'The Re-education of the Adult: The Neurasthenic in War and Peace', *Sociological Review*, 10 (Summer 1918) 25–40, repr. as *Papers for the Present*, no.4 (Sociological Society, ?1918); 'The Occupation Cure in Neurasthenia', *Edinburgh Medical Journal*, May 1923; 'Dreams, Folklore and our Present Spiritual Distress', *Hibbert Journal*, 87 (Apr 1924) 487–500.

Geddes was in India in 1917. The few Outlook Tower Association members available were struggling to keep the Tower open. Miss Wyer was standing in for AJB as secretary of the Open Spaces Committee. AJB and others tried to revive the place after the war but it seems never to have regained its old vigour. The Victorian *camera obscura* is still open to the public but Geddes's elaborate interiors have long since gone. Sources: Philip Boardman, *Patrick Geddes: Maker of the Future* (N. Carolina, 1944); Amelia Defries, *The Interpreter Geddes: The Man and His Gospel* (Routledge, 1927); Tower records and Geddes MSS (National Library of Scotland); Sociological Society papers (University of Keele).

5   ROBERT BALDWIN ROSS (1868–1918)

Art-dealer (ran the Carfax Gallery with More Adey, 1900–8) and critic (*Morning Post*, 1908–12), benefactor of the National Gallery, patron of many young artists and writers. His housekeeper, Mrs Burton, kept rooms above the flat in Half Moon Street, Mayfair, where friends, including WO, were welcome to stay. As Honorary Adviser, Imperial War Museum (from Dec 1917) and British War Memorials Committee (from Mar 1918), he played a significant part in setting up the Museum's art collection although his grand ideas for it were never realised (M. and S. Harries, *The War Artists*, 1983). He must have been deeply hurt when quietly relieved of these posts after the 1918 trial. For affectionate portraits of him, see SS (1945) 30–2 etc., and OS (1950) 98–101. Margery Ross, *Robert Ross: Friend of Friends* (Cape, 1952) contains letters to him from SS, RG, Nichols and many others. Friendship 'was the chief business of his life' (*Times* obituary, 7 Oct 1918; cf. *CL*, 585).

The many letters and biographies of Oscar Wilde shed light on Ross's earlier life. Accounts of the Ross-Douglas feud are often unreliable. For Douglas's side, see his *Autobiography* (1929); W. Sorley Brown, *Life and Genius of T. W. H. Crosland* (1928); R. Croft-Cooke, *Bosie* (1963). Ross's supporters gave him £700 and a public testimonial after the 1914 trials. *Times* trial reports: Apr, July, Nov, Dec 1914 (Douglas); 30 May – 5 June 1918 (P. Billing); 11 Jan, 4 Mar 1918 (Millard).

6   CHARLES KENNETH SCOTT MONCRIEFF (1889–1930)

Scholar at Winchester, where he rashly published a homosexual story (1908). Read Law at Edinburgh (MA, 1914). Reservist before the war. Mobilised, Aug 1914. Captain, 1915. Repeated illness. Severe leg wound, 1917; awarded Military Cross and put on administrative duties at War Office. Lived in Italy after the war but never recovered his health. Outstanding translator and imitator, his talent evident in his solutions to *Saturday Westminster* competitions; reviews and verse in *New Witness* (harsh comments on SS's war poems, 28 June 1917); translations of *Song of Roland* (Chapman & Hall, 1919), *Beowulf*, Proust, etc. Pre-war friendships with Millard, Philip Bainbrigge and others confirmed his sexual tastes. He and Bainbrigge used to exchange scholarly but obscene verses and parodies (d'Arch Smith, 1970). His many letters to Vyvyan Holland (Tex) show a similar ingenuity.

According to his own account he met WO at RG's wedding, Jan 1918, after a day giving evidence 'ineffectively, at a Police Court'. This may refer to Millard's trial (at Bow Street? – cf. 'Bow St. cases', MS of 'Who is the god of Canongate'), which would have been a topic of conversation at Ross's flat that evening. In his one surviving letter to WO (26 May 1918, OEF, part-quoted in *CL*, 553 n.1, and not destroyed by HO, who may

not have seen its implications), CKSM says he is writing sonnets
out of 'passion' as a means of both 'vivisecting' their relationship
and discovering how Shakespeare felt(cf. OS's later reference to
CKSM's 'ghoulish process' of trying to inhabit Proust's mind –
Pearson (1978) 211). With the letter is a Shakespearian sonnet
written by CKSM on Half Moon St paper and dated 19 May
1918 by WO, who was staying above the flat for a few days then;
the poet records that he had fallen in love with WO and had
tried to 'draw thy heart to me', but had been found 'unworthy'.
Despite this rejection, he praises WO's poetic 'merit' and hopes
to share in his future fame. On 7 June the *New Witness* published
another 'Sonnet' by 'C.K.S.M.':

> Thinking Love's Empire lay along that way
>   Where the new-duggen grave of Friendship gaped,
> We fell therein and, weary, slept till day.
>   But with the dawn you rose, and clean escaped,
> Strode honourably homeward . . . .
>   . . . you were gone from sight
> To Honour in an honest House of Shame . . .

The faint traces of 'To Eros' here seem to link this 'coded' poem
with WO. One more 'Sonnet' appeared in the same periodical
on 10 January 1919:

> When in the centuries of time to come
>   Men shall be happy and rehearse thy fame,
> Should I be spoken of then, or they grow dumb –
>   Recall thy glory and forget thy shame?

The poet goes on to say he does not care, since neither fame nor
'any breath of scandal' could shake him if he were 'in Heaven
with thee . . . Where two contented ghosts together lie'. The echo
of 'I am the ghost of Shadwell stair' ('I with another ghost am
lain'), a poem with which CKSM was somehow associated, is
unlikely to be coincidence. It seems probable that both the *New
Witness* sonnets were among those addressed to WO in May 1918.
A revised version of the second one, entitled 'To W.E.S.O.' and
dated 1918, was published as one of the three dedicatory poems
to friends killed in the war (Bainbrigge, WO, Ian Mackenzie)
which preface CKSM's *Roland*, in place of the dedication to 'Mr
W.O.' which he had drafted in 1918. Changes in wording include

a more discreet fourth line, 'Recall these numbers and forget this name?', as well as 'envy' for 'scandal' and 'stay' for 'lie'. The cryptic references to shame in these sonnets may allude to WO's alleged cowardice or, more probably, to his relationship with CKSM. There have always been rumours about this relationship, some of them emanating from RG, a notoriously unreliable source (cf. Fussell, 1975, 216, on the 'fatuous, erroneous or preposterous' material in *Goodbye to All That*). Hearsay at several removes is at best doubtful evidence, but there seem to be some grounds for supposing that CKSM at least tried to seduce WO. RG repeated to Martin Seymour-Smith a story from Ross that CKSM had not only tried but succeeded, having got WO drunk, and that WO had been deeply distressed. At any rate, something seems to have happened to cause 'scandal' in Ross's circle, and that may help to explain why SS (1945, 82–3) and OS loathed CKSM. RG also said that WO himself told him in 1918 that he had picked up young men in Bordeaux but had never overcome guilt feelings sufficiently to form any lasting relationship; since RG had by then repudiated the fervent Uranianism he had expressed to Carpenter in 1914, his enthusiasm for WO cooled a little after this confession. Mr Seymour-Smith, who has kindly told me about his conversation with RG on these matters, is convinced that for once RG can be trusted. There is no evidence, incidentally, that there was any physical relationship between WO and SS; SS told RG in a letter that he had never been physically attracted to WO.

C. K. Scott Moncrieff: Memories and Letters, ed. J. M. Scott Moncrieff and L. W. Lunn (1931), includes some verse. CKSM's published memories of WO: letter, *New Witness*, 2 Jan 1920, 117; 'The Poets there are. III – Wilfred Owen', *New Witness*, 10 Dec 1920, 574–5; 'Wilfred Owen' (letter), *Nation and Athenaeum*, 26 Mar 1921, 909–10.

7   HAROLD OWEN AS FAMILY HISTORIAN

By all accounts a charming man of complete integrity, HO knew nothing of literature or scholarship and was in some ways highly eccentric. *JFO*, his absorbing memoirs of the Owen family, is written with considerable artistic licence. Similarities between his memories and WO's poems *may* illuminate the poems or may reflect his long years of guarding and puzzling over WO's MSS, which he hoped one day to edit. Even his one specific gloss on a poem is suspect: *JFO*, I, 176–7, describes his remark that he had feet of gold after a family walk through buttercups, and WO's ensuing comment that 'Harold's boots are blessed with gold'. This has been accepted by all commentators (including myself) as giving the origin of the 'blessed with gold' image in 'Spring Offensive'. But in his 'Working Copy' of EB (OEF) HO wrote against the image, 'The family walk home from Uffington on early summer evenings through the water-meadows when we would look with delighted wonder on our shoes and stockings flushed with gold – unconscious recollection?' This note is convincing, with its intuitive understanding of poem and poet, but the published account may be an imaginative version which quotes from the poem rather than from an actual conversation. The way in which HO's remark is ignored by the family and is taken over by WO, who rewords it to sound priggishly 'poetic', is symbolic, like so much of *JFO*, of the psychological burden that HO laboured under as the obscure, cruelly uneducated younger brother.

HO, and SO before him, did their best to control public perception of WO. It was on WO's own instructions that SO burned 'a sack full' of his papers (EB, 3) but HO thought later that she had probably destroyed more (HO to EB, 13 Oct 1947, Tex). Certainly, remarkably few letters to WO survive. As the portrait of WO the archetypal soldier poet in EB's 1931 memoir is SO's, who supplied all the key information for it, so that in *JFO* of the idealist indifferent to immediate human ties is HO's. Always insisting that only 'members of the family' were able to understand the poet, HO turned away researchers (notably DW) in his efforts to prevent all discussion of WO's personal life. He destroyed some of WO's letters and mutilated others. One of his chief motives was dread that someone might raise the 'frightful implication' of homosexuality. He wrote in 'desperate anxiety' to EB (16 Oct 1950, Tex) for help when some enquiry was made, saying he had 'taxed' WO on the subject but WO had denied all personal involvement although admitting 'abstract' interest because homosexuality seemed to attract so many intelligent people. At a late stage of writing *JFO*, HO inserted a 1918 conversation between the brothers (III, 163–6) in which he revises the information he had given EB. Instead of 'taxing' WO, HO innocently asks for enlightenment about goings-on between sailors, whereupon WO says he had intended making a similar request but since they both knew nothing they might as well talk of other things. No doubt this represents the truth as far as HO saw it. A friend of CKSM and Ross would not have asked for illumination from a puritanical younger brother, but WO's words can be read as regret that HO's evident disgust had prevented confidences. By contrast, as WO may have known, SS's younger brother, Hamo, had eased SS's pre-war worries by admitting to being contentedly homosexual.

If WO did not confide in HO he seems to have done so in ELG, giving him the 'key' to many poems (CL, 508 – the omitted words refer to one of ELG's girlfriends) and even referring mysteriously to '*mon petit ami*' in Scarborough (*CL*, 544). But ELG says these things were too long ago for him to remember.

Appendix B: Owen's Manuscripts and their Chronology

The odd history of WO's MSS has yet to be unravelled, but almost all of them are now at last in libraries and available to researchers. There are three main collections.

1   BL. Two volumes of verse MSS bought by public subscription in 1934. It seems possible that some or all of these folios were never returned to the family after being lent to Edith Sitwell and SS for *Poems* (1920). Add. MS 43720 (bound for SS before 1930?) consists of the drafts treated as more or less definitive for *Poems* (1920). Add. MS 43721 is a larger volume of further drafts and fragments, read by EB for *Poems* (1931); many of these MSS remained unpublished until *CPF*, and researchers were not supposed to quote from them.

2   OEF. A large, apparently random collection of loose MSS retained by the family. There seems to be no record of why these were kept or what state WO's papers were in at his death, but the family's storage methods were haphazard and these MSS may simply have been overlooked before Mary Owen's death in 1956 (cf. *CL*, 1). Accompanying notes show that HO began to assemble them at about that time in the hope of preparing an edition. It seems that only Patric Dickinson and CDL saw this material before Jon Stallworthy was appointed official biographer and editor. The verse MSS have now been arranged in *CPF* order and numbered by Professor Stallworthy. Other MSS include *Hydra* editorials, some notes on Keats and Dickens (for talks at Craiglockhart?), and fragments of letters (see Lett). ELG has generously added all the WO MSS and books in his possession. OEF also has WO's library and many other

202

items to do with WO and HO.

3   Tex. MSS (described in detail in Lett) of WO's letters home, unfortunately sold after *CL* was completed. Also a curious assortment of letters, photocopies, MSS, etc., connected with WO and other war poets, gathered in the 1950s by Joseph Cohen for a projected 'Wilfred Owen War Poetry Collection'. Whether this was a serious project or just a means of persuading donors to present material is not clear. The more interesting WO items are mentioned in my chapter notes. Tex also has EB's papers, many letters from SS to friends, and other relevant material.

I have not seen the originals of WO's letters to SS, recently bought by Columbia University, nor the few MSS still in private hands.

My own edition, *WPO*, which was put together in 1968–70 during HO's lifetime, had to be based on BL MSS only. I had no access to the material now in OEF, nor was I permitted – by the publishers – to make the substantial changes to the EB–CDL text that were clearly desirable. I was even forbidden to restore WO's 'moans in man' in 'Insensibility' on the grounds that EB and CDL had preferred his cancelled 'mourns' and they as poets knew best. The *WPO* text was an advance on CDL, but *CPF* goes much further.

In general *CPF* and *CL* are meticulously accurate. A few amendments to *CL* are suggested elsewhere (Lett) and to *CPF* in my quotations and chapter notes. Professor Stallworthy and I exchanged many letters but I have not attempted to check his text against MSS except for the passages I have quoted. His generosity and kindness over the years seem poorly rewarded if I criticise any aspect of *CPF*; his edition is a splendid achievement and a great asset to scholarship. But researchers need to bear some caveats in mind. *CPF* cannot show all draft workings, and the conversion of heavily revised manuscript into uniform print has sometimes obscured the sequence of WO's alterations or the relationship between one word and another. Some drafts may still be sundered from the poems they were written for, and a few may be yoked mistakenly together (perhaps I should take responsibility for the debatable separation of 'Purgatorial passions' from 'Mental Cases' and 'Earth's wheels' from 'Strange Meeting'). The arrangement of poems in the order of their final drafts must never be mistaken for the order of composition; WO copied out a lot of early verse in 1917–18 so *CPF* has to put juvenilia among mature work. Finally, researchers should be aware that dates ascribed to MSS are often open to argument, although *CPF* is far more accurate than any previous authority.

One representative problem may illustrate the complexity of the evidence involved in establishing dates for MSS. MS letters show that a particular watermark is invariably limited to one brief period; it follows that any verse MSS with that watermark probably belong to the same period. But this rule is less reliable from mid 1917 because WO bought special paper for his 'job' as a poet and only used odd sheets for letters. Society Bond paper (SB), for example, is used for only four letters (late Jan–28 Feb 1918) but for drafts of eleven poems. *CPF* ascribes the drafts of 'My Shy Hand' (final draft), 'Sunrise' (final) and 'I am the ghost of Shadwell Stair' (first) to January–February, presumably on the basis of the watermarked letters. Puzzlingly, the only drafts of 'Schoolmistress' and 'The Letter', and the final drafts of 'A Tear Song', 'Strange Meeting', 'Dulce' and 'Conscious' are ascribed to January–*March*; the evidence for extending the limit to March in these cases is not clear. (The watermark is also taken as proof that 'The Letter' was 'written' in January–March; that may well be the date of the surviving draft, but the Sassoonish style suggests the poem was first composed at Craiglockhart. There must have been at least one preliminary draft before the SB fair copy; WO certainly fair-copied many of his 1917 poems in 1918.) Moreover, *CPF* implies that the final (SB) draft of 'Insensibility' may be no later than January. (The eleventh of the SB drafts is the first complete version of 'Exposure' but *CPF* does not need to date this because there is an earlier one-line fragment which can itself be dated.) But all these dates seem undermined by one more SB MS, the first of WO's two tables of contents (*CPF*, 538), which lists several poems ascribed by *CPF* to April–May (i.e. late March – early June, the Borrage Lane period). It would be inconsistent with the principles governing *CPF*'s use of watermarks to suggest that this was a stray sheet or that the April–May titles were additions to the list. *CPF* is uncertain in the date it ascribes to the two lists of contents ('March–June', 186, '20 May – 30 July', 537) but it does not suggest that the SB list could be as early as January–February. The strong probability is that both lists of contents were drawn up at Borrage Lane. If WO used one sheet of SB there, *any* otherwise undated SB MS *may* belong to that period. I think one has to conclude that SB was one of the several types of paper which he had available between January and early June 1918. It is a matter of instinct where one places most SB drafts within that period, but I incline to March–June (Borrage Lane) in general. WO composed new poems there and 'realized many defectuosities in older compositions' (*CL*, 543), so one would expect the drafts to include revised 1917 work as well as fresh material. He had been giving time at Scarborough to 'Sonnets' (hence, perhaps, the numerous fair copies of sonnets and lyrics on Pompeian Parchment) but then the war changed. 'Strange Meeting', in particular, seems to me to be a response to the carnage that began on the Western Front on 21 March.

Appendix C: *The Hydra*

OEF has a complete set of nos 1–12 (fortnightly, 28 Apr – 29 Sep 1917) of *The Hydra: Journal of the Craiglockhart War Hospital* and six of the monthly New Series (Nov 1917 – Jan 1918, May–July 1918). WO seems to have become editor for no. 7 (21 July). No. 10 (1 Sep) includes a version of two 'Dead-Beat' stanzas, and 'Song of Songs'; no. 12 (29 Sep) includes 'The Next War'. All three pieces are anonymous, like WO's editorials and prose contributions (which are in the whimsical style then considered appropriate). The November issue reports Mr Owen as being on the Debating Society Committee and as having given an 'interesting' lecture on the classification of soils on 1 October. Many other hospital activities are recorded. WO was succeeded in October by J. B. Salmond, whose experience as a professional writer is evident in the great improvements in design and organisation introduced for the New Series. Salmond said later (letter in Edinburgh University Library, microfilm in Tex) that WO acted as his sub-editor and was responsible for recruiting November contributions from G. K. Chesterton and John Drinkwater. The editorial says Wells and Bennett were also approached (presumably at SS's suggestion). Drinkwater's MS, a poem called 'Reciprocity', is still in OEF; the poem was included in his *Tides* (1917), which is no doubt why WO asked for the book as a Christmas present. The May–July 1918 issues were presumably sent to WO by AJB, who contributed an article in each of them on Edinburgh Regional Survey. AJB marked these articles (with a characteristic cross often found in his own MSS) and a poem which laments that Craiglockhart patients, who apparently had to wear an identifying tab, were stared at in Princes Street ('all people think us mad'), an experience WO may well have suffered in 1917. The January issue announces that Mr Owen's poem on Antaeus will appear in February, but no copies of the missing 1918 issues have yet been traced.

See also DH, 'Some Notes' (1982) and 'A Sociological Cure' (1977).

Notes

The following abbreviations are used in the Notes and Appendixes.

AJB	Arthur John Brock
BL0,1; BL1, 4v; etc.	British Library Additional Manuscript 43720, folio 1; 43721, folio 4 verso; etc.
BNY	Berg Collection (Marsh correspondence), New York Public Library
Casebook	*Poetry of the First World War: A Casebook*, ed. Dominic Hibberd (1981)
CDL	C. Day Lewis *or* his edition of *The Collected Poems of Wilfred Owen* (1963)
CKSM	Charles Kenneth Scott Moncrieff
CL	*Wilfred Owen: Collected Letters*, ed. Harold Owen and John Bell (1967)
CPF	*Wilfred Owen: The Complete Poems and Fragments*, ed. Jon Stallworthy (1983). Numerals refer to *pages*, not poems
DH	Dominic Hibberd
DW	Dennis Welland
EB	Edmund Blunden *or* his edition of *The Poems of Wilfred Owen* (1931)
ELG	E. Leslie Gunston
Geo	Dominic Hibberd, 'Wilfred Owen and the Georgians', *Review of English Studies*, 30 (Feb 1979) 28–40
HO	Harold Owen (brother)
JFO	Harold Owen, *Journey from Obscurity*, 3 vols (1963–5)
JS	Jon Stallworthy, *Wilfred Owen: A Biography* (1974)
Lett	Dominic Hibberd, 'Wilfred Owen's Letters: Some Additions, Amendments and Notes', *Library*, 4 (Sep 1982) 273–87
OEF	English Faculty Library, Oxford (Wilfred Owen Collection). Numerals refer to MSS in this collection
OS	Osbert Sitwell
RG	Robert Graves
SO	Susan Owen (mother)
SS	Siegfried Sassoon
Tex	Harry Ransom Humanities Research Center, University of Texas at Austin
TO	Tom Owen (father)
WO	Wilfred Owen
WPO	*Wilfred Owen: War Poems and Others*, ed. Dominic Hibberd (1973)

Page references to *CPF* are not given when the title or first line of the poem or fragment is obvious.

Books and articles are identified only by surname of author/editor and date of publication (of the edition referred to), except that an abbreviated title is given when confusion might otherwise result: AJB (1924); Bäckman (1979); DH, 'Rival Pieces' (1976); etc. All items thus referred to are listed in full in the Bibliography or Appendixes.

A reference to '*CL* [Lett]' indicates that a quotation from a letter involves an amendment or addition to the *CL* text.

In some cases several consecutive references are contained in one note and identified where necessary by key words.

Readers unfamiliar with *CPF* may find the following list of fragments helpful; it includes those I have discussed as P1–6 under 'Perseus'.

P1: 'What have I done, O God, what have I done', BL1, 155–6v (*CPF*, 467–70).
P2: 'The cultivated Rose', BL1, 153 (*CPF*, 465–6).
P3: 'Shook and were bowed before embracing winds', OEF 216 (*CPF*, 464).
P4: 'The sun, far fallen in the afternoon', OEF 209 (*CPF*, 449).
P5: 'About the winter forest loomed', OEF 208 (*CPF*, 448).
P6: 'Speech for King', BL1, 154 (*CPF*, 470).
The fragment about Broxton: 'Instead of dew, descended on the moors', OEF 203 (*CPF*, 433–4).
The fragment about touching a boy's hand: 'We two had known each other', OEF 204v (*CPF*, 437–8).

The following titles used (and in some cases invented) by earlier editors may still be more familiar than the more accurate versions given in *CPF*:

'All sounds have been as music', *now* 'I know the music' (*CPF*, 485).
'Antaeus', *now* 'The Wrestlers' (*CPF*, 520).
'Bold Horatius', *now* 'Schoolmistress' (*CPF*, 139).
'Bugles sang', *now* 'But I was looking at the permanent stars' (*CPF*, 487).
'Shadwell Stair', *now* 'I am the ghost of Shadwell Stair' (*CPF*, 183).
'Sonnet to a Child', *now* 'Sweet is your antique body, not yet young' (*CPF*, 129).
'To a Comrade in Flanders', *now* 'A New Heaven' (*CPF*, 82).
'To my Friend', *now* 'With an Identity Disc' (*CPF*, 96).
'Voices': *see* 'Bugles sang'.

CHAPTER 1. THE ORIGINS OF A POETHOOD [1893–1911]

1 WO's surviving books and other possessions may be assumed to be in OEF unless otherwise stated.
2 *CL*, 271.
3 'Instead of dew ...' (*CPF*, 433–4). This 1914 fragment says 'ten ye[ars]' have elapsed since Broxton. For another reference to the Bagnères moon, see *CL*, 464.

4   *CPF*, 68 (Welsh blood). *CL*, 581 (forefathers), 256 (uncle, aunt).
5   *CL*, 68, 186 n.1. Cf. 'Perversity', line 14.
6   *CPF* ascribes 'To Poesy' to 1909–10 but without clear evidence, and 'Written in a Wood, September 1910' to 1910 on the basis of the title. The wood seems to be Hampstead, visited by WO in September 1911, but '1911' in the title would have required 'ninety-one' in the poem, ruining the metre.
7   Quoted by Patric Dickinson in a BBC talk, 17 Aug 1953 (script at Tex). HO showed Dickinson some early MSS, now lost, including verse (*CPF*, xxv) and another prose sketch (for a sonnet on poetic desire). CDL, 14–15, quotes a third sketch.
8   In WO's Keats, I, 10–11 and 14–15.
9   *CL*, 273 n.1.
10 JS, 78, and *CPF*, 409n., need slight amendment. Keats's house was not yet open to visitors. The Dilke Collection, including MSS and a lock of hair (*CPF*, 447) was presented to the Hampstead library in 1911 and displayed there. (Information from Assistant Curator, Keats House.)
11 *CPF*, 447, 409.
12 T. S. Eliot, 'Little Gidding'. Eliot's many sources may include 'Strange Meeting', a poem he admired (see his contribution to Walsh, 1964).
13 *CL*, 112.
14 For titles, inscriptions, etc., see JS, 308–23, and DH, 'WO's Library' (1977). When WO's books were transferred to OEF, the Librarian, Miss M. Weedon, shelved them as HO had listed them in 1920.
15 Keats to Taylor, 24 Apr 1818; cf. *CL*, 325, and 'To Poesy'. *CL*, 150 (King). Lett, 286–7 (Collingwood).
16 SO's (Anglican) Evangelicalism is shown by her attendance at the Evangelical St Julian's rather than her parish church, the Abbey, despite a steep walk. WO's name is on war memorials in both churches, in the former as a 'member of the congregation', in the latter as a parishioner. Among the notes in SO's Bible is 'Feb 22nd 1909 Wilfred spoke at Frankwell', perhaps a record of his first public talk or prayer. See also Alec Paton in Walsh (1964).
17 *CPF*, 447.
18 *CL*, 118, 150, 106.
19 *JFO*, I, 120. Unlike HO, WO praised SO's self-sacrifice (*CL*, 32, 479).
20 SO to EB, n.d., *c*.1930 (Tex).
21 JS, 27; *CL*, 68. JS tends, I think, to overrate SO's intellectual ability and interests.
22 *CL*, 99. I assume WO is answering a comment from SO.
23 JS, 28; *JFO*, I, 103. Here as elsewhere JS seems a little too ready to accept HO's record. HO was writing years later about an event which occurred in his absence when he was five.
24 See DH, 'Images' (1974).

CHAPTER 2. THE FIRST CRISIS: RELIGION [1911–13]

1   'O World of many worlds'.
2   *CL*, 75.

3 *CL*, 102. *CPF*, 394 (CPF may be mistaken in treating this fragment as part of 'Spring not . . .'; the two fragments differ in form and subject). WO was worried about his heart at Dunsden (cf. *CPF*, 36; *CL*, 131, 271).

4 *CL*, 273 n.1.

5 *CL*, 123.

6 *CPF*, 16, 397. *CPF* follows my suggestion in identifying the first piece as a 'chorus' rather than a hymn. Evangelical choruses are informal verses on scriptural themes, to be chanted at meetings; this one is on God's promises. WO composed another in February (*CL*, 115).

7 For details see DH, 'Rival Pieces' (1976).

8 *CL*, 122, 73 (Sept 1912: letter misdated in *CL*, see Lett, 278). Beckett (1879); WO's copy has a Reading bookshop label.

9 Rossetti (n.d.) 158.

10 *CL*, 93.

11 'I simply daren't be sincere' (*CL*, 155).

12 Black: see Appendix A1.

13 *CL*, 174. Cf. n. 38 below.

14 'O hard condition!' (*CL*, 181).

15 JS, 85–6. I quote from MS (OEF 259v) with slight simplification. The Keats quotations are from 'Eve of St Agnes' and 'Lamia'.

16 'Spring Offensive'.

17 *CL*, 131. WO misquotes from Magnus (1902) 146; his copy has a Reading bookshop label. With 'wolfish', cf. Dickens, *Hard Times*, ii.vi ('Reality will take a wolfish turn, and make an end of you'); 'press upon' echoes Keats (to Dilke, 21 Sep 1818).

18 *CPF*, 387. Symonds (1878) quotes Mrs Shelley's note on *The Revolt*.

19 *CL*, 96, 109–10. For details see Geo, 28–9.

20 *CPF*, 396.

21 Monro (1911) 130. *CPF*, 372.

22 Geo, 30 (quoted, *CPF*, 107).

23 Monro was friendly with Carpenter in 1910–11. See Grant (1967).

24 *CL*, 167. Cf. Lord Henry Wotton's typically Decadent thought, 'One could never pay too high a price for any sensation' – Wilde (1948) 55.

25 *CL*, 171.

26 *CL*, 206 (phantasies), 212 (horrors), 235 (phantasms: here he records that they began in late February 1913).

27 The scene of Perseus showing Andromeda the head's reflection in a well or pool occurs several times in ancient and Pre-Raphaelite art. With WO's description of Medusa, cf. his feeling 'helpless' before the 'thrilling eyes' of a girl in a painting, 1912 (Lett, 278).

28 *CL*, 135, 137 (Borrow); 153 (bicycle). Discussed in DH, 'Images' (1974).

29 *CL*, 339.

30 *CPF*, 12. JS, 71, seems to overstress the conclusion to this poem; WO is not 'in the throes of an almost Christlike suffering' but enjoying the pleasure of *recovering* from indigestion. 'Bliss' may imply sexuality or it may refer to religious ecstasy (with 'dangerous air', cf. 'madness-giving air' in 'Unto what pinnacles').

31 *CPF*, 228 (mother-arms), 378 (rapture).

32 *CL*, 175 ('On my Songs', Jan 1913).

33 *CL*, 186 n.1.

34   *CL*, 161. Tom Coulthard suggests a parallel with Trelawny's snatching the heart out of Shelley's burning corpse, a story WO had read in Symonds (1878).

35   *CL*, 536.

36   *CL*, 137.

37   *CPF*, 403.

38   *CL*, 400 (eyes, school), 123 (tea), 119–20 (piano), 118 (ramble), 174–5 (furor). *CPF*, 437 ('We two had known each other'; *CPF* follows my suggestion in identifying the subject as Rampton). The Librarian of Reading School tells me V. C. Rampton (born 23 Nov 1899) entered the school in September 1915 and left in July 1917 (for war service?).

39   Draft refers to pines, beeches, one June/August day (*CPF*, 207). I am not quite convinced by *CPF*, 58, that a parish outing is referred to since it was over before dusk and WO was not alone with one child (*CL*, 145).

40   *CL*, 181.

41   *CPF*, 269 (draft). Cf. the sermon WO read to an old parishioner: 'Not more frail are flowers ... or more fleeting meteors than Human Life' (*CL*, 154).

42   *CPF*, 413. *CL*, 400; Simon Wormleighton recently met an elderly Dunsden resident who remembered how Wigan used to 'preach and preach and preach'.

43   *WPO*, 31 (quoted, *CPF*, 72). Other sources of WO's imagery may include Tennyson, 'The Silent Voices', and Keats, 'The Poet'.

44   DH (1975), 10; Geo, 30 (both quoted, *CPF*, 74). With the image in the poem, cf. E. Vedder's Decadent painting, *Superest Invictus Amor* (P. Jullien, *Dreamers of Decadence*, 1971, pl. 71).

CHAPTER 3. AESTHETE IN FRANCE [1913–15]

1   *JFO*, III, 52–6. Typically, HO says WO did not start the 'Sir Thomas' rumour. But who else could have done? It paid off. The Légers lived in one of the grandest streets in town. Miss de la Touche had been governess to princesses, her sister was a baroness, her nephews went to an English public school; WO told her he was 'preparing for Oxford', more hopeful than true (*CL*, 299,305).

2   WO had affected purple ink, tie, slippers (*CL*, 111, 127). His early Aestheticism also appeared in 'Wilfred's Church' and burying SO in flowers (*JFO*, I, 77–8, 150–1). (Flower burials occur in Morris, 'The Wind', and d'Annunzio, *The Triumph of Death*.)

3   I use the word loosely to include more specific movements such as Symbolism and Decadence. For more substantial accounts, see Carter (1958), Gaunt (1945), Kermode (1957), Nordau (1895), Praz (1970), Stephan (1974), and many others.

4   Tailhade: see Appendix A3.

5   Baudelaire, 'Le soleil s'est noyé dans son sang qui se fige' ('Harmonie du Soir'); Flaubert, 'Legende de St Julien l'Hospitalier'; d'Annunzio, see Praz (1970) 279; Wilde, 'Panthea'; Hardy, *Tess*, xxi; Dickens, *Our Mutual Friend*, IV.i. Cf. SS, 'Last Meeting'; Rossetti, 'A Last Confession'; Flecker, *Hassan*; etc.

6   Nordau (1895) 2.

7   'Hymn to Proserpine'.

8   Locke (1905) 77. The best and easiest way to get bloodshed, says the character, will be to challenge and 'exterminate Prussian Lieutenants'.

9   Brooke to St John Lucas, Oct 1907 (MS at King's College, Cambridge). Seven years later Brooke really did see fire and men hurt in Antwerp, finding 'release there' ('Peace') as he had lightly hoped in 1907.

10 Quoted in Gaunt (1945) 119.

11 Wilde (1948) 90.

12 But it was also to have had type like *Before Dawn* (BL1, 163v).

13 The 'chasms' round mad eyes in 'Mental Cases' began as 'blue chasms' in 'Purgatorial passions'. Such marks usually signify sexual excess in Decadent imagery.

14 *CL*, 441.

15 *CL*, 431.

16 *CPF*, 465, a 'Perseus' MS. A list of pararhymes headed 'Nocturne' on a draft of 'The Imbecile' (JS, 105) matches pararhymes in 'Has your soul sipped' (1917), a poem WO might well have intended to call 'Nocturne', and may thus be much later than 'The Imbecile' itself (1913). For WO's rhymes and their French–Welsh origins, see Bäckman (1979), DH (1978), Masson (1955), DW (1950), etc. It would be pleasing to prove us all wrong by finding that his pararhymes originated as a solution to a literary competition, as 'The Imbecile' did – DH, 'Problems A' (1980).

17 *CL*, 234 [Lett, 280].

18 *CL*, 243–5. WO quotes the 'One remains' line from *Adonais* in this letter, which reinforces the link between Henriette and the sonnet. He had recently been given a Shelley for his twenty-first birthday.

19 *CL*, 280, 295.

20 Léger family: see Appendix A2.

21 *CL*, 271.

22 Appendix A3.

23 *CL*, 271–2. WO's playing of Chopin was admired at home (*JFO*, II, 756; cf. ELG, 'Music'). The *Marche* was one of his favourites.

24 The book cover in OEF corresponds with the one in the photograph.

25 *CPF*, 435–6. 'Ballade élégiaque pour le morose après-midi' (*morose* was a favourite Decadent adjective).

26 WO seems to have lent the first volume to SS, the second – now lost – to Monro (*CL*, 361, 493), which implies he was proud of the friendship despite HO's later attempt to play it down (*CL*, 2).

27 Tailhade (1913). For sources for this paragraph, see Appendix A3.

28 *CL*, 282.

29 Kolney (1922) 32.

30 Lett, 286.

31 Cf. Tailhade's titles: 'Lundi de Pâques', 'Vendredi-Saint'.

32 *CPF*, 433–8.

33 *CL*, 286, 291.

34 *CL*, 350, 348. I assume WO read *Bovary* in 1915. A phrase from it is echoed in 'A Palinode' (Oct); see *CPF*, 78, but the point was first made by Bentley (1970).

35 Steiner (1971) 25.

36 *CL*, 347.

37   *Pace* DW (quoted, *CPF*, 70), the poem seems nearer Tailhade and Flaubert than Wilde and Swinburne. Cf. Salammbô's prayer to the moon (*Salammbô*, iii).

38   *CPF*, 444.

39   *CL*, 322 (novelists); 320 [Lett, 281] (soundings); 285 (hospital).

40   See Appendix A6.

41   *CL*, 333 (writings, trees), 325 (boy), 352 (heather), 295–6 and 300 (useful to England), 342 (Vigny quotation).

42   *CPF*, 471, 545, gives some dating-evidence. P1–2 could as well be 1914 as 1915, I think. P3–5 have the same watermark as the 'certain writings' letter (Apr 1915). *CPF* groups under 'Perseus' only P3, P2, P1, P6, in that order. P1 is a large double sheet, of which *CPF* treats the first and second sides as separate pieces, breaking the fragment in mid stanza. The third side is blank. The fourth has four isolated lines about Pluto (*CPF*, 470). Other MSS which may be 'Perseus' work include *CPF*, 528 (winged sandals); 492, interesting for its links with 'Strange Meeting'; 463; and possibly 'Purgatorial passions' (455).

43   *CL*, 439. ELG has no recollection of 'Icarus' but it may have begun as a tribute to Gustav Hamel, a celebrated aviator who had given displays from the racecourse in front of the Owens' house and who crashed into the sea in 1914. There is a poem to his memory in ELG's book. ELG once told me the cousins liked H. J. Draper's painting *The Lament for Icarus* (Tate).

44   My readings differ from *CPF*. In addition to those shown in my reconstructions, I read 'no heat', 'soul' 'drenched the fires', for *CPF*'s 'us heal', 'rough', 'drenched the flowers'. I think WO wrote 'I touch the god', altered it in stages to 'I hav touched the godess', and then left it unrevised. MS shows that the two lines in *CPF* about Persephone's body and soul were afterthoughts.

45   Quoted from BL1, 11, the earlier draft.

46   For Decadent hermaphrodite imagery, see Busst (1967).

47   MS appears to read 'feel' but I assume WO meant 'fell'. With WO's details in 'Perseus', 'Storm' and 'Supposed Confessions', cf. Morris, 'The Doom of King Acrisius' (*Earthly Paradise*), including: Zeus as sunlight, thunder; Perseus a 'fair' child; 'aged crone'; Gorgon's head reflected in a pool. WO was reading about Morris in 1915 and thinking of poetic creation as flight (*CL*, 315–16). With Danae's captivity, cf. Henriette's 'plaints of captivity ... she is watched and warded everywhere' (*CL*, 244).

48   *CL*, 408. *CL*, 334, mentions Perseus as a sailor, referring to HO's tribulations in the Navy.

49   *CL*, 348, 353.

50   *CL*, 458.

51   *CPF*, 463.

52   Endymion falls from love into hell, where he meets Age. The blank verse of P1 and P6 reads like a crude imitation of speeches in *Atalanta*. Pluto and Persephone: appear in P1 only, perhaps dropped thereafter.

53   WO mentions buying 'Casts ' to ELG in 1915 (*CL*, 355). The statuette, presumably a copy of a Roman votive figure, came to ELG after WO's death. Cf. 'By Hermes, I will fly' (*CL*, 408). When ELG first showed me this bronze, I understood him to say that it came from Bordeaux, but on two later

occasions he assured me that WO bought it in Scarborough. It may well have been one of WO's purchases in the Scarborough antique shops in early 1918. Bainbrigge would have liked it.

54   Brown (1934) 95.
55   *CPF*, 76. See also Lett, 287.
56   *CL*, 332.
57   *CL*, 341. WO knew the area, his uncle having been a doctor there.

CHAPTER 4. PREPARING FOR WAR
[Oct 1915 – Dec 1916]

1   Chapple (1970) 262 (war scare). A. Paton to Joseph Cohen, 7 Feb 1954 (Tex) (toy soldiers). *Christian*, 22 July 1912, 15.
2   Pound (1964) 55.
3   Beerbohm (1911) 291. Before the sacrifice the *Marche funèbre* was played and grey morning clouds 'massed themselves . . . an irresistible great army' (207). Cf. Ch.3, n. 23, above, and dawn 'massing . . . her . . . army . . . in ranks on . . . ranks of grey' ('Exposure').
4   ' . . . the highest moral act possible, according to the Highest Judge' (1916 – *CL*, 387).
5   Murry (1921), in *Casebook*.
6   *CPF*, 151, 487–8. *CPF* suggests July 1917 for the 'Artillery' sonnet without clear evidence; content seems pre-SS but the poem could have been written in France, possibly in response to Sorrell's challenge in March (*CL*, 441).
7   *CL*, 304. Cf. 160 (Severn), 310 ('the End'; quoting the prophecy).
8   *CPF*, 500–9. WO probably meant this poem when he referred to 'that War Ballad' in 1916 and the 'Ballad' in 1917 (*CL*, 416, 476). A 1915 version was called 'The Ballad of Purchase Moneys' (cf. Acts 20:28).
9   *Christian*, 3 Sep 1914. WO probably saw this number (*CL*, 279).
10 *CL*, 484. For examples of soldier-Christ poems and paintings, see DH and Onions (1986).
11 *CL*, 388.
12 *CL*, 381.
13 *CL*, 393. *CPF*, 114, and JS, 139, identify the addressee as Johnny de la Touche, but I suspect Professor Stallworthy attaches rather too much importance to him. If the poem refers to old memories, there are several more likely addressees: if it refers to 10 May 1916, de la Touche is unlikely to have been able to visit London from his Somerset boarding-school on a termtime weekday.
14 *CPF*, 223. Draft version quoted for its botanical metaphor ('die back') and the detail that war memorials were beginning to appear even in 1916. Another draft substitutes 'brows' (cf. 'Anthem') for 'breasts', more appropriate to 'Acropole'. Another title: 'To a Comrade in Athens' (hence 'Acropole') or 'To a Comrade in Flanders'. The comrade may have been Lt Briggs, 'my closest chum', who was warned he would be sent out in September (*CL*, 391, 407). The many current rumours may have included talk of being sent to Athens, where the Navy was involved in a local crisis, but in the event Briggs no doubt went to Flanders.
15 *CPF*, 228.

16 Draft last line. With 'mock', cf. 'mockeries' in 'Anthem'.
17 *CL*, 492.
18 See DH, 'Rival Pieces' (1976) and 'The End' (1983).
19 Kyle: Monro MSS (BL Add. 57739); *The Times*, 17–23 Mar 1922. 'Macdonald' was a major publisher of 'soldier-poetry'. For WO's Little Books, see Lett, 287, and Geo, 32. 'Minor Poems' MS (OEF) includes:

> Points to note about my Sonnets
> 1 They are 'correct' as regards rime, but the system of rime is not necessarily classical.
> 2 [They range from the simplest to the] Such as I send you here conform to the essential   of a sonnet unity of idea – and a solemn dignity in the treatment. At the same [time I] The only corrections I have made have been [on the side] by way of simplification. I can write nothing simpler, & at the same time, poetical than the Sonnet 'The One Remains'.

'Sonatas': *CL*, 454; BL1, 163v. In the numbered sequence of MSS (all watermarked 'Pompeian Parchment', so second half of 1917 or first half of 1918), 1–9 are the sonnets numbered 100–8 in *CPF*; 10–14 are missing (but *CPF* nos 95 and 113–16 have drafts on Pompeian); 15–20 are *CPF* nos 109–11, 154, 112, 119 ('The End', a tailpiece).

20 Quotations selected from OEF 333 (partly reproduced in *CPF*, 323–6). With drums and trumpets in 'The End', cf. 'The thunder of the trumpets of the night' ('Laus Veneris', last line).
21 *CL*, 354.
22 Joergens's 'Attar of Roses' is subtitled 'Tannhäuser" and begins 'Great Venus, at thy knees I am lying low' (ELG has MS). No doubt the three sonneteers knew Wagner's opera and perhaps that attar was his favourite scent.
23 'I looked and saw' ('1915'), 'Afar off, afar off' ('Ideal Love'). The latter has stresses marked; perhaps WO was looking for a drumbeat to go with his 'drummers drumming'.
24 BL1, 31v (mentioned, *CPF*, 496). Several images from BL1, 31–2, reappear in 'Purple'.
25 *CPF*, 457. With 'my heart has failed . . . in sleep', cf. *CPF*, 36 (Dunsden hypochondria) and 'Mental Cases', lines 8–9.
26 BL1, 127.
27 WO confused Inferno and Purgatorio, using both terms to describe the war zone (*CL*, 425, 440). He knew the *Inferno* (*JFO*, II, 71) but may not have read further. Purgatorio is not subterranean.
28 *CL*, 383–4. For Monro's influence on WO, see Geo, 28–34. They may have met more often than *CL* records; WO may even have been at one of Monro's drinking-bouts (*CL*, 501).

CHAPTER 5. THE SECOND CRISIS: SHELLSHOCK [Jan– July 1917]

1 DH, 'Concealed Messages' (1980).
2 *CL*, 427. *Tale of Two Cities*, closing lines.

3   *CL*, 431 ('unnatural' is underlined in MS).
4   *CL*, 424, 449, 427, 425, 429.
5   *CL*, 432.
6   *CL*, 424, 427.
7   *CL*, 95, 211, 285, 322.
8   *CL*, 429.
9   *CL*, 428. Several sharp comments on the Gunstons, usually Gordon, are omitted from *CL*, ranging in date from July 1916 to October 1918. Cf. Ch. 11, n. 4, below.
10   The battalion felt 'bitterly towards those in England who might relieve us, and will not' (*CL*, 453). And see other comments in these letters (e.g. on Somme films, 429, 440).
11   *CL*, 425 (Poems), 434 (Leslie's), 482 (Tennyson). Benson (1912) 79. 'Adolescens' appears on BL1, 3, and Bodleian draft of 'Happiness'.
12   'Happiness' draft and notes: *CPF*, 88, 227–30.
13   *CL*, 444. JS, 171n.
14   *CL*, 448 (rest), 452 (hole). WO ascribed his shellshock to the shellburst and the proximity of Lt Gaukroger's shattered, presumably decaying corpse (cf. JS, 182n.). *CL*, 475 (coming-to). JS, 183, presumably paraphrases from a reliable source in saying the Medical Officer found 'him to be shaky and tremulous and his memory confused' (he was still slightly concussed from the cellar fall). The gaps and confusions in his letters about Savy Wood, which he always called 'Feyet', seem to confirm the Medical Officer's view.
15   Sources for this paragraph: CKSM (1920, 1921); SS to SO, '20 February' (OEF); E. Sitwell (1921); RG (1929), end of ch. xxiv (modified in later editions); EB's draft memoir and correspondence (Tex). See also DW (1978) 159–60.
16   Brown (1934) 88. My information on shellshock comes mainly from Brown, Brock, MacCurdy and Rivers (see Bibliography). Brown worked in France in 1916–17, treated WO in May 1917 (*CL*, 455–6), and was put in charge of Craiglockhart in 1918 (presumably in the upheaval there referred to in *Sherston's Progress*).
17   Brown (1934) 91. Pre-war stress emerged in war dreams which had 'their source in the patient's earlier life'; treatment was by recall and analysis of pre-war and war experience.
18   The notebook (BL1, 157–64) includes drafts of 'With an Identity Disc' and 'Sunrise'; the fragments 'When on the kindling wood', 'It was the noiseless hour', 'Hide ah! my Flower'; notes for 'Sonatas in Silence'; and a Shakespeare sonnet.
19   *CPF*, 460–1. *CL*, 234.
20   Craiglockhart registers, Public Records Office (MH 106.1887).
21   *CL*, 473, 478. A mock ballad? Cf. his hospital parodies of the Bible (458–60), medical chart (465) and litany (557–8 – 8 June 1917, misdated in *CL*).
22   (*a*) *CL*, 389, (*b*) 406–7; (*c*) *CPF*, 353, (*d*) 351, (*e*) 487, (*f*) 175. For other dock imagery, cf. 'A Palinode' and 'I know the music'. Drafts of 'Disabled' show that WO imagined the soldier sitting 'between his crutches in the park' (no longer a 'movable' body).
23   *CL*, 502.

24 The only surviving MS has a July–August 1917 watermark; however, *CPF* does not add that EB seems to have seen another draft of unknown date (CDL, 113–14). CDL suggested 1916 because MS has a note of Sorley's 1916 book, but perhaps SS or RG recommended Sorley, whom they both admired, in 1917 as an antidote to the lushness of WO's poem.

25 RG made this mistake several times. Cf. Seymour-Smith (1982), 97, 113.

26 AJB: see Appendix A4.

27 WO found Outlook Tower people held 'my own – almost secret – views of such things as sculpture, state-craft, ethics, etc. etc.' (*CL*, 491).

28 *CL*, 476. MS notes in OEF (I amend slightly). WO acquired one Tower publication, Branford (1913), an essay by a leading Geddesian which AJB's book recommends as a study of poetic idealism–hysteria.

29 *CL*, 497, 481 n.3.

30 AJB to Geddes, 17 Mar, 16 Apr 1920 (National Library of Scotland, Geddes MSS 10546).

31 *Hydra*: see Appendix C. AJB (1911) first defines Ergotherapy. 'Effort and work are the instrument of all attainment' (WO, Tower essay).

32 Bergson (1915) argued that Germany had become a machine and would therefore be defeated.

33 *Health and Conduct*, 278.

34 *CL*, 473 (Mother). Lett, 283 (indigoes). *CL*, 479 (Sacrifices), 475 (activities), 479 (energy). The Craiglockhart letters repay careful reading. SO would have heard AJB's theories and WO promised her 'an hour on the Tower' (477), so subsequent letters to her assume a knowledge of Ergotherapy.

35 Article possibly by WO. The January one, signed 'Acturus' (he who is about to act), probably by AJB.

36 MS in OEF. *CL*, 478. WO had studied at a more advanced level than Cassell's, marking passages about plant behaviour in Keeble (1911).

37 See Ch. 3, n. 16, above.

38 *CL*, 255.

39 *CL*, 482 (WO first wrote 'wicked' for 'weary'), 484, 488, 490, 496.

40 *Health and Conduct*, 150.

41 *CL*, 536.

42 *CL*, 377, 427. Bäckman (1979) 103.

43 AJB (1924) 494.

44 SO to EB, n.d., *c.*1930 (Tex). EB, 29.

CHAPTER 6. SASSOON [Aug–Oct 1917]

1 RG (1930) 56. Seymour-Smith (1982) gives RG's side of the protest affair. There are many other published and unpublished sources.

2 SS's letters in Tex unless otherwise stated. These quotations: to Morrell, 26 Apr 1917 (snapping), 13 Nov 1917 (nightmares); to Meiklejohn, 25 Feb 1918 (horrors); to Morrell, 30 July 1917 (dotty), 21 Aug 1918 (cursed), 8 July 1917 (B. R.). Information on Rivers: SS to Meiklejohn, 1 Aug 1917; to Morrell, 30 July 1917.

3 SS to Meiklejohn, 25 Mar 1917, quoting Hardy, 'Night in the Old Home'.

4 For some criticism of SS's war poems, see *Casebook*.

5 SS to Morrell, 21 May 1917.

6 SS to Nicholson, 21 June 1917 (Harvard). For Nicholson, see SS, *Diaries* (1983) 33, and d'Arch Smith (1970). Beauty, compassion and friendship were Uranian values.

7 SS to EB, 4 Aug 1965.

8 *CL*, 486.

9 *CL*, 484.

10 SS (1945) 58.

11 *CL*, 567. SS's first mention of WO in a letter seems to be in answer to RG's enthusiasm in October (*Diaries*, 191). Evidence suggests that RG admired WO's verse before SS did. SS's silence about WO contrasts with his 1918 excitement to many correspondents about Frank Prewett (cf. *CL*, 482).

12 *CL*, 489 [Lett, 283]. Proof, were any needed, that WO wrote no trench poems before meeting SS.

13 For poem, drafts and associated comments, see *CPF*, 144, 298–300; *CL*, 485–6; *Hydra*, 1 Sep 1917; *CL*, 436 (*Mirror* photographs). WO was probably mistaken in marking the Belloc line as old; he should have marked the line from 'Perseus' about the Kaiser (*CPF*, 449).

14 OEF 249 (*CPF*, 255–6, 372). *CPF* treats both fragments about civilians as work for 'Six O'clock', although that poem is different in tone and style.

15 *CPF*, 137, ascribes 'The Letter' to 1918 but 1917 seems more likely. See Appendix B.

16 *CL*, 493. 'Hymne à Dionisius', Tailhade (1907) 105. DW (1960), 127, suggests this echo.

17 *CL*, 497. BL1, 60.

18 *CL*, 489 (Aug 1917).

19 1917 edition. SS reshuffled the sequence in his *Collected Poems*.

20 SS to EB, 12 Nov 1930 ('2.30 a.m.').

21 RG (1929, 3rd impression) 326. DW (1951), 349, disagrees with RG; the argument seems to depend on what date one gives to 'Exposure'.

22 *CL*, 494 (parallel). SS to Morrell, 19 Aug 1917 (Keats). *CL*, 487 (Hardy). *CL*, 492 (the folders may have included work by ELG). SS (1945) 59 (some confusion here: SS quotes from a 1918 poem as an example of 'lusciousness'). *CL*, 486 (Sonnets), 491 (condemned). *CL*, 490 [Lett, 283] (doubtful); 'Song of Songs' was published anonymously. The annotated *Hydra* is among SS's letters to Morrell; another, unmarked copy is in BNY, presumably sent by him to Marsh. He also sent a copy to Ross and promised one to Meiklejohn (SS to Meiklejohn, 9 Sep 1917).

23 See DH, 'Some Notes' (1982) for references for this paragraph. 'The Triumph' is last in the notebook, eighth from the end in SS (1918) and omitted from SS (1961).

24 *CL*, 494.

25 MS in OEF. JS, 212, says WO echoes SS, but WO's letter shows his poem preceded SS's.

26 Tennyson, 'Merlin and the Gleam'. Tennyson explained 'the gleam' as 'the higher poetic imagination' but it may have had some further meaning for WO and SS. Cf. *CL*, 512, and 'follow gleams more golden and more slim' (*CPF*, 184).

27 For example, Silkin (1972) 210–11.
28 An echo pointed out by David Perkins in a perceptive discussion of WO's Romanticism (*Casebook*).
29 Binyon, 'At the going down of the sun ... We will remember them' (1914). Brice, 'the voice of monstrous guns' (1916). WO liked Binyon's poem in 1915 (*CL*, 355).
30 *CL*, 248 (funeral); liturgical details in 'Anthem' drafts are Roman Catholic, so in calling them 'mockeries' WO was in line with his Evangelical past and not necessarily condemning all formal observance. Craiglockhart fragments: 'But I was looking' (*CPF*, 487–8), 'I know the music' (*CPF*, 485–6). Bells: cf. *CL*, 205, and 'passing-bell' (*Hyperion*, I. 173). Acolytes: favourite Decadent–Uranian subjects; cf. 'Maundy Thursday', and *CL*, 311.
31 Final draft marked '*Nation*'. Cf. *CL*, 506–7 (despite the footnote, WO could not have been referring to 'Miners' so early); SS (1945) 59–60. Two of RG's letters are in *CL*, 595–6; a third is in OEF.
32 Lang (1906) 172 (my italics). WO bought a copy of this edition (OEF), inscribing it 'W. E. S. Owen   Scarborough   Dec: 1917' (cf. *CL*, 520 n. 3).
33 For *CPF's* 'small ... vile ... kissed' and 'turning ... turning ... writhing', I read 'sin ... vice ... kissed' and 'turning ... twinging ... turning ... writhing' (*CPF*, 293).
34 Cf. Shelley's image of the face of war, eyes starting 'with cracking stare', tongue 'foamy, like a mad dog's hanging' (*Revolt*, lines 2476–83).
35 RG to Gosse, 24 Oct 1917 (Brotherton Collection, Leeds).
36 Gleam: *CL*, 512 (cf. n. 26, above), 505 (comet). 'Storm': cf. WO's early phrase 'Consummation is Consumption' (*CPF*, 384). *CL*, 492 (light the darkness), 567 (candle).
37 *CL*, 492.

CHAPTER 7. NEW INFLUENCES: GEORGIANS AND OTHERS [Nov 1917–Jan 1918]

1 *CL*, 513 (3 December), 520 (revise), 525–6 (art), 535 (bluebells), 533–4 (duty).
2 RG to Marsh, 1 Jan 1916 (BNY).
3 ''Orace Cockles' in draft (cf. Horatius Cocles).
4 For details, and references for this and preceding paragraph, see Geo, and DH, 'WO's Library' (1977).
5 'It would be a crime to exhibit the fine side of war, even if there were one!' – Barbusse (1929 edn) 342.
6 Nichols to SO, 10 Dec 1920 (OEF).
7 OEF 284. This hurriedly amended version of the last stanza is arguably later (but not better) than that given in *CPF*. Cf. imagery of gods, laughter and the exclusiveness of art in 'Sweet is your antique body' (*CPF*, 129).
8 *CL*, 510. Second sentence from a passage about ELG's book omitted from *CL*, 516. In May WO had sent ELG a roundel, having found it easy to

write without 'either emotion or ideas' (*CL*, 250; for date see Lett, 281); there are five roundels in ELG (1917). So the cousins had been in disagreement about truth to experience before WO met SS.

9   Keats to Haydon, 8 Mar 1819. Cf. 'Nothing ever becomes real until it is experienced' (to George Keats, 19 Mar).

10   *CL*, 520, 526, 531.

11   *WPO*, 122. *CL*, 416. 'L'Amour' first published in *YM*, the YMCA magazine, 10 Nov 1916, reprinted in ELG (1917).

12   'Tears, idle tears'. WO was reading Tennyson again with pleasure despite this gibe (*CPF*, 127). Cf. *CL*, 152.

13   WO got SS to autograph at least two copies (ELG still has one), having read the poem before meeting him (*CL*, 494).

14   *CL*, 514.

15   *CL*, 521. Cf. 'I suffer a temptation . . . to remain a poet's poet!' (*CL*, 520).

16   Two *Battle* poems were in the *Nation* in October 1914. See DH and Onions (1986). Possible traces of Gibson in WO's poems include a parallel between 'Makeshifts' (*Livelihood*, 1917) and 'Disabled' and 'A Terre'.

17   WO first wrote in his Preface that his subject was war and 'the pity of it' (the phrase originates from *Othello*). With another Hardy title, 'The Souls of the Slain', cf. phrases in WO's list of contents, 'The Soul of Soldiers' and 'The Women & the Slain'. WO described Hardy as 'potatoey' in 1917, perhaps alluding to the potato conversation in *Under the Greenwood Tree*; he may not have been aware of Hardy as a poet before meeting SS (*CL*, 487).

18   In Hardy's 'To the Moon', the moon sees life as a 'show'. *Dynasts* quotations: iii.iv.1, ii.vi.1, iii.i.9, Fore Scene. Some of these parallels are noted by Cohen (1957).

19   CL, 406. Tennyson, 'The Lotos-Eaters', line 159.

20   Wells (1914), 113–14. Cf. 'the last War of the World' (*CL*, 274).

21   SS to Morrell, 19 Aug 1917. *CL*, 520. Glover (1980) shows how Fitzwater Wray's spirited translation influenced WO's diction and imagery in 'Dulce' and other poems.

22   *CL*, 545 ('A Terre'). EB, 134 (Nicholson). Ross used such photographs (OS, 1950, 96); perhaps he got them from a senior civil servant such as Meiklejohn, who could also have given some to SS. But if SS avoided talking of horrors to WO, 'knowing they were bad for him', he is unlikely to have given him nightmarish pictures.

23   *CL*, 509.

24   *CL*, 512.

25   *CPF*, 514–17, gives three of five drafts, treating them as fragments. *Poems* (1920) included one as an 'Alternative Version' of 'Strange Meeting'. I assume they represent a complete, independent poem. I quote a fourth draft (BL0, 5) and fragments from others.

26   Cf. WO to SS: 'my principles and your mastery' (*CL*, 582). Cf. 'Had such a mastery of his mystery' (Tennyson, 'The Last Tournament', line 327).

27   Ezekiel 25:10: 'the noise . . . of the wheels, and of the chariots . . . as men enter into a city wherein is made a breach'. Joel 2:5, 7: destroyers shall come like 'the noise of chariots' and 'they shall not break their ranks'. Wells of water: John 4:14; Isaiah 12:3 (cf. also 58:11, one of WO's 1903 'promise texts').

28   See Mrs Shelley's note on *The Revolt of Islam*. Shelley's Prometheus is referred to as 'One' several times and has 'Drops of bloody agony' on his white brow (*Prometheus Unbound*, 1.564). Cf. 'the wise and free / Wept tears, and blood like tears' ('Ode to Liberty', line 269). Bäckman (1979) suggests other interesting parallels, including war's red chariot wheels in *Queen Mab*.

29   'Letter to Maria Gisborne', line 121. With 'wise', cf. 'wise and free' (preceding note) and 'We wise' ('Insensibility').

30   *Revolt of Islam*, line 1042.

31   Wells (1916) 22. *CL*, 527.

32   *CL*, 504. A visit to Lady Ottoline would have established a contact with Russell.

33   Russell, *Principles* (1916) 10. Cf. 'Those who are to begin the regeneration of the world . . . must be able to live by truth and love, with a rational unconquerable hope; they must be honest and wise, fearless, and guided by a consistent purpose' (246).

34   RG's poem, in his unpublished book *The Patchwork Flag* (1918), and associated letters are in BNY. SS to Morrell, 28 May 1918. *CL*, 530 (quotation unidentified – it is perhaps nearer Wells or even Wilde than Russell), 498 (washy), 521 (blindfold).

35   *CL*, 520 (propaganda). SS to Forster, 1 May 1918, quoted in P. N. Furbank, *E. M. Forster* (1978) 47. SS to Marsh, 18 July 1918 (BNY); SS (1945) 70–1.

36   'My soul looked' – still a trace of Joergens ('I looked and saw') as well as Shelley.

37   *CL*, 458, 544.

38   *CL*, 521.

CHAPTER 8. THE PITY OF WAR [Mar–May 1918]

1   *CL*, 535 (catastrophic), 538 (Farewell), 543 (get fit, cottage-window). WO, *CL* and JS spell 'Borage'; WO probably never saw it written down but then as now it was apparently 'Borrage' (from 'burgherage', not 'borage'). 'No one here knows of my retreat' (*CL*, 547): JS, 259, may be mistaken in implying WO actually lived in the cottage, since he would presumably have had to inform his superiors. He was free daily from about 3 p.m. to Lights Out. Borrage Lane compositions probably include 'As bronze', 'Arms and the Boy', 'The Send-Off', 'Strange Meeting', 'Futility', 'Mental Cases', 'Elegy in April' (four stanzas). Many poems begun earlier were extensively revised, including 'Exposure', 'A Terre' and 'Insensibility'.

2   Wordsworth, Preface to *Lyrical Ballads* (1802). *CL*, 544 (public attention was on the German breakthrough at St Quentin), 524 (Hamel), 542 (Wordsworth wrote three poems to the 'small' celandine). Murry (1921). *CL*, 543 (numbs), 581 (charred).

3   *Pace CPF*, 147, 'Insensibility' is hardly a reply to Wordsworth's 'Happy Warrior', which is not a Pindaric ode and contains little that WO would have disagreed with (cf. Ch. 11, n. 23, below).

4   'Happy': a key word in Keats's odes but also in wartime propaganda (cf. 'Smile, Smile, Smile', and Plates 13, 14).

5   The 'wise' in 'Insensibility' are poets (cf. Ch. 7, n. 29, above, and *Adonais*, line 312). WO does not imply any comment on class or education as some critics have supposed.

6   Symons, 'To the Merchants of Bought Dreams', *Poems* (1906) II, 175. BL1, 15.

7   The first draft (*CPF*, 304) refers to frozen bodies underfoot. Cf. *CL*, 542.

8   *English Elegies*, etc.: titles on verso of first list of contents (*CPF*, 538; cf. JS, 265, and *CL*, 561 n. 3, an inaccurate note). Lang: see Ch. 6., n. 32, above. Milton, Gray: Bäckman (1979) 40–3. Tennyson: DH, *WO* (1975) 32–3; *In Memoriam*, cxxix.

9   *CL*, 511 (acid). EB, 125 (sour).

10 MSS, like 'An Imperial Elegy', have musical annotations. At one stage entitled 'Ode for a Poet reported Missing: later, reported killed', this may be the 'Ode' in the 1918 lists of contents (*CPF*, 147 headnote). Subtitle, '(jabbered among the trees)', from SS, 'Repression of War Experience'. In stanzas 1–4 (Apr) the poet seeks a companion missing in spring; in stanzas 5–7 (Sep) he laments him killed in autumn. Presumably the poem was meant for SS, but who is its subject?

11 *CL*, 526. Gosse in *Casebook*, 44–5.

12 Some careless remarks of mine years ago on protest and elegy have been rebuked. DW (1978) 167–8 rescues me from an absurd position with characteristic generosity. Silkin (1980) seems less persuasive.

13 '[And have I shut the last book I shall read?]' OEF 339–42, not in *CPF*.

14 For both lists, see *CPF*, 538–9. They are of war poems only, but WO continued to plan work unconnected with war, assembling sonnets at Scarborough ('Farewell . . . Sonnets' – *CL*, 538). A note of 5 May 1918 (OEF; *CL*, 551 n. 1) lists four 'Projects': verse plays on old Welsh themes, like Tennyson's English and Yeats's Irish dramas (so WO envisaged a role as a Welsh poet); *Collected Poems* (1919); 'Perseus'; 'Idylls in Prose'. But he would have been less old-fashioned than this if he had lived; he had yet to encounter Modernism.

15 For the date of Society Bond MSS, see Appendix B.

16 MS reproduced as frontispiece to *CPF* and in DH (1975).

17 CKSM to WO, 26 May 1918 (OEF).

18 *CL*, 554.

CHAPTER 9. TO SUFFER WITHOUT SIGN [1918]

1   See also Appendix A 5–7.

2   Carpenter Collection, Sheffield Central Library, has letters from SS, including five in 1917, and one from RG (1914). SS sent a copy of his protest to Carpenter, who satirised the official response to it in 'Lieutenant Tattoon, M. C.' (*Three Ballads*, 1917). SS considered taking a factory job near Carpenter in August 1918. See also Tsuzuki (1980), Grant (1967).

3   'The Intermediate Sex' – Carpenter (1911) 130 – an essay which SS said had 'opened up a new life' for him in 1911. Cf. 'We wise who with a thought besmirch / Blood over all our soul' ('Insensibility').

4   WO and SS seem to have shared confidences; see WO's allusions to Sodom, beautiful bodies, 'exposed flanks' (*CL*, 506, 512, 582). Cf. n. 31

below. Homosexuality was on SS's mind at Craiglockhart – R. Gathorne-Hardy, *Ottoline at Garsington* (1974) 230.

5 For Sherard, see JS, 242, 320. Ross wrote to Gosse about covering up SS's protest (19 July 1917, BL Ashley MSS). SS often asks after Ross in 1918 letters. For other sources for this paragraph, see Appendix A5, and *CL*.

6 SS (1945) 29.

7 OS (1949) 115. WO sent him a crudely spelled copy of 'Long ages past', perhaps to show that his own juvenilia had included *Salomé*-like verse.

8 'Armchair', prefatory poem in *Wheels 1917*. For 'old men', see 'Adelyne More', 'The one thing needful', *Cambridge Magazine*, 7 (19 Jan 1918) 315–17.

9 TO's war scrapbook (OEF) shows the simple patriotism for which WO sometimes criticised him. It contains reports of actions fought by WO's regiment; the report of WO's death is accompanied by news of Victoria Crosses won at the time and some consolatory newspaper poems.

10 WO ordered *Wheels 1917* in late July (cf. S. Sitwell's 'Barrel-Organs' in this volume with 'The roads also'). CKSM gave him S. Sitwell (1918) in August (JS, 320). For an assessment of what WO might have achieved, see P. Hobsbaum in *Casebook*.

11 Details in this paragraph from OEF MSS and sources given in Appendix A6.

12 TS in BNY, n.d.

13 The evidence for *CPF*'s 1917 date for the first draft of 'Reunion' is not clear; the second draft is almost certainly early 1918. The three ballads are on Clarence Gardens paper (like Nov–Feb letters). WO may have sent them to a *Bookman* competition, announced in January; the results (May) list him (Clarence Gardens address) and ELG as runners-up in the ballad section ('Song of Songs' won a consolation prize and was printed in the lyric section). In February WO saw much of P. Bainbrigge, skilled author of 'ballads of a . . . private kind' (CKSM, 1920).

14 Cf. WO's poems about the 'god' Eros, and the 'phantoms' seen in his 1913 illness.

15 *CL*, 461 (cf. Tailhade, 'Ne tue pas!'); WO gave his religion as 'Primitive Christian' in June (467). *CL*, 468 (Oxenham), 483–4 (*CL* gives 'lad's cheeks'; MS is ambiguous; plural seems more likely).

16 *CL*, 536 ('as one says'). WO misquotes from Wilde (1905), 116. The style of *De Profundis* is evident in this letter.

17 Wilde (1905) 88–9. *CL*, 521, 573, 580. WO, 'The Calls'.

18 But see Breen (1974) 177, Silkin (1972) 234–5, etc. My comments on and quotations from 'Greater Love' drafts are based on MSS, not *CPF*. Unlike *CPF*, I take BL1, 40 (a version of lines 1–18) and OEF 269v (lines 19–24) to be the earliest surviving draft. These and BL1, 38v and 39v, were scribbled out on the backs of fair-copied sonnets at some point in the period summer 1917 to spring 1918. Two later drafts are probably summer 1918.

19 But 'greater love' also occurs in Keats ('Isabella', xl).

20 WO imitates OS's epigram 'Ill Winds' (*CL*, 561–2). He copied it for his family, adding, 'I need not show unto you this Jesus. For myself, I have seen him with my eyes, and touched his blood with my hands. / I am now engaged in teaching him to lift his cross by numbers, and inspecting

his feet, that they be worthy of the nails. WEO' (MS in OEF). R. Nichols described WO himself as a crucified, gospel-bringing Christ: 'You owe it to his wounds, received for your sake, . . . to try and realise what war is, as he has shown it, and, having learned, to teach others' (cutting dated 1920, OEF). WO might have preferred N. Royde-Smith's comparison of his work to 'the torso of some unimaginably beautiful antique marble, mutilated, unrestorable, as splendid as dreams' (*Time and Tide*, 13 June 1931).

21 With the walls and terraces in 'The Kind Ghosts', cf. the terrace in *Salammbô*, xiii, so heaped with bodies that it seems to be made of them.

22 Wilde (1948) 56.

23 *CL*, 563. *CPF* perhaps ascribes too many drafts to Scarborough (July–Aug); WO had much less time there than at Ripon (Mar–May).

24 SS to WO, c. 7 Aug 1918 (OEF); cf. *CL*, 567.

25 Mrs Gaskell defined the Victorian author's role as spokesman in saying she wrote *Mary Barton* 'to give utterance to the agony which, from time to time, convulses the dumb people' – K. Tillotson, *Novels of the 1840s* (1961), 205.

26 *CL*, 568, 570. WO misquotes 'Testament' (see DH and Onions, 1986).

27 WO was willing to go out again but it is not clear whether he volunteered or awaited orders. Recorded events as follows: 9 August: draft order issued (OEF). 11th: taken off draft by Medical Officer. What happens next is uncertain (*did* the War Office intervene?) but he gets a few days' leave (*CL*, 16, says 12–18 Aug but is probably guessing). 17th (Saturday): returns to Scarborough after a London afternoon with OS (1950, 108–9) and SS (1945, 71–2). 26th: embarkation order (OEF) issued at Scarborough, the Medical Officer having now certified him fit to proceed overseas (how could a cardiac irregularity have disappeared in a fortnight?). 30th: in Hastings with SO and Colin, returns Victoria alone late, met by CKSM. 31st: 7.35 a.m. train, Victoria-Folkestone; embarks later that day. WO's letter about his last evening in London has been mutilated by HO. I have not seen WO's War Office file.

28 SO to Alec Paton, 27 May 1919 (Tex).

29 *CL*, 570–1.

30 *CL*, 589. A fellow officer in October later recalled that WO 'mentioned that he had the highest regard for the Poet A. C. Swinburne and very often I found him studying a volume of his works' (Lt J. Foulkes, unsigned notes used by EB, *c*. 1930, Tex). For the volume, see JS, 321.

31 DW (1978) 158–9. *CL*, 580–1. WO told SS he was glad Jones was 'happily wounded: and so away from me. He had lived in London, a Londoner.' This may be a coded message, and 'Who is the god of Canongate' may be relevant to an interpretation of it; Jones seems in fact to have come from Hereford (*CL*, 582, 560).

CHAPTER 10. 'STRANGE MEETING' [1918]

1 *CL*, 560.

2 DH, 'The Date of . . . ' (1976). MS of 'Cramped' has draft lines used in 'Exposure' and 'Mental Cases' (*CPF*, 513).

3 'A New Heaven' (*CPF*, 223). *CL*, 424.
4 Reconstructed from BL1, 28v. For other workings, see *CPF*, 367.
5 *CL*, 126.
6 Cf. Ch.4, n.27, above.
7 Revelation 7:13–14 – Sinfield (1982) also makes this point. Dante, *Vision of Hell*, tr. Cary (1814) v.50–1; 'Mental Cases' is reminiscent of many passages in this translation, which both WO and Keats knew.
8 *CPF*, 492. I reconstruct a little. MS has features in common with those of 'Spells and Incantation' (late 1917) and possibly 'Mental Cases'. It ends '[or clo]', another echo of Keats ('emperor and clown').
9 An ambiguity first pointed out by Gose (1961).
10 I prefer a Ripon (March – very early June) date to *CPF*'s January–March (see Appendix B).
11 *EB*, 128.
12 Bäckman (1979) 101–2. As elsewhere, HO's memory may have been influenced by WO's poetry but Bäckman's suggestion is striking.
13 MS in BNY.
14 *Faerie Queene*, I.i.41–2. The preceding stanza is quoted in the Hazlitt volume WO read in January. He summarised *Faerie Queene*, I, at school.
15 *Endymion*, IV.512–42.
16 'Ode to Liberty', lines 256–7. *CL*, 571.
17 A recurrent theme in *Health and Conduct*.
18 SS to Forster, 30 June 1918 (King's College, Cambridge), records his respect for and disagreement with Russell's views. SS's political understanding of the war was now that German militarism was preventing peace.
19 The *alter ego* reading was proposed by DW (1960) and subsequently widely accepted (but see Silkin and DH, *Stand*, 1980). Earlier examples of doubles include Shelley, *Prometheus Unbound*, I.192–9; Dickens, *Our Mutual Friend* (Headstone–Riderhood), *Dombey* (Phiz illustration to ch. xl). See Miyoshi (1969). The sexuality in 'Strange Meeting' is noticed by John Bayley (*Casebook*, 156–7).
20 Ser Brunetto is the obvious example, familiar from 'Little Gidding' (cf. Ch. 1, n. 12, above).
21 *Fall of Hyperion*, I.462–3 (an echo suggested by Bäckman); I.260–3.
22 Murry (1919) 1284. Murry deals admirably with borrowing: 'no danger to the real poet. He is the splendid borrower who lends a new significance to that which he takes . . .'.

CHAPTER 11. 'SPRING OFFENSIVE' [Sep–Nov 1918]

1 *CL*, 571 (serenity); 580–1 (nerves); 573 (sorrows); 575 [Lett, 284] (hawked); 580 (Strictly); 583 (circulation); 591 (band of friends; cf. 'band of brothers', *Henry V*, IV.iii.60, and *Revolt of Islam*, line 2407).
2 *CL*, 572.
3 *CL*, 574. The Harrow boy was 'the best piece of Nation left in England' (570). By chance, in February WO had read Vachell (1905), a Harrow story, in which a boy redeems past failings by dying a hero in South Africa. The scene possibly influenced 'Spring Offensive': troops are 'halted

at the foot of the hill, halted in . . . a storm of bullets. Then the word was given to attack . . . fire from invisible foes . . . He ran . . . as if he were racing for a goal', dying at 'the highest point', 'smiling at death' (295). Harrow Hill reminded WO of Broxton, his own place of education (*CL*, 535).

4   *CL*, 589 [Lett, 284]. An ensuing sentence about Gordon Gunston is silently omitted from *CL*, one of three similar comments omitted from late-1918 letters.

5   *CL*, 585.

6   'Smile, Smile, Smile'.

7   *CL*, 589 (atrocities), 583–5 (Peace Talk).

8   WPO, 134.

9   'Happy': cf. 'Insensibility', and Ch. 8, n. 4, above. Anger against newspaper optimism had been an Army grouse for some years; WO is acting as spokesman.

10   *CL*, 498 (reputation), 582 (confidence), 580 (pleader).

11   The *CL* wording (580 n. 2) follows a carbon TS preserved by HO (OEF), who may have thought it genuine. The TS contains at least one error and has no official mark. The type of carbon looks early. Is it a forgery by SO? John Bell cannot explain the discrepancy between it and the apparently official wording given (with slight variations) in TO's scrapbook, Higham (1922), Walsh (1964), and JS, 279 (where the puzzle is not remarked on).

12   *CL*, 580. Cf. 'one seraphic lance corporal' (582).

13   Two reports by William Beach-Thomas (cf. *CL*, 584).

14   *CPF*, 540, 193.

15   *CL*, 458.

16   Years before WO had asked for a poetic vision like those granted to 'old dreamers on May Morn' (*CPF*, 385).

17   *WPO*, 135, gives the time as late afternoon, an inexcusable error.

18   *CL*, 510 (croziers; quoted from ELG, 'From the Caradoc in June'). Appendix A7 (HO's memory of buttercups). *CL*, 588 (Haughmond), 581 (forefathers; HO perhaps remembered this when saying he too thought of them in times of danger – *JFO*, ii, 77).

19   *CL*, 381.

20   *CPF*, 12, 378.

21   Cf. imagery of men as trees in Monro's 'Trees' (*Strange Meetings*, 1917), a poem strongly influenced by Carpenter.

22   Cf. 'Stand ye . . . Like a forest . . . With . . . looks which are / Weapons of unvanquished war' (Shelley, *Mask of Anarchy*, lines 319–22).

23   Matthew 17:2; Revelation 1:16. Cf. *CPF*, 194 n. 2, and Wordsworth, 'Happy Warrior', lines 51–2 (the soldier called to fight is 'happy as a Lover; and attired / With sudden brightness, like a Man inspired').

24   Wells (1917) 77–8, 123. *CL*, 487 (WO had deplored a passage from the book when it was published in May – 461).

25   'Fearfully flashed', 'the whole sky burned': cf. rape of Dânae ('the whole sky fell / In . . . flashings' – P3); and 'Fitfully flash' (*Dynasts*, iii.iii.2).

26   Cf. 'the last sea' ('Insensibility'). Men thought the end of the world had come at Roncesvalles.

27 WO was not erudite enough to have been aware of the arguments about what is really meant by the many Old Testament references to 'high places', and to Moloch (Molech) and child sacrifices by fire; but he would have remembered Milton's description of Moloch, and such biblical phrases as 'They have built also the high places of Baal, to burn their sons with fire' (Jeremiah 19:5), 'no man might make his son or his daughter to pass through the fire to Molech' (2 Kings 23:10), 'Then did Solomon build an high place . . . for Molech' (1 Kings 11:7).

28 Moloch's priests wear scarlet (cf. WO's battledress soaked in Jones's blood and his simile for it, 'crimson-hot iron'). Another *Salammbô* chapter WO may always have remembered is the last, which brings together key images of scarlet, blood, fatal love and stare, passivity, sacrifice, etc., in a ghastly climax.

29 Cohen (1965) 259–60.

30 *CL*, 431.

31 *Paradise Lost*, I.746–7, referring to a 'Sheer' fall (James McLaverty suggests this parallel). WO's line is persistently misread. In Arthur Bliss's war oratorio *Morning Heroes* (1930), it is intoned as Christian consolation. But Bliss did see the heroic element in the poem. Rutherford (1978) is one of the few modern critics to write perceptively about heroism in this and other Great War poems.

32 *CL*, 458.

33 *CL*, 432.

Bibliography

A select list of published works consulted (for MSS, see Notes and Appendixes).

Space prevents a complete list of Owen articles. William White, *Wilfred Owen (1893–1918): A Bibliography* (Kent State University Press, 1967) gives a near-complete record to 1965. The following books, in addition to others listed in (3) below, contain discussion of Owen's work: A. Banerjee, *Spirit above Wars* (1975); Bernard Bergonzi, *Heroes' Twilight* (1965); D. J. Enright, essay in *Pelican Guide to English Literature*, v, ed. Boris Ford (1961); Paul Fussell, *The Great War and Modern Memory* (1975); John H. Johnston, *English Poetry of the First World War* (1964); Philip Larkin, *Required Writing* (1983); David Perkins, *A History of Modern Poetry* (1976); C. H. Sisson, *English Poetry 1900–1950* (1971); Hilda D. Spear, *Remembering, We Forget* (1979); C. K. Stead, *The New Poetic* (1964). *Poetry of the First World War: A Casebook*, ed. Dominic Hibberd (1981), reprints studies by Sassoon, Blunden, Murry, Day Lewis, Welland, Bayley, Johnston, Perkins and others.

For 1914–18 poets, see Catherine W. Reilly, *English Poetry of the First World War: A Bibliography* (Prior, 1978), and the biographical–bibliographical notes in *Poetry of the Great War: An Anthology*, ed. Dominic Hibberd and John Onions (1986).

Books marked * are a small selection of those known to have been read and/or owned by Owen (in the edition shown, where this can be ascertained). (*) indicates probability rather than certainty. Standard works by major authors are usually omitted.

Abbreviations: *ELT– English Literature in Transition 1880–1920*; *N&Q– Notes and Queries*; *RES– Review of English Studies*. Place of publication (books) London unless otherwise stated.

1 POEMS PUBLISHED DURING OWEN'S LIFETIME

'Song of Songs', *Hydra*, 1 Sep 1917, and *Bookman*, May 1918.
'The Next War', *Hydra*, 29 Sep 1917.
'Miners', *Nation*, 26 Jan 1918.
'Futility' and 'Hospital Barge', *Nation*, 15 June 1918.

2 EDITIONS OF OWEN'S WORKS

Poems by Wilfred Owen, ed. Siegfried Sassoon [actually mainly ed. Edith Sitwell] (Chatto & Windus, 1920).

The Poems of Wilfred Owen, ed. Edmund Blunden, with a memoir (Chatto & Windus, 1931).

The Collected Poems of Wilfred Owen, ed. C. Day Lewis, with an introduction and notes (Chatto & Windus, 1963).

Wilfred Owen: Collected Letters, ed. Harold Owen and John Bell (Oxford University Press, 1967).

Wilfred Owen: War Poems and Others, ed. Dominic Hibberd, with an introduction and notes (Chatto & Windus, 1973).

Wilfred Owen: The Complete Poems and Fragments, ed. Jon Stallworthy, 2 vols (Chatto & Windus, The Hogarth Press and Oxford University Press, 1983).

Wilfred Owen: Selected Letters, ed. John Bell (Oxford University Press, 1985).

The Poems of Wilfred Owen, ed. Jon Stallworthy (Chatto & Windus, The Hogarth Press and Oxford University Press, 1985).

3 GENERAL

Bäckman, Sven, *Tradition Transformed: Studies in the poetry of Wilfred Owen*, Lund Studies in English 54 (Lund: C. W. K. Gleerup, 1979).

*Barbusse, Henri, *Under Fire: The Story of a Squad*, tr. Fitzwater Wray (Dent, 1917). Originally *Le Feu* (Paris: Flammarion, 1916).

Bateson, F. W., 'The Analysis of Poetic Texts: Owen's "Futility" and Davie's "The Garden Party" ', *Essays in Criticism*, 29 (Apr 1979) 156–64.

Bebbington, W. G., 'Jessie Pope and Wilfred Owen', *Ariel* (Leeds), 3 (1972) 82–93.

*Beckett, Sir Edmund, *On the Origin of the Laws of Nature* (SPCK, 1879).

Beerbohm, Max, *Zuleika Dobson, or an Oxford Love Story* (Heinemann, 1911).

*Bennett, Arnold, *The Pretty Lady* (Cassell, 1918).

*Benson, A. C., *Tennyson* (Methuen, 1912 edn).

Bentley, Christopher, 'Wilfred Owen and Gustave Flaubert', *N&Q*, 17 (Dec 1970) 456–7.

Bergson, Henri, *The Meaning of the War: Life and Matter in Conflict* (T. Fisher Unwin, 1915).

*Branford, Victor, *St Columba: A Study of Social Inheritance and Spiritual Development*, illus. John Duncan (Outlook Tower and Chelsea: Patrick Geddes and Colleagues, 1913).

Breen, Jennifer, 'Wilfred Owen: "Greater Love" and Late Romanticism', *ELT*, 17 (1974) 173-83.

———, 'Wilfred Owen (1893–1918): His Recovery from "Shell-Shock" ', *N&Q*, 23 (July 1976) 301–5.

Brock, A. J.: *see* Appendix A 4.

*Brooke, Rupert, *1914, and Other Poems*, 13th impression (Sidgwick & Jackson, May 1916).

Brown, William, *Psychology and Psychotherapy*, 3rd edn (Arnold, 1934).

Busst, A. J. L., 'The Image of the Androgyne in the Nineteenth Century', in *Romantic Mythologies*, ed. Ian Fletcher (Routledge, 1967).

Carpenter, Edward, *Love's Coming-of-Age: A Series of Papers on the Relations of the Sexes*, 7th edn (George Allen, 1911).

———, *Three Ballads (an Intermezzo in War Time)*, pamphlet (Oct 1917).

Carter, A. E., *The Idea of Decadence in French Literature 1830–1900* (Toronto: University of Toronto Press, 1958).

Chapple, J. A. V., *Documentary and Imaginative Literature 1880–1920* (Blandford Press, 1970).

Cohen, Joseph, 'Wilfred Owen's Greater Love', *Tulane Studies in English*, 6 (1956) 105–17.

———, 'Owen's "The Show" ', *Explicator*, 16 (Nov 1957) item 8.

———, 'Owen Agonistes', *ELT*, 8 (Dec 1965) 253–68.

*Colvin, Sidney, *Keats* (Macmillan, 1907 impression)

Cooke, William, 'Wilfred Owen's "Miners" and the Minnie Pit Disaster', *English*, 26 (Autumn 1977) 213–17.

*Dante, *The Vision, or, Hell, Purgatory, and Paradise of Dante Alighieri*, tr. H. F. Cary (Warne, n.d.).

d'Arch Smith, Timothy, *Love in Earnest: Some Notes on the Lives and Writings of the English 'Uranian' Poets from 1889 to 1930* (Routledge, 1970).

Das, Sasi Bhusan, *Wilfred Owen's 'Strange Meeting': A Critical Study* (Calcutta: Firma KLM Private, 1977).

———, *Aspects of Wilfred Owen's Poetry* (Calcutta: Roy & Roy, 1979).

(*)Flaubert, Gustave, *Madame Bovary* (first published 1857).

*———, *Salammbô* (first published 1862).

*———, *La Tentation de Saint-Antoine*, Collection Gallia (Paris: Crès; London: Dent, n.d.).

Freeman, Rosemary, 'Parody as a Literary Form: George Herbert and Wilfred Owen', *Essays in Criticism*, 13 (Oct 1963) 307–22.

Gaunt, William, *The Aesthetic Adventure* (Cape, 1945).

Geddes, Patrick: *see* Appendix A 4.

*Gibson, Wilfred Wilson, *Battle* (Elkin Mathews, 1915).

———, *Collected Poems, 1909–1925* (Macmillan, 1926).

(*)———, *Fires*, 3 vols (Elkin Mathews, 1912; repr. as one vol., 1915).

Glover, Jon, 'Owen and Barbusse and Fitzwater Wray', *Stand*, 21, no. 2 (Spring 1980) 22–32.

———, 'Whose Owen?', *Stand*, 22, no. 3 (1981) 29–31.

Gose, E. B., 'Digging in: An Interpretation of Wilfred Owen's "Strange Meeting" ', *College English*, 22 (Mar 1961) 417–19.

Graham, Desmond, *The Truth of War: Owen, Blunden and Rosenberg* (Manchester: Carcanet, 1984).

Grant, Joy, *Harold Monro and the Poetry Bookshop* (Routledge, 1967).

(*)Graves, Robert, *Over the Brazier* (Poetry Bookshop, 1916).

(*)———, *Goliath and David* (Chiswick Press, ?Dec 1916). [Owen referred to Graves's 'books' in Oct 1917.]

*———, *Fairies and Fusiliers* (Heinemann, 8 Nov 1917).

———, *Goodbye to All That: An Autobiography* (Cape, 1929, extensively revised in later editions).

———, *But It Still Goes On: An Accumulation* (Cape, 1930)

———, *In Broken Images: Selected Letters 1914–1946*. ed. Paul O'Prey (Hutchinson, 1982).

*Gunston, E. Leslie, *The Nymph, and Other Poems* (Stockwell, Nov 1917). [Dedicated to Owen.]

(*)Hardy, Thomas, *The Dynasts: An Epic-Drama of the War with Napoleon* (Macmillan, 1903–8).

*———; *Under the Greenwood Tree* (Macmillan, 1907).

———, *Moments of Vision* (November 1917).

Hassall, Christopher, *Edward Marsh: A Biography* (Longman, 1959).

———, *Rupert Brooke: A Biography* (Faber, 1964).

Hibberd, Dominic, 'Images of Darkness in the Poems of Wilfred Owen', *Durham University Journal*, 56 (Mar 1974) 156–62.

———, *Wilfred Owen*, Writers and their Work 246, (British Council and Longman, 1975), repr. in *British Writers, VI* (New York: Scribner, 1983).

———, ' "Rival pieces on a chosen theme": A Note on Some of Wilfred Owen's Minor Poems', *Four Decades of Poetry 1890–1930*, 1 (Merseyside, Jan 1976) 70–5.

———, 'The Date of Wilfred Owen's "Exposure" ', *N&Q*, 23 (July 1976) 305–8.

———, 'A Sociological Cure for Shellshock: Dr Brock and Wilfred Owen', *Sociological Review*, 25 (May 1977) 377–86.

———, 'Wilfred Owen's Library: Some Additional Items', *N&Q*, 24 (October 1977) 447–8.

———, 'Wilfred Owen's Rhyming', *Studia Neophilologica*, 50 (1978) 207–14.

———, 'Wilfred Owen and the Georgians', *Review of English Studies*, 30 (Feb 1979) 28–40.

———, 'Some Contemporary Allusions in Poems by Owen, Rosenberg and Sassoon', *N&Q*, 26 (Aug 1979) 333–4.

———, 'Silkin on Owen: Some Other War', *Stand*, 21, no. 3 (Spring 1980) 29–32.

———, ' "Problems A": The Solution to Wilfred Owen's "The Imbecile" ', *N&Q*, 27 (June 1980) 232–3.

———, 'Concealed Messages in Wilfred Owen's Trench Letters', *N&Q*, 27 (Dec 1980) 531.

———, 'Some Notes on Sassoon's *Counter-Attack and Other Poems*', *N&Q*, 29 (Aug 1982) 341–2.

———, 'Wilfred Owen's Letters: Some Additions, Amendments and Notes', *Library*, 4 (Sep 1982) 273–87.

———, ' "The End": Wilfred Owen and Leslie Gunston', *N&Q*, 30 (Aug 1983) 325–6.

———, and Onions, John (eds), *Poetry of the Great War: An Anthology* (Macmillan, 1986).

Higham, S. Stagoll (ed.), *Regimental Roll of Honour and War Record of the Artists' Rifles*, 3rd edn (Howlett, 1922).

*Housman, A. E., *A Shropshire Lad* (Grant Richards, 1915).

Hydra: *see* Appendix C.

*Joergens, Olwen A., *The Woman and the Sage, and Other Poems*, Little Books of Georgian Verse (Erskine Macdonald, Apr 1916).

*Keats, John, *Complete Works*, ed. H. Buxton Forman, 5 vols (Glasgow: Gowans & Gray, 1900–1).

*Keeble, Frederick, assisted by M. C. Rayner, *Practical Plant Physiology* (Bell, 1911). [Owen knew Miss Rayner, and probably Professor Keeble, at Reading in 1912.]

Kermode, Frank, *Romantic Image* (Routledge, 1957).

Keynes, Geoffrey, *A Bibliography of Siegfried Sassoon* (Hart-Davis, 1962).

Lane, Arthur E., *An Adequate Response: The War Poetry of Wilfred Owen and Siegfried Sassoon* (Detroit: Wayne State University Press, 1972).

Lang, Andrew: *see Theocritus, Bion and Moschus*.

*Larronde, Carlos (compiler), *Anthologie des écrivains français morts pour la Patrie*, 4 vols (Paris: Larousse, 1916).

*———, *Le Livre d'Heures* (Paris, 1913).

Léger, Albine: see Appendix A 2.

*Locke, J. W., *The Morals of Marcus Ordeyne* (Bodley Head, 1905).

MacCurdy, John T., *War Neuroses*, with a preface by W. H. R. Rivers (Cambridge University Press, 1918).

MacDonald, Stephen, *Not About Heroes: The Friendship of Siegfried Sassoon and Wilfred Owen*, play (Faber, 1983).

*Magnus, Laurie, *Introduction to Poetry* (Murray's School Library, 1902).

*Marsh, Edward (compiler), *Georgian Poetry 1916–1917* (Poetry Bookshop, November 1917).

Masefield, John, *The Everlasting Mercy* (Sidgwick & Jackson, 1911).

Masson, David I., 'Wilfred Owen's Free Phonetic Patterns: Their Style and Function', *Journal of Aesthetics and Art Criticism*, 13 (Mar 1955) 360–9.

Miyoshi, Masao, *The Divided Self: A Perspective on the Literature of the Victorians* (New York University Press and London University Press, 1969).

*Monro, Harold, *Before Dawn: Poems and Impressions* (Constable, 1911).

*———, *Children of Love* (Poetry Bookshop, 1914).

*———, *Strange Meetings* (Poetry Bookshop, 1917).

*Morris, Sir Lewis, *The Epic of Hades* (first published, 1877).

Murry, John Middleton, 'The Condition of English Poetry' (review), *Athenaeum* (5 Dec 1919) 1283–5.

———, 'The Poet of the War', *Nation and Athenaeum*, 28 (19 Feb 1921) 705–7.

*Nichols, Robert, *Ardours and Endurances* (Chatto & Windus, 1917).

Nordau, Max, *Degeneration*, tr. from the German (Heinemann, 1895).

O'Riordan, Conal, 'One More Fortunate', in *Martial Medley: Fact and Fiction* (Scholartis Press, 1931). [Contains a reminiscence of Owen.]

Orrmont, Arthur, *Requiem for War: The Life of Wilfred Owen* (New York: Four Winds Press, 1972).

Owen, Harold, *Journey from Obscurity*, 3 vols (Oxford University Press, 1963–5). *See also* Appendix A 7.

———, *Aftermath* (Oxford University Press, 1970).

*Oxenham, John, *The Vision Splendid: Some Verse for the Time and the Times to Come* (Methuen, Mar 1917).

Pearson, John, *Façades: Edith, Osbert, and Sacheverell Sitwell* (Macmillan, 1978).

Pound, Reginald, *The Lost Generation* (Constable, 1964).

Praz, Mario, *The Romantic Agony* (Oxford University Press, 1933; reissued, 1970).

*Renan, Ernest, *Souvenirs d'enfance et de jeunesse* (Paris: Nelson, n.d.).

Rivers, W. C., 'Mr Yeats Analyses his Soul', *Cambridge Magazine*, 7 (19 Jan 1918) 315–17.

Rivers, W. H. R., *Instinct and the Unconscious: A Contribution to a Biological Theory of the Psycho-neuroses* (Cambridge University Press, 1920).

———, *Conflict and Dream* (Kegan Paul, 1923).

Ross, Robert: see Appendix A 5.

Ross, Robert H., *The Georgian Revolt: Rise and Fall of a Poetic Ideal 1910–1922* (Faber, 1967).

*Rossetti, W. M., *Life and Writings of John Keats*, Great Writers (Walter Scott, n.d.).

Royal Academy Pictures (being the Royal Academy Supplement to *The Magazine of Art*), annual volumes (1888–1915).

Russell, Bertrand, *Justice in War-Time* (Chicago and London: Open Court, 1916).

———, *Principles of Social Reconstruction* (Allen & Unwin, 1916).

Rutherford, Andrew, *The Literature of War: Five Studies in Heroic Virtue* (Macmillan, 1978).

*Sassoon, Siegfried, *The Daffodil Murderer* by Saul Kain (John Richmond, 1913).

*——, *The Old Huntsman, and Other Poems* (Heinemann, May 1917).

*——, *Counter-Attack, and Other Poems* (Heinemann, June 1918).

——, *Sherston's Progress* (Faber, 1936).

——, *Siegfried's Journey 1916–1920* (Faber, 1945). [Ch. vi describes Owen.]

——, *Collected Poems 1908–1956* (Faber, 1961).

——, *Diaries 1915–1918*, ed. Rupert Hart-Davis (Faber, 1983).

——, *The War Poems*, ed. Rupert Hart-Davis (Faber, 1983).

Scott Moncrieff, C. K.: *see* Appendix A 6.

Seymour-Smith, Martin, *Robert Graves: His Life and Works* (Hutchinson, 1982).

*Shelley, P. B., *Complete Poetical Works*, ed. T. Hutchinson (Oxford University Press, 1912).

*Sherard, Robert H., *The Real Oscar Wilde* (T. Werner Laurie, n.d.).

*——, *Oscar Wilde: The Story of an Unhappy Friendship* (Greening, 1909 edn).

Silkin, Jon, *Out of Battle: The Poetry of the Great War* (Oxford University Press, 1972).

——, 'Owen: Elegist, Satirist or Neither; A Reply to Dominic Hibberd', *Stand*, 21, no. 3 (Spring 1980) 33–6.

Sinfield, Mark, 'Wilfred Owen's "Mental Cases": Source and Structure', *N&Q*, 29 (Aug 1982) 339–41.

*Sitwell, Edith (ed.), *Wheels 1917: A Second Cycle* (Oxford: Blackwell, 1917).

——,(ed.), *Wheels 1919: Fourth Cycle* (Oxford: Blackwell, 1919). [Dedicated to Owen, contains seven of his poems.]

——, *Selected Letters*, ed. John Lehmann and Derek Parker (Macmillan, 1970). [Contains ten letters to Mrs Owen, 1919–21.]

Sitwell, Osbert, *Argonaut and Juggernaut* (Chatto & Windus, 1919). [Includes war poems.]

——, *Laughter in the Next Room* (Macmillan, 1949).

——, *Noble Essences or Courteous Revelations* (Macmillan, 1950). [Ch. iv is about Owen.]

*Sitwell, Sacheverell, *The People's Palace* (Oxford: Blackwell, 1918).

(*)Sorley, Charles Hamilton, *Marlborough, and Other Poems* (Cambridge University Press, 1916).

Stallworthy, Jon, 'W. B. Yeats and Wilfred Owen', *Critical Quarterly*, 11 (Autumn 1969) 199–214.

——, 'Wilfred Owen', Chatterton Lecture on an English Poet, repr. from *Proceedings of the British Academy*, 56 (Oxford University Press, 1970).

——, *Wilfred Owen: A Biography* (Oxford University Press and Chatto & Windus, 1974).

Steiner, George, *In Bluebeard's Castle: Some Notes towards the Re-definition of Culture* (Faber, 1971).

Stephan, Philip, *Paul Verlaine and the Decadence 1882–90* (Manchester: Manchester University Press, 1974).

*Swinburne, A. C., *Poems and Ballads: First Series*, Golden Pine edn (Heinemann, 1917).

*Symonds, John Addington, *Shelley*, English Men of Letters (Macmillan, 1909 edn).

Tailhade, Laurent: *see* Appendix A 3.

Theocritus, Bion and Moschus, tr. Andrew Lang (Macmillan, 1906 edn).

Thorpe, Michael, *Siegfried Sassoon: A Critical Study* (Leiden: Leiden Universitaire Pers; London: Oxford University Press, 1966). [Reprints *The Daffodil Murderer*.]

Tsuzuki, Chushichi, *Edward Carpenter, 1844–1929: Prophet of Human Fellowship* (Cambridge University Press, 1980).

*Vachell, H. A., *The Hill: A Romance of Friendship* (Murray, 1905).

Walsh, T. J. (compiler), *A Tribute to Wilfred Owen* (Birkenhead: Birkenhead Institute, 1964).

Welland, Dennis S. R., 'Half-rhyme in Wilfred Owen: Its Derivation and Use', *RES*, I (July 1950) 226–41.

———, 'Wilfred Owen: The Man and his Poetry', unpublished PhD thesis (University of Nottingham, 1951).

———, 'Wilfred Owen's Manuscripts', two articles, *Times Literary Supplement*, 15 and 22 June 1956.

———, 'Sassoon on Owen', *Times Literary Supplement*, 31 May 1974.

———, *Wilfred Owen: A Critical Study* (Chatto & Windus, 1960; repr. with a postscript, 1978).

*Wells, H. G., *The Passionate Friends* (Macmillan, 1913).

*———, *The Wife of Sir Isaac Harman* (Macmillan, 1914).

———, *The World Set Free: A Story of Mankind* (Macmillan, 1914).

*———, *What is Coming? A Forecast of Things after the War* (Cassell, 1916).

(*)———, *God the Invisible King* (Cassell, May 1917).

White, Gertrude M., *Wilfred Owen*, Twayne's English Authors (New York: Twayne, 1969).

*Wilde, Oscar, *De Profundis*, ed. Robert Ross (Methuen, 1905).

*———, *Poems* (Methuen, June 1916).

———, *Works*, ed. G. F. Maine (Collins, 1948).

General Index

Owen, Wilfred: attitudes to war, 49,
55–9, 66, 71–2, 103–4, 110–11, 114,
180–1, 191; cowardice, alleged,
76–7, 133, 135, 162–3, 199; and
the Decadence, 16, 29–34, 38–41,
47–8, 53, 56–7, 68–9, 74, 78, 82–3,
90, 112, 114, 122, 156, 158, 160–
1, 170, 211; dedication to poetry,
x, 2–6, 25, 43, 119; 'Disabled and
Other Poems', 146–8, 152; 'Do
Plants Think?', 88–9, 188, 217;
dreams, 17–19, 41, 48, 68, 71, 77–
8, 82, 89, 91–4, 119, 134, 138, 166,
168, 170, 216; as educator,
spokesman, x, 13, 25, 42, 61–2,
71–2, 128–31, 158, 162, 175, 182,
191, 226; elegy, 140–2, 149, 191;
and enlisting, 37, 43, 49–50, 57–
8; and Ergotherapy, 84–90, 93–4,
187; and the Georgians, 15, 70,
119–26, 134–6, 143, 154; imitating
other poets, 3–4, 6, 15, 39–41, 98–
9, 101, 123–6, 153–4; library, 4,
41, 202, 209; lists of contents
(1918), 136, 146–8, 159, 204, 222;
manuscripts, ix, 202–5; Military
Cross citation, 182–3, 226; 'Minor
Poems', 63, 215; pararhyme, 34,
79, 82, 88, 90, 106, 123, 125, 129–
30, 134, 140, 155, 212; photographs
of casualties, 129, 220; 'the pity of
war', 126–8, 139–42, 147–9, 174,
220; poems, *see separate Index*;
Preface (1918) and allusions to,
33, 88, 122–3, 127, 136, 139–41,
146–9, 153, 183–4, 186, 191–2, 220;
and religion, 5–7, 9–13, 23–6, 58–
·, 156–60, 189, 209, 219, 223–4;
return to the front, 161–5, 180–3,
224; work at Ripon (Borrage
Lane), ix, 136–49, 169, 172, 204–
5, 221; sexuality, 19–23, 26–7, 34–
5, 38–9, 43, 46–8, 150–3, 155–6,
192, 197–201; shellshock, 17–18,
75–9, 91–4, 134–5, 216; and
Shropshire hills, 1–2, 8, 36, 89,
186; 'Sonatas in Silence', 33, 63,
78, 215–16; sonnets, 41–2, 62–3, 73,
78, 107, 123, 205, 215, 222;
'sympathy for the oppressed', 1–2,

13–15, 53, 125, 128, 158; in
trenches, 71–6; Welsh blood, 2–3,
186
Oxenham, John, 7, 157, 223, 232; *The
Vision Splendid*, 121

Paris, Owen in, 54, 64, 82, 194–5
Patmore, Coventry, 74
Paton, Alec, 209, 214, 224
Pearson, John, 198, 232
Pemberton Billing, Noel, 152, 197
Perkins, David, 219, 228
Perseus, *see Index of Owen's Poems*
Poitou, Henriette, 34–5, 43–6, 50,
187, 212–13
Pope, Alexander, 138
Pope, Jessie, 114–15
Pound, Ezra, 70, 195
Pound, Reginald, 214, 232
Praz, Mario, 53, 211, 232
Prewett, Frank, 218

Rampton, Vivian, 21–3, 27, 34, 68,
187, 211
Reilly, Catherine W., 228
Renan, Ernest, *Souvenirs d'enfance et de
jeunesse*, 39–40, 232
Rivers, W. C., 177, 232
Rivers, W. H. R., 84, 95–6, 116, 132,
195, 216–17, 232
Ross, Robert, 29, 106, 151–2, 154–6,
163, 181, 196–7, 199, 201, 218, 220,
223
Rossetti, Dante Gabriel, 30–1;
'Aspecta Medusa', 18; 'A Last
Confession', 211
Rossetti, William Michael, *Life and
Writings of John Keats*, 10, 20, 210,
232
Ruskin, John, 1, 5–6
Russell, Bertrand, 95, 103, 116, 119,
132–3, 175–6, 221, 225, 233
Rutherford, Andrew, 227, 233

Salmond, J. B., 206
Sassoon, Siegfried, 15, 26, 29, 33, 41,
55, 58, 70, 76, 81, 84, 90, 94–5,
114, 116–26, 128–34, 136, 141–2,
148, 150–4, 158, 160–5, 169, 173,
175, 179–84, 189, 197, 199, 201–
4, 206–7, 211–12, 216–19, 220–5,

Index of Owen's Poems

241

THE RISING TIDE

Molly Keane
(M. J. Farrell)

Introduction by
Polly Devlin

A *Virago* Book

Published by Virago Press 1984

Reprinted 1984, 1985, 1988, 1990, 1994, 1997

Copyright © M. J. Farrell 1937
Introduction copyright © Polly Devlin 1984

First published in Great Britain by Collins 1937

The moral right of the author has been asserted

A CIP catalogue record for this book is available
from the British Library

ISBN 0 86068 472 5

Printed and bound in Great Britain by
Clays Ltd, St Ives plc

Virago
A Division of
Little, Brown and Company (UK)
Brettenham House
Lancaster Place
London WC2E 7EN

INTRODUCTION

Molly Keane started to write in the 1920s, for "pin" money when she too was in her twenties, using the pseudonym M.J. Farrell which name she borrowed from an Irish public house as she clattered by on her Irish horse near her Irish home in County Wexford after a hard day's hunting. She needed a pseudonym since in her Anglo-Irish and somewhat philistine world, writing was no occupation for a young woman. "For a woman to read a book, let alone write one was viewed with alarm, I would have been banned from every respectable house in County Carlow." So, apparently casually, did her remarkable career begin, a career that has spanned half a century, though her creative powers quivered, almost broke and certainly lay mute in her middle years, stilled by the double blows of the sudden death of her beloved and dashing young husband and, later, the failure of her fourth play in London's West End—a play out of kilter with the new atmosphere in the theatre of that time, since epitomised by *Look Back In Anger*.

Then in 1981 she published *Good Behaviour* under her own name. It was shortlisted for the Booker Prize and she has written steadily ever since, to the enormous pleasure of a new generation of readers and indeed television viewers since *Good Behaviour* was adapted for a three part "classic" serial. The story of the emergence of Molly Keane, a delightful, witty woman in her late seventies from the ghost of the anonymous, sexless M.J. Farrell is charming and arbitrary, as so much connected with this diffident author seems to be. She wrote *Good Behaviour* in her late sixties and sent it to Sir William Collins, who had published most of her earlier works. He turned it down on the grounds that it was too black a comedy—"Couldn't I make the characters more pleasant? Imagine a black comedy with charming characters. I couldn't and wouldn't do it—the bravery—and put it away in a drawer and thought 'I'm too old. I've forgotten how to write plays and now I can't write a novel any more.'"

There the matter literally rested until Dame Peggy Ashcroft came to stay with Molly Keane in her house perched on a cliff-top in Waterford and, ill in bed with 'flu, wanted something to read. She gave her the manuscript to read and Dame Peggy loved it—her enjoyment and encouragement spurred Molly Keane to send it to André Deutsch, who found they had a new classic on their list, greeted with popular and critical acclaim. It was no new thing for Molly Keane—she had had success from her earliest days, both with her novels and with two early plays written in collaboration with John Perry, directed by John Gielgud, and, in *Spring Meeting*, introducing the inimitable Margaret Rutherford to delighted London audiences. Hugh Walpole wrote,

I think that Miss Farrell is one of the best half-dozen younger women novelists now writing in England [sic]. She has in the first place a beautiful gift of description ... She has, secondly, a real sense of drama ... Thirdly, she can create character. There is beauty here with understanding and an original mind.

The myth is that she shut herself up for a time whenever her dress allowance—some £30 a year—ran out and she herself seems to subscribe to this explanation of her artistic motivation. But her writing has no feeling of expediency, her novels are never journey-work—indeed they contain few of the ingredients necessary for writing as a commercial venture. She writes of narrow horizons, élitist occupations, the preoccupations of a moneyed, hunting, curiously dislocated class of people, floating as it were over the political, angry geographical reality that was Ireland. In *The Rising Tide* there is no mention of political turmoil though at the time in which it is set the issue of Home Rule was tearing the country apart.

vi

VIRAGO
MODERN CLASSICS
137

Molly Keane

Molly Keane was born in Co. Kildare, Ireland in 1904, into 'a rather serious Hunting and Fishing Church-going family', and was sketchily educated by governesses. Interested in 'horses and having a good time', Molly Keane wrote her first novel, *The Knight of Cheerful Countenance*, when she was seventeen in order to supplement her dress allowance. She used the pseudonym M. J. Farrell 'to hide my literary side from my sporting friends'. Between 1928 and 1961 Molly Keane published ten novels under her pen name, novels in which she brought acuteness and good-tempered satire as well as affection to her portrayals of the ramshackle Anglo-Irish way of life. She also wrote several successful plays. But the untimely death of her husband brought a break in her career which ended only in 1981, when *Good Behaviour* appeared under her own name, triggering a revival of interest in and respect for her work. Molly Keane died in 1996.

In this disregard for the outside world she is akin to Jane Austen; in concentrating on the two inches of ivory of one Edwardian family, in her feeling for the minutiae of human behaviour, she gives an unforgettable picture of a vanished world, the world Home Rule was threatening. These people are sustained by an absolute sense of their own superiority, by a certainty about the appropriate social response to every crisis, including tragedy. It is a book about, among other things, heartlessness. It is also about houses, hunting and horses, her central themes.

It is remarkable in its conjuring up of the lust some women have for hunting: literally mad for it, finding in the chase everything lacking in their ordinary lives—glamour, muscle, music, passion, thrills, admiration, spills and exhaustion. Horses were the great pursuit in Molly Keane's Ireland—that other Ireland. Everyone rode, even the most timid and shrinking of daughters. It was mandatory for acceptance in that world and in *The Rising Tide* she paints a shocking picture of the heroine Cynthia's utter belief in the "rightness" of riding.

She did not love her children but she was determined not to be ashamed of them. You had to feel ashamed and embarrassed if your children did not take keenly to blood-sports, so they must be forced into them. It was right. It was only fair to them. You could not bring a boy up properly unless he rode and fished and shot. What sort of boy was he? What sort of friends would he have?.

And was it really like that?

"When I was young" Molly Keane says, almost with disbelief "I really disapproved of people who didn't ride, it was the only thing that counted. I've had this feeling for horses all my life. It's a frightening excitement that takes the place of something else. Of course my father provided the horses I rode in my early days, but he was as distant about hunting as he was about most things. He was a beautiful horseman but never instructed us. It was all long before things like the pony club existed. Then you just were, or were not any good at it, there was no question of your being taught. It was your occupation, it was supposed to be hereditary, you just knew how."

She *did* know; she always loved and enjoyed hunting, as did all her family, save her mother who did not ride (one brother, Walter Skrine, rode in the Grand National although he had been badly injured in the war). But this mastery of the saddle never precluded her imaginative leap into the minds of her compatriots who were terrified of riding. "When I started to write there were lots of books about hunting and awful hunting romances. . . *The Queen of the Chase* and all that kind of thing. I thought it much more interesting to write about fright. . .It's natural to be frightened." She describes brilliantly the horror of children rising in the cold damp chill of a November morning, when the very hot water bottle is frozen in the bed, to drive in an ancient car to the meet, to get onto their ponies. Driven also by the fear of their parents, by the social shame of refusing.

They looked at the fields and the fences as they drove past them, seeing them with a curious relationship of fear and not fear. The fields were not ordinary fields as in summer. They were places you had to get out of, that you were inexorably carried over. That field now, with green plover waddling and pecking about on its dark, sheep-bitten turf, was a dreadfully unkind field with wire on its nice round banks and as cruel a coped stone wall across one side of it as a frightened child could face. The very young may be sick with cowardice about a fence but they are more afraid not to jump it than to jump it . . . It was not soft falls in water or bog that they dreaded, but that shameful, hurting falling off and the moments before you fell, their agony seeming to endure in interminable uncertainty before you went with a sort of sob and the ground hit you from behind, strangely like a house falling on you, not you falling on a house.

As children Molly, her sister and her three brothers were much neglected and left to their own and the servants' devices—and from her earliest days she was fired with a fierce

viii

determination to lead her own life away from the influence of her mother, who lived in an atavistic world of her own and didn't pay the least attention to her children who were educated by default. Her mother was a writer, who wrote under the pseudonym Moira O'Neill and was known as the Poet of the Seven Glens. She had great success with her "folkloric" poems which are often magical and touching although they incline to show the peasantry in a moral, amiable and trusting attitude which seems out of order with the real thing like pious children in Victorian genre-painting. "I fought her every inch of the way" Molly Keane says, "She really didn't know how to treat us. You can't think how neglected we were, by our parents. I mean they didn't do anything with us at all, they simply didn't bother. They were utterly reclusive. My mother had great taste but was totally oblivious to comfort. Life was much more stringent then, there was no such thing as hot water or central heating. There were fires but they went out and I remember the deadly cold of the school room and the blue cold coming off the wall. I never remember a fire in my father's library or in the dining room, although my father was perhaps a bit more warmth conscious." One memorable day when her father had forbidden her to ride a horse she'd been given she was discovered weeping behind a fox covert by a Major Perry who lived in a beautiful Queen Anne house, Woodrooff, in the centre of the glamorous Tipperary hunting country. She went to stay with him, his wife Dolly "the dearest woman who ever lived" and their two children Sylvia and John, who also hunted like lunatics. (In later years John Perry encouraged and helped her write her first play, Spring Meeting, and sent it to Gielgud.) Woodrooff became her home—"I almost lived there for six or seven years, mostly in the winter months, when I hunted three days a week on horses largely provided by Woodrooff, although a few others chipped in. There were so many horses in those days of the late twenties and early thirties that if you were lightweight and a moderately useful rider your fun was endless."

. "My mother disapproved of Woodrooff—she was frightened by the idea of it. She belonged to the nineteenth century and didn't change, and I think I must have been a hideous worry and an anxiety to her. When she was very old she said to my sister Susan 'Oh the mistakes I made about Molly.' Rather a sad comment. I expect we all make mistakes about our children and always will. She feared that I would get into a fast set and fall into bad ways—she was as worried as a mother would be now about her child going on drugs. There was a woman there who'd been divorced and some what *she* would have called dirty talk which I didn't know a thing about, but I soon found out about and was rather good at. My mother was alarmingly prudish and old-fashioned in those ways. In fact everyone there was wonderfully kind to me. I met marvellous people, and it was only when I went to stay in London with some of the women I'd met there, one especially who was very sophisticated and on the fringe of literary and intellectual life in London and taught me what to read and to go to the theatre, that I began to be in any way educated. I was utterly ignorant."

"I remember once one of my books got into one of the bestseller lists and I was asked to a literary party at the Piccadilly Hotel and I thought 'what awful, ghastly boring people—still I'd better go'. I travelled over on the boat with this hunting gent—really rather mad and in fact deadly boring, but how I wished I was with him instead of at this dreary party. I met Elizabeth Bowen and I thought 'what a strange lady', not what I considered attractive, and she introduced me to David Cecil and I thought he was ghastly. Just shows my state of ignorance. Elizabeth Bowen became my greatest friend in life and when I met David Cecil again I realised he was the most divine person who ever breathed. But the hunting people had their own style. It was like a club with its own language and rules."

In fact her whole life belonged to a club—an exclusive club

x

called the Anglo-Irish. The meaning of the term—though most people seem to accommodate imaginatively what it means—is elusive, full of heft and emotion, and covers anything from a cool description of a class to an indictment, depending on who is using it and how they are using it. It is for a start a comparatively new term. Before the breakaway from England many of those people now described as Anglo-Irish would certainly have described themselves as Irish though, (and this is where things thicken) the native Irish would not have allowed them the privilege of their Irishness, as a kind of revenge. It is an angry refusal, perhaps the only refusal that was left to them since the Anglo-Irish were powerful and believed themselves to be socially superior. In the time when M.J. Farrell was writing—a generation before Molly Keane and a generation after *The Rising Tide*—Ireland was a place of three countries, the newly created states of Eire, Northern Ireland and this older, floating notional world of Anglo-Ireland. The "real" Irish (the peasant Catholic Irish) seem only to enter this world—and thus Molly Keane/Farrell's faithful rendering—as a sub-species, good for opening gates and giving amusing, barely subservient lip service, the words and rhythms of which were recounted as hilarious anecdotes with broguish emphasis on the Oirishness of it all. Perhaps in no other country has simply and artlessly belonging to that country been made such an inherently ridiculous thing, an object for mirth.

Indeed Oirishness was the Anglo-Irish interpretation of how the Irish who lived outside their club and class behaved and spoke: there was no question of admittance. "No-one would have thought of marrying someone not of their own class," Molly Keane said. "It would have been more than death. It simply wasn't an idea. Those things were completely part of the code." These co-existing cultures had nothing in common except their country and that extraordinary Irish climate, which as George Bernard Shaw said "will stamp an immigrant more deeply and durably in two years apparently than the English climate will in two hundred".

It worked the other way round too—even those great families who had been in Ireland for centuries and regarded themselves as more Irish than the Irish, when they "turned" Protestant for political expedient and earthly survival, became aliens to the common mass of the people, indistinguishable from the English. There is a famous story of how Oscar Browning rushed up to Tennyson and said "I am Browning" and Tennyson said "You are not." The Anglo-Irish have said over and over again "We are Irish" and the Irish have said "You are not." That the class that had endowed them with a sense of social inferiority and, as they believed, taken their land, should claim the same national identity whilst despising it in its purest form was an anathema to the mass. De Valera said that to know what the Irish people wanted he had only to examine his own heart. There was little room in his heart for the Anglo-Irish, though he could far less easily claim to be Irish than those he excluded.

There is little point to saying that all this is history and should be forgotten for, as Terence de Vere White wrote in his book on the Anglo-Irish, "The Irish mind hops back to the flood when discussing a leaking tap." In many ways the world about which Molly Keane writes is like that of the Raj in India. W.B. Yeats was once asked was Oscar Wilde a snob; "No" he said "I would say that England is a strange country to the Irish. . . to Wilde the aristocrats of England were like the nobles of Baghdad." The Anglo-Irish were like the nobles of Baghdad to the Irish people and the gradations of class within the Anglo-Irish so cunningly observed, so slyly mapped out by M.J. Farrell would certainly not have been apparent to the "common people". Many of the Anglo-Irish weren't rich or grand, but appeared so since it cost next to nothing to keep horses and a groom then. "The very rich did much more about being more English and their sons went to Eton and Oxford. Everything was part of the code . . ." Molly Keane says, and when she is talking one hears bewilderment that the thing has irrevocably gone and astonishment that it should have existed. There is none of that guilt that the native Irish hope the Anglo-Irish should suffer, if only because they had so much enjoyed the country in which they lived and from which enjoyment the natives were excluded—as Indians under the Raj suffered an acute sense of inferiority whilst serving their masters with loyalty, a loyalty which was largely a myth. But it was never a question of anything so subtle as gradations between the Irish and the Anglo-Irish. An unbridgeable chasm lay between them. "Those who know Ireland," wrote an observer about the time in which *The Rising Tide* is set, "need not be told that the feeling of the average Irish Protestant towards Roman Catholics was a repugnance, instinctive rather than reasoned, based on social and racial as much as religious antipathies."

In the South of Ireland that world shuddered to an end in the Troubles, in the same way that Edwardian England ground to a halt with the First World War and in *The Rising Tide* M.J. Farrell makes a spectacular imaginative leap back into that lost Edwardian world, that mysterious generation that has more irrevocably vanished and which just preceded her youth. *The Rising Tide* was her seventh book and one of her most accomplished: in it she shakes out the crushed memories of an age lying pleated like parchment maps at the bottom of a century's memory, and snatches back the looks and feel and textures of that time, its mixture of meanness and voluptuousness, its sumptuousness and sternness, its scent and meanings, and displays the protocol of its discretions—and indiscretions. And then she juxtaposes this world against the chic gloss of the world of the twenties and watches as they turn each other upside-down.

The title is layered with meaning, as closely lapped and integrated as feathers on a wing: for the tide that is rising engulfs not only Lady Charlotte, the old tyrant who rules Garonlea, the grand house at the centre of the book; Cynthia, her daughter-in-law and her foe who inherits the house; but also engulfs an age, a class and that mythical country in which the Anglo-Irish lived with such apparently endless cocooned luxurious security. The tide rose silently, they seemed to hear no tocsins. "People simply didn't visualise any change coming. They believed life would go on like that for all time, and for some it did, because they were rich enough to remain insulated from everything that was happening—except of course for those who had the rude shock of seeing their houses with their entire contents burned around them."

Which is what happened to her father. And when her husband died she had to leave their beautiful house, Belleville. Her acute feeling for the secret life of houses, the mark that the years have laid on them, the turmoil, the passions, the domestic histories that have seeped into the very fabric, the way the colours of ages past have run into the spirit of a later age, all these intangibles play a particular and powerful part in her plots and especially in *The Rising Tide*. This is as much the story of the house, Garonlea, a great crenellated Gothic mansion, full of malevolence and power, lying brooding at the bottom of its secret valley as it is the story of two indomitable women pitted against each other like primeval forces.

Garonlea seems a convincingly Irish house, but in fact it is Warleigh Manor, lying in a wooded valley outside Bath. Her description is superb. Walking round the house—now a school—one gets the measure of her gift for evocation, of her extraordinary apprehension of atmosphere and spirit.

What is there that can be told about Garonlea and the evil that can be on a place through want of happiness. Or even a will towards happiness. Family tragedy is brief and sudden in comparison to this that lies like the breath of mould in old clothes on the people who live in such a place. It seems as though nothing could ever dissolve such mists and ill vapours, or only for such a little while. So inexact, so dim is such a gloom, it is hard to say whether it is the effect of place on character or character on place.

To this house comes Cynthia, the new daughter-in-law to

be presented to her mother-in-law, Lady Charlotte, its chatelaine and its evil genius. Cynthia acknowledges that both the house and its ruler are her enemies and that she must fight and defeat them, and the Old Queen, unscrupulous herself, knows well what the new pretender is like. "She took the utmost from everyone around her far more than she gave. She failed them and charmed them to her again. She leaned upon them and queened it over them at the same time."

The portrait of Cynthia is magnificently done: and what makes it so artful is that one never quite knows how mendacious Cynthia actually is, how unscrupulous, how fine. She is like a woman of one's own acquaintance whom one cannot help admiring, yet about whom one has grave doubts which are at times assuaged, at others nourished with foreboding. She is an early Superwoman, bearing children, running a house impeccably, hunting every day if possible, keeping her courtiers entranced and always looking glorious.

She removes herself to the altogether sunnier dower house across the valley, and there as liege lady she holds a permanent tournament. The forces are drawn up on the other side, and it is a formidable battle, a crash of styles and ways of living, a collision of hatred as well as a clash between two queens of indomitable will, and Molly Keane doesn't miss a trick in recording the fight to the finish. She knows whereof she writes. Lady Charlotte's genesis might lie in the nineteenth-century attitudes of Molly Keane's mother, but the character is based more closely on one of her aunts "who was far grander and far more deadly than my mother". Her every wrinkle and quirk, her every meanness and pretension is limned in with dreadful precision.

Lady Charlotte French-McGrath mounting the stairs in her daughters' wake was a shocking despot, really swollen with family conceit and a terrifying pride of race. She had a strange sense of her own power made real indeed by a life spent at Garonlea with her obedient husband, frightened children and many tenants and dependants. Here she loved and suffered and here she was supreme.

Her moral dereliction is delivered in one deadly sentence. "She was mean, although not so mean as her husband whom she had taught to be mean."

The Edwardian details are impeccable and irresistible.

Lady Charlotte rang for her maid. She then washed her hands in buttermilk soap, folded the neck of her combinations down towards the top of her corsets (those corsets which propped so conscientiously the bosoms like vast half-filled hot water bottles) and thus prepared stood while her evening dress was put upon her and sat while her hair was fiddled not redone. Her hair was never washed but it did not smell of anything but hair. The switches and curls of false hair were drier and frizzier in texture than her own.

The portrait of Cynthia too was taken from life—based on a beautiful Irish woman (a Fitzgerald) who was married to the local glamorous M.F.H. (who came from Warwickshire). "We all worshipped her," Molly Keane recalls, "thought she was marvellous. She had great chic and we tried to look like her and dress like her and felt that we, too, would like, if at all possible, to marry the same kind of man. Looking back I feel she was extraordinarily kind and generous to us younger generation. It was a little clan."

The Rising Tide did not just engulf an age, that clan, a woman and then another woman. It engulfed a way of life and in the book you watch it happen. As Molly Keane says, "A way of life that people thought would go on for ever, ended. It was another way of life and one treats it, must treat it as, in a much bigger way, people who are expelled from Poland, say, treat their exile—looking back as though it had never been. The place existed, but you can no longer return there."

It is gone, as you might say, for good. In Molly Keane's books it lives on.

Polly Devlin,
Bruton, Somerset 1983

I

WHAT don't we know about the Early Nineteen Hundreds. 1901 and 2 and 3 and up to '14 we can feel about only very dimly. Leisure and Richness and Space and Motor-veils and the bravery of those who flew in the earliest aeroplanes impress us. But we can't feel about those years really. Not in the way we feel about the War. There we are conscious.

It requires an effort to realise the necessities, pleasures, colossal bad taste, Romance, trust, suspicion, pride of those years. The War is forced on us, horror is so actual. But those years, the years of our cousins' youth, avoid us and will not be known. We almost forget how deeply that youth was influenced by the generation that got it. Influenced and prescribed for in a way we can't know about. So much and such nearly complete power was in those elder hands. Over the trivialities or fatalities of Life our cousins and aunts accepted so much and really managed it with admirable smoothness and dignity.

Pain they endured and accepted.

Endless Chaperonage.

Supervision of their correspondence.

The fact that Mother Knew Best.

That Father Says So.

That there is no more to be said on the subject, they accepted.

They accepted their leisure without boredom.

They accepted having occupations found for this leisure.

5

They accepted trivialities and treated them with that carefulness and detail which rounds such perfect smallness and makes it an acceptable part of life.

With all this acceptance they could preserve a death-like romantical obstinacy where their hearts were concerned—they had a true romantical outlook, infinitely less destructible than the quick love encounters we so often know. On absence their Romances throve. They were not afraid of sentimentality—they were not afraid of being thought girlish. They never needed to explain their emotions to themselves or to their friends. The indecency of knowing what it was all about would have been appalling to them; they didn't want to know—the mystery and the thrill enough and most secretly their own. Hence much rapture and much failure, and a certain dignity too. This outward smoothness of Life which at all costs they struggled to achieve was a politeness of living which we may envy them.

"Eleven o'clock—more than bedtime!" Lady Charlotte French-McGrath had four daughters, and at these words Muriel immediately folded up her work. Enid ceased tracing a picture of a stag's head into her album called Sunlight and Shadow. Violet gathered up the cards with which she and her father had been playing picquet, and only Diana, Little Diana, showed no speed in closing her book. Really, Mother might not have spoken——

"Bed-time, I think, Diana."

Diana shut her book guiltily and was the first of the four to kiss her father good-night.

"Ha," he said. "Ha-Ha. Bed-time. Bed-time, I suppose. Good-night, my dear. Candles, now, let me see, candles." He crossed the room to that small, dark

6

table where immemorially the candlesticks were set out and lit the five candles. Giving each daughter a kiss and a candlestick and the same to his wife, he followed them with a very satisfied eye as they went out of the door, crossed the hall one behind another and mounted the stairs in the same pretty succession.

Muriel first with her fluffy brown hair, thin neck and little birdy body. Poor little Muriel—time she was finding a husband. Twenty-four and nothing satisfactory turning up yet. Enid then, with her purple eyes, deep voice and dark hair. She would have been a beauty if Violet had not come after her, and Violet was an Edwardian classic. Skin like shells and peaches, bosom like the prow of a ship, smooth thighs, features of bland and simple beauty and a head crowned by obedient golden hair and unhampered by brains. A satisfactory daughter. And then Diana—little Diana—there was not much of the Edwardian classic about Diana. Her mother could not find her very satisfactory, since she had neither the charm nor the biddable disposition of her elder sisters. And none of their beauty. Small and dark and angular and inclined even at the age of seventeen to a dark and downy growth of hair (ignored by her family, for what could be done about it?) Diana was hardly due for success in 1900.

Lady Charlotte French-McGrath mounting the stairs in her daughters' wake was a shocking despot, really swollen with family conceit and a terrifying pride of race. She had a strange sense of her own power, made real indeed by a life spent chiefly at Garonlea with her obedient husband, frightened children and many tenants and dependants. Here she had lived and suffered and here she was supreme.

Married at the age of eighteen to a man of good family

7

and one who owned moreover the best woodcock shoot in the west of Ireland, Lady Charlotte had borne six children in the first eight years of her married life— four daughters and two sons; one son unhappily died of convulsions when an infant. She was mean, although not so mean as her husband whom she had taught to be mean. She ran Garonlea like a court— her daughters like the ladies in waiting.

"Muriel, write my notes for me——"

"Enid, how about your little job of washing the china in my boudoir this afternoon——"

Such awful little employments—the walks and messages to the needy tenantry, bestowals of charities and reprimands, piano practising and "Lying Down," that cure for all ills.

God should have chastened Lady Charlotte with one malformed or unsatisfactory child. But they were all miracles of aristocratic good looks—inclined to anæmia perhaps, except for the divine Violet.

One by one the Lady Charlotte French-McGrath presented her daughters to their sovereign. But as far as entering the social contest and finding them husbands went, God or Garonlea and its famous woodcock shoot might do the rest. So far, whatever God might send them in that lonely countryside, the woodcock shoot produced for the most part only their father's friends and contemporaries. Determined as she was that the girls should marry men of Property and Title, Lady Charlotte did nothing at all about collecting these mythical and appropriate husbands. She showed such marked disapproval of any young friend that her son Desmond ever brought to stay, should the friend fulfil neither of these requirements, and such embarrassingly obvious tactics should he fulfil either or both that Desmond desisted in disgust,

8

for he was a charming creature and entirely free from his mother's influence. His sisters adored him and it was his firm intention to do all in his power for the girls when he should marry. Until then he looked on them as rather boring princesses set for a time in a castle beyond a wood.

Meantime it would be unfair to her not to allow that Lady Charlotte loved her daughters with a passion none the less genuine if it demanded first their unquestioning obedience, and fed itself on a profound jealousy of any interest in their lives other than those she might herself prompt or provide. She felt that her children owed to her as a mother, not as a person, love, confidence and obedience. She felt this tremendously. It was a true thing with her. Among all the travails and secret adventures of her own life both mental and physical she had endured, raising no manner of complaining, shyness as much as stoicism helping her here. She absolutely required that her children should prove a justification, as she should see it, of herself.

So they were—they were practically all that she required. Her son Desmond and Violet were the two at the top of her estimation but she trusted herself to keep this sacred maternal secret safely from the other three daughters. "I love all my children equally," she was very fond of saying. Although vaguely impressed by the proper feeling she thus displayed, the three less-favoured daughters were never slow to employ either Desmond or Violet as their intermediary if for any reason matters should be strained between themselves and their mother.

To-night at the stairhead they parted from her.

"Good-night, Mother dear."

A gold head bobbed for a kiss in the candle-light.

"Good-night, my child. Don't let me hear you and Muriel talking at one o'clock as I did last night on my way to—the bathroom. Please, Violet."

"Good-night, Mother dear."

"Good-night, Muriel. Now remember."

"Good-night, Mother dear."

"Good-night, Enid. Have you been taking your senna regularly? I see you have another spot on your chin."

"Oh, but I have been taking it."

"Always remember, darling, a bad skin is most unattractive to gentlemen."

Enid flushed up to her beautiful brow. Oh, the shame of those spots, the shame and the recurring horror.

"Couldn't we perhaps ask Doctor Maxwell if he knows of a cure," she had once asked her mother; but the answer, "Some things are best left to *Nature*, Enid," had quelled the ardour of her vanity.

"Good-night, Mother."

Diana as usual trying to be a little different from the rest.

"Good-night, Diana—*dear*."

The quick brush and escape of lips. Strange the lack of confidence in that child. Sad for her. A pity.

Lady Charlotte trailed the length of her oyster satin skirt down the passage to her bedroom and with the help of her maid undressed so far as those black satin corsets which had been in her trousseau. Then she slipped the fine white flannel nightgown over her head and fumbled under it for a long time before she thrust her arms into its sleeves.

In the blue-and-white bedroom at the other end of the house the three sisters were gathered solicitously round Violet, who lay on her bed in a state of pretty severe pain. Now that the long formality of the evening was over she could collapse and leave go that curious control which worked somewhere outside herself because it must.

"Let me help you with your stays, Violet dearest. You'll feel so much more comfy in your nightie."

"Oh, don't touch me, please, I'm in such pain." Violet sank her face into her pillow for a moment, then sat up bending herself together taut and convulsive with pain. The hair on her forehead and on the back of her neck was wet and sticking to her flesh. She was entirely in pain and moaned helplessly.

"I must go and tell Mother," Enid said in a frightened voice. "She'd give you a glass of ginger."

Violet signed to her desperately not to go.

Muriel was crying quietly for she adored Violet.

Diana said, "Mother ought to see how bad Violet is."

They all felt quite desperate and quite helpless as they stood round that neat brass bedstead where Violet lay suffering so horridly at the hands of Mother Nature. Each of them knew that the glass of hot water and ginger, their mother's sovereign remedy in such times of stress, was calculated rather to make the sufferer vomit than to relieve her pain. The only thing it did relieve was their mother's sense of responsibility towards sickness. That and "Lying down," were her two invariable specifics for all female ills.

Violet whispered, "I'm so dreadfully cold."

Muriel sobbed, "I'll sleep with you darling, and warm you."

This not very hygenic plan these sisters often followed, for fires and hot-water bottles alike were considered vaguely sybaritic influences and seldom appeared in either the Blue and White or the Pink and White bedroom. To-night the Blue and White room was full of October air, cold and hollow as an October mist. Between the window panes that flattened the outside night and the white curtains it was present and in the room too, circling bluely the candle flames when the girls had put down their candlesticks here and there at unequal levels and distances. They shivered a little in their low dresses and put a flannel dressing-gown tenderly over their poor sister at whom they continued to gaze in helpless concern.

"But where has Diana gone?" Enid whispered to Muriel. "To Mother's room?"

"Or do you think the bath-room?"

"Oh, dear, shall I go and see?"

"She's not in our room and she's not in the bath-room," Enid reported, important with omen. "Can she have gone to Mother?"

"Well, if she has and Mother catches you in here there'll be trouble. Oh, dear, do go to your own room."

With many a backward and pitying look at Violet, Enid retired as advised. She had plaited her hair in its two soft dark plaits and put on her frilled nightdress and blue padded silk dressing-gown before Diana came in.

Diana was so sharp and aggressive, tearing off her dress and her corsets and doing her Swedish exercises in her black silk stockings and her chemise. Really she didn't seem to mind at all . . . Enid couldn't always look. Embarrassing.

"Where did you go to ?" she asked when she had poured her senna out of the window and hopped into bed. "Poor Violet, wasn't it awful !"

"If you did drink that senna you wouldn't have so many spots," Diana told her, still exercising with energy.

"Oh, but I simply can't. I would if I could. I hate

12

deceiving dear Mother. But where did you go, Diana?"

"I went down to the kitchen and got some hot milk and brandy for Violet and a hot-water bottle."

"Brandy—Diana, but what would Mother say? And a hot-water bottle. You'll catch it if she finds out."

"Why should she find out? Anyhow Violet's better now. The hot milk acted like a charm."

"The hot milk? Oh, Diana, I'm afraid it may have been the brandy." Enid said this sadly without a ray of amusement. She saw nothing either funny or comforting in her sister relaxing into a drunken stupor. The thought of such pain as she had seen assuaged hardly touched her.

"I don't care whether it was the milk or the brandy or the hot-water bottle, she's better now than if we'd left it all to Nature," Diana said this in the hard unappealing way which Enid admired while secretly feeling repelled by it. Now she said as she tested the almost damp cold of her bed with her thin blue feet:

"Oh, Diana, it is cold. Shall we sleep together?"

Diana said quickly, "Oh, no it's not cold enough for that." She had a dreadful inner feeling about sleeping with Enid and never did so if it could be avoided. Enid slept with her only for warmth as sheep sleep and huddle together, but Diana was afraid about this. Afraid only of herself. Obscurely obstinate in her avoidance of such contacts.

AMBROSE FRENCH-MCGRATH who had begotten these daughters was a gloomy and nervous man who infinitely preferred the company of his inferiors to that of his equals by birth and station. He had suffered all his life from being the son of a man famous for his wit and renowned for his recounting of those tales of Ireland's merry peasants which found such delighted audience in those simpler and heartier days. Always as compared to his nimble-witted parent Ambrose had been accounted but a sad dog and a dreary one, indeed had it not been for the woodcock shoot Lady Charlotte's father would hardly have looked on him as much of a pretender for the hand of his daughter. However, as things turned out he made little Char a most suitable husband. Suitable indeed he was for he interfered with his wife in nothing, her moderate good sense was far more obvious to him than her complete tyranny. She looked to the running of Garonlea with a competent discretion that every one except the land agent felt to be a miracle of insight and almost supernatural power in a woman of those days. The agent knew too well how pig-headedly obstinate and reactionary Lady Charlotte could be, how unfair in her preferments and disposals. He alone could know how many awkward situations she had created and left him to deal with, but he knew too that good agencies were not too many, and as he was an old and tactful man, he was able to contain himself and agreed heartily with all those who so often said that Lady Charlotte was truly wonderful.

Ambrose her husband really thought so. He was implicit in his loyalty and belief in her and grateful indeed for the leisure her activities left him to pursue his own mild pleasures and businesses. His pleasure in the chase had been great though mild. There was no hunting now near Garonlea. It could never have been described as an active pursuit, but his own cowardice annoyed him not at all, so fox-hunting had been really a pleasure and not a scourge in his life. He was only a moderate shot, which of course was a pity, but a fine fisherman, and had, moreover, a sound knowledge of forestry which he was able to put into practice over many acres of his estates. His green felt hats were dented always in the same precise manner and each morning he could scarcely wait for the post to come in, so anxious was he to be off to the woods to grub up elders, or to the fields to wage his lifelong war, spud against thistle. He took his duties as a magistrate and a churchman seriously and discharged them with a punctiliousness which took no account of his own personal convenience. He was tremendously and rather touchingly proud of his house and his lands, touchingly because his pride was so truly that of a tenant for life of these possessions. He had inherited from so many before his father and the place would go down, he hoped, to as many beyond his son.

There were so few things really to lay hold of about Ambrose, that dim, gloomy, kind man. The man was so hidden beyond his circumstances—"Son of old Desmond McGrath—most amusing fellow that ever was." "His wife's a wonderfully able woman." "Best cock shoot in Ireland." These were the phrases used to describe Ambrose. Nothing about him as a person. Nobody knew that he had eaten boiled eggs for breakfast until his youngest child was ten, simply

15

because he fed the salty top to each child in turn at breakfast time. Nobody knew how he hated elders and loved ash trees though so much of his life was spent in uprooting one species and planting the other. Nobody knew how pure a pleasure it had given him to see hounds hunting a fox, or his setters in the heather or a young hound slipped on a hare. He had more connection with these things than with his daughters' beauty, or with his satisfactory soldier son, or with his lands. His wife had made all these things too much her own.

Garonlea where these McGraths had lived for a long time had its share in the forming and making of that sadness in their natures which so few of the family seemed entirely to escape. Now and then the sadness would miss a beat, as in the case of old Desmond and Ambrose's own son Desmond, and there was a character ready for pleasure and unvexed in mind, but the effect such owners had upon the place itself was as uncertain and as quickly dispersed as was the stamp they left on their successors who seemed the heavier for that brief shifting of gloom.

What is there that can be told about Garonlea and the evil that can be on a place through want of happiness. Or even of a will towards happiness. Family tragedy is brief and sudden in comparison to this that lies like the breath of mould in old clothes on the people who live in such a place. It seems as though nothing could ever dissolve such mists and ill vapours, or only for such a little while. So inexact, so dim is such a gloom, it is hard to say whether it is the effect of place on character or character on place. Thus was Garonlea affected beyond its native melancholy by these gloomy McGraths who had lived there such a dreadfully long time.

16

One side of a deep valley lay Garonlea with its rich lands and its woods of Craiga and Laphonka and Gibbets Grove, and its village of Garonlea. And on the farther side of the river that ran down the valley length, on this opposite side of the valley there is a gayer, lighter air to breathe. It is really and tangibly a better place altogether.

The house of Garonlea is built so near the river that the terraces of its garden drop down from the windows to the lush banks of the river. And above the house the comely embowering woods fulfil the act of clothing the mild valley in a manner more suited to an English manor than to a sad Irish place. The House was best described by some lady whose gifted pen wrote of it in a weekly paper as "an elegant castellated mansion." Such a house indeed was built by the McGrath who pulled down a Georgian dwelling house and built himself this habitation of glorious Gothic, out of a rich wife's money. Well, they had let themselves go and probably enjoyed the result.

But with all its vulgarity of architecture some curious eternal line of beauty remains to the house. Perhaps because its awful Gothic mass is built of the local gold-grey stone. But colour is not line. Vaguely the house is the right height for its width and has been built in the right attitude exactly for the shallow wooded height of the valley behind it. Below, the terraces dropping flight by flight, lion-guarded at each flight, towards the river, were well and elaborately cultivated. Crimson ramblers threading an intricate pergola, much bedding out of lobelia and geraniums in their proper season.

Here the girls would walk and sit and sew together and read the novels of Sir Walter Scott aloud to their mother. And if their brother Desmond was at home he would come out and have a joke with them now and then. They thought Desmond was marvellous. So daringly frivolous. So of a farther world. So brave, so sweetly scented when he kissed them in the morning. So full of jokes he might not say to them. So languorous and dreamy on some summer night when the valley was heavy in mist and white jasmine scented the air, its sprays leaden white near a dark window.

III

Iᴛ was in November that Desmond McGrath brought to Garonlea his love and Bride-to-be. It was her first visit to his home. Nor had she met Lady Charlotte before nor Desmond's father nor one of the four sisters.

It had been one of the quick things in her life, this engagement to Desmond, and her father saying, "Well, you have only known him six weeks—— But the Garonlea shoot is famous, really famous. Besides there is something I like about the boy. You must please yourself, my pet, and remember I hate to lose my little girl."

"You mean you hate to lose your little housekeeper," Cynthia had said with a practicality unusual at her date. "You'll marry again, Daddy. You're a very attractive man. I've had a hard time keeping off prospective stepmothers. Much worse than you've had with my suitors." It was the wit of the day, and ran in rather flowing periods. Nobody resented this.

"A naughty little girl, I regret to say," he answered.

One of those admirable filial while maternal kisses and she had gone.

At Garonlea Lady Charlotte was saying to her husband, "Her family is all one could wish. But we have never had anything to do with Racing. I could have hoped that was different."

Ambrose answered, "Well, she has seen her father lose a lot of money—what about the Cambridgeshire when his horse started a red-hot favourite and finished down the course."

"Let us hope the child knew nothing about the matter. Anyhow it's good blood, remember, and I was presented the same season as her mother. So really one can't help feeling dear Desmond has made a good choice."

The girls were told:

"Desmond has written to his father and to me saying he wishes to be engaged. He is coming over for the first woodcock shoot and hopes to bring his—his bride-to-be with him so that we may all get to know her and to love her."

"Mother! Do we know her?"

"Next week?"

"Oh how exciting!"

"Who is she?"

"Well, Diana, if you will give me your attention for a moment and try not to interrupt I will tell you. She is a daughter of Colonel George Holland-Mull and her mother, who was presented the same year that I was, was a Hamish, a daughter of the last Coolcullen."

"But they're all as mad as hatters."

"Di—ana?"

"Oh, but it's true. Half the Hamishes are locked up."

"My child, to repeat such foolish gossip is both vulgar and dangerous. Please never mention the subject again to any one."

"All right. But it is true."

"Diana, I don't wish to hear another word from you. The idea. Are you aware that you are being extremely impertinent to Mother?"

So Cynthia Holland-Mull (such good blood, remember) came to Garonlea for the first time one November evening when there was a high star in the sky and a light frost and the moon as clear as glass above the

20

high hedges and faintly patterned fields and the thin distant line of mountains. She and her lover bowled along in a dashing high-wheeled dogcart behind a brisk chestnut horse. Desmond, though his unqualified happiness was almost unendurable, did not touch her hand with his once, not her gloved hand beneath that fur rug as he drove the seven miles from the station to Garonlea. After all there was a stable boy sitting with his back to them in the rear seat of the dog-cart.

" You're certain you're not cold?" he asked her every mile or so and she smiled at him through her veil, for how could she be cold in her dark green driving coat lined with fur and its three-tiered cape like a highwayman's, and her feet in their pointed glacé kid shoes were warm as toasts in a blue fur-lined foot muff. And though she would scarcely have said so, how could she feel cold in the glow and the thrill of Desmond beside her.

"Oh, no, I am deliciously warm," she said.

"We should have a good day to-morrow, they've had nearly a week's frost, Dempsey told me——" Desmond could hardly have felt happier. His Love and his Bride by his side and the prospect of two excellent days' shooting in front of him.

"Are we near Garonlea yet?" Cynthia asked.

Desmond said, "Yes, this is Garonlea now."

But they drove on for miles before they came to an avenue gate, and always down hill. Down hill with woods sloping down to the road on their left hand and dropping thickly below them down the side of the valley.

"What a long way down hill," Cynthia said brightly. Through all her insensitive happiness a faint feeling of hostility and coldness was at her heart which she could not explain. Perhaps she was a little chilly after all. Perhaps it was nerves at the thought of meeting Desmond's relations. Whatever it was it did not leave her when the dog-cart turned in at a vast and over-powering gateway, an impressive cross between the entrance to a mosque and a street lavatory, and bowled (that was what the best dog-carts did) smoothly down the smooth falling slope of the avenue.

Cynthia was not shy when she first met people. She was too certain of herself. Quite sure of her success. So that this first encounter with Desmond's relations was hardly the reason for the constricted chill that was causing her to feel so unreal to herself; on the defensive and alone when they stopped before the great wan house sprawled at the bottom of the valley with the mist rising round it off the river below. Soon, in a moment almost, the door was opened and Desmond had lifted her down and she was real and breathing easily again, feeling as well and free as she only did near him. Her head swam a little as they went together through the strange light halls and into the library where Desmond's mother and all his pretty sisters kissed her with eager embarrassed welcome. All but one kissed her so, the little dark one with a stiff white collar up to her young soft chin, she shook hands nervously and with a blushing determination that evaded graceful readiness for a sisterly salute.

"And how cold you must be!"

"Such a journey!"

"Did you have a long wait at the junction?"

"Muriel, my dear, you may take Cynthia up to her room."

"Yes, I'm rather a dirty girl, I think," said Cynthia, blinking like a cat, a gold cat in the warm light room where white chrysanthemums smelt antiseptically and a majestic silver tea service glittered on an elaborately clothed table.

"There will be some fresh tea for you and Desmond by then," Lady Charlotte passed over Cynthia's little joke as perhaps a pity. "We're all rather smutty after that journey," she added.

Cynthia, knowing she looked superb in her new green facecloth coat and skirt with a dark flat velvet cap skewered through the pale puffed wings of her hair, agreed and followed the shy eager Muriel upstairs.

She had been put into the most important bedroom —very high and papered in white and silvery white stripes with a dado of brisk pink roses. Rosebuds on the chintzes too and fresh frilled covers on all the cushions. Two brass bedsteads shrouded in starched elaborate white. A red carpet on the floor and a greataunt's faded water colours on the walls. Malmaison carnations in a silver vase on the writing table and a small coal fire in the grate. The rest of the furnishings by Mr. Maple.

Cynthia changed with speed and total disregard of cold into a mauve chiffon blouse, dislocating herself cleverly to fasten its back buttons and the nine minute hooks that fastened the wire-supported net collar, which finished in a wee Toby frill under her chin. An embroidered yoke and deep soft folds of tulle, long big sleeves gathered in ten different directions and very tight on the wrist—that was how this romantic blouse was made, and worn with a dark purple skirt and bronze kid slippers with bronze bead buckles. Cynthia tucked a bunch of silk violets into her waistband and was ready when Muriel knocked at her door again.

That was the start of the shooting party at Garonlea

23

in 1900; the party that was to make known to the county the next Mrs. French-McGrath.

Cynthia enjoyed it a great deal. She understood the organised formal entertainment perfectly and played her part as star guest with tact and an exquisitely right sense of drama.

Only Desmond knew something of her under that perfectly constructed facade towards which his mother could not deny her approval and with which his father and his sisters were soon half in love. She excited them all. He saw that and it pleased if it disturbed him. He knew she was a powerful and secret person and it delighted him profoundly that this creature should with him be moved by so strong a current that she was as helpless as a body in a smooth flood. He liked her efforts at restraint or dignity and would uphold her in them with a certain subtlety until she abandoned them because she was entirely and urgently in want of love. He was not grand about this to himself. It did not make him feel either as masterful or protective or indifferent as it might have done. As it certainly should have done according to the "never give all" theories of the date. Not that it was possible, or even thinkable, to give or take what they called "all" at such a party, but moments of practically unendurable ecstasy and severe strain were permissible and considered both romantical and seemly under the auspices of that magic word—Engaged. To such moments Cynthia brought a wild whole life and an entire lack of austerity, very touching if it had not been for the complete freedom with which she regained her balance when they were over.

A DAY and a night from such a party, what were they
really like? Not like the jolly reminiscencing of
some countess who had tremendous success with
King Edward—— "We joined the ' guns ' for luncheon
in a keeper's cottage, all very jolly and informal.
I remember old Lord X having five helpings of Irish
stew." . . . No. Probably depressingly the same as
the same shoot to-day—except that the women didn't
drink so much or talk dirt. Anyhow there is no truth
to be got from the queens of these past festivities. How
vainly we seek it. Why can't we get hold of any truth
about them ? They are defensive now—the Enids and
Muriels and Cynthias. One asks them some absurd
question—clothes, for instance!

Instead of saying, "My dear, you should have seen
our buttoned boots, they were too nice. And you
really were a dashing girl if you put black ribbon in
your camisoles." Or, "How well I remember my first
ball dress—White. You practically had to wear White.
White satin, the bodice cut down to the bust line
and my bosoms propped and supported, all most
pneumatic and attractive to the gentlemen. Puff
sleeves and white kid gloves turning the elbow. And
such a waist, darling—twenty-three inches without
going black in the face. Then a full flowing skirt
(bodices and skirt were not attached, you wore a folded
silk belt pointed back and front). But petticoats—there
was drama for you!—Flounces and lace and insertion
and ribbon bows among the lace. It all gave one such

a sensation of one's own glamour. Glamour I think was what we had. Glamour was the thing——"

No. They say instead, "Well, I think our clothes weren't so very unlike what you wear now. Long skirts in the evening, you know. Really I think we were very much the same—not so very different. We used to have tremendous fun. Perhaps we wore more elaborate clothes in the afternoon."

And they were so different. Really they had everything we haven't got. Why don't they boast about it more? Why aren't they prouder of their glamour and their chastity and their dainty boots and their rich leisured lives. Why don't they see that they lived in a definite Period? Not in a moment of transition between the Victorian and the present age. Why don't they face the Edwardian Practical Jokes and insist that many of them were fine jokes, much funnier than simply breaking things?

Perhaps it is because they are too Spartan ever to face the truth. One only gets at the truth by admitting a lot of pain. By not minding being a fool for a moment. By forgetting one's natural tremendous vanity. By coming off one's poses. But they cannot speak a lessening thing of themselves, a thing that leaves them open to wounding, anything near the truth. Still we must allow them their glamour.

Breakfast time at Garonlea. The ladies were all down to breakfast—very neat. Shirts and fringes and little watches ticking on their left bosoms, pinned there by a silver, a gold or a jewelled bow. Sunlight poured in through the long windows and an enormous breakfast was on the sideboard. Undefeated by the hour, the wit of the party, Colonel Stagg (who had not missed a Garonlea shoot for ten years) made his

26

jokes and the ladies all laughed merrily. He told Lady
Charlotte what an Irishman called Pat had said, which
put the table in a roar. After this he ate a dish of
kedgeree, a snipe and some cold ham, and got to the
lavatory first. He was a very civilised man and knew
the technique of these parties well.

Lady Charlotte was having a talk about St. Bridget
anemones and the new rhododendron, Pink Pearl,
with his wife on her right. This was her mate for the
party. Enid was sitting silently beside a young man
called Poor Arthur, with whom she had surprised herself
by her success the night before. "He has far more in
him than I ever thought," she told Diana when they
went to bed. Violet and Muriel maintained strained
but bright conversation, one with her Uncle George
who told her in sentences not longer than nine words
how well his daughter Phyllis was getting on, and the
other with Lord Jason Helvick, who only cared about
Bird Life and was destined to disappoint Lady Charlotte's
hopes that he might care about Muriel too. After all
he was forty-eight and it was time he settled down.

Diana sat between her father and Muriel so that
she need not talk. Cynthia on Ambrose's other side
delighted him with details of moors he had shot over
in Scotland and men he knew, but had been too shy
to make friends with. She had an entirely accurate
and reliable memory. Desmond sat on her other
side in a live union of silence, delighted because his
father was so impressed by his able young Love.
Across the table Diana with frequent shyness raised
her eyes to Cynthia's gold head and dropped them
quickly on her plate. Puffed and back combed, braided
and whorled, Cynthia's hair could not quite lose the
yellow extreme softness of willow flowers—the yellow
that is almost silver.

27

After breakfast the men disappeared for the morning and the ladies sat about in the drawing-room chattering amiably. They looked out through the window down the drop of the terraces and across the sleekly flooded river to the high bright rise of woods on the opposite side of the valley. The morning was damp and full of sunlight, still sweet air as soft as a cat's fur.

But inside the drawing-room the day seemed far off and hostile to the creatures and furniture and ornaments enclosed. There was a good deal of furniture in the drawing-room and the ladies, poised about on gilded chairs, their heads bent over pieces of needlework, looked like faintly dusty stuffed birds—the sun in their fuzzed hair made a dusty effect and their bosoms and padded behinds were birdlike. In the crowded groups of Dresden china the sun picked out dust in the raised wreaths of flowers and struck high lights off the smooth bosoms of those other china ladies and fat behinds of china Cupids. The lovely chandelier looked dark and sinister as dark glass can look, the chairs were slippery with sun, and the fire strove against the light till Muriel, at her mother's nod, drew down a blind.

As though she had been waiting for this, Diana jumped up and said to Cynthia:

"Wouldn't you like to come out? It's such a lovely morning."

Her words sounded distressingly prearranged, she had been rehearsing them for nearly ten minutes, and brought them out now with unnatural speed and boldness.

Lady Charlotte looked up from the note she was writing. She paused a second, raised pen in hand, and in that second in that room streaked by long moted beams of sun, she seemed to swell with strange arrogance among those swords of sunlight before she said in her voice of commanding benignancy:

"My dear, I think you and Diana will be quite tired enough by to-night without going for one of her long rambles now. Besides which, Diana"—the voice lost some of that benignancy—"you must remember Desmond might like to take Cynthia round himself."

Cynthia smiled faintly in good-humoured acquiescence. It was not till after Diana's blush of fury had died its last beneath her high collar, and the ripple of embarrassment had widened to oblivion through the room, and Lady Charlotte's pen pursued once more its dignified course, that she lifted her eyes to send Diana a look of such complete understanding that Diana's heart was stilled for a suffocating moment before it went galloping off on its pledged way. She was exquisitely, powerlessly committed to Cynthia. It was one of the most important moments in her life.

To Cynthia the incident was chiefly important as a mark for her when the faint hostility she had felt last night towards Garonlea became through that enclosing room, that quiet arrogant voice, that helpless blush of rage, a real thing which she must hide from every one except the little Diana. But most especially from Desmond.

V

THE ladies and the luncheon set off at 12.15, to meet the guns. A phaeton with curling springs and blue cushions carried most of the party. A side-car with a red crest on its back and food in its well the remainder. Gloved in tight dog-skin and neatly veiled the ladies drove along the little roads that climbed and twisted and shelved their way about the valley side of Garonlea.

Then greetings at a keeper's cottage high above a wood of hazel and rhododendron. Chaff and laughter, always bursts and ready tinkles of laughter, and an enormous quantity of rich hot food served in the dark parlour of the cottage where pictures of Popes and stags and Jesus (with little faded crosses from the last Palm Sunday stuck in their frames) looked on at the gay and social scene.

There were more men now to pay compliments if not court the girls. Violet's smooth blush rose and fell at things that Colonels said. And Muriel's sweet rather deprecating giggle was drowned in the hearty male voices of men who were men.

Poor Arthur had been shooting well (for him) and sought Enid's company in a gay and confident manner that did not escape Lady Charlotte's notice. A pity he was staying in the house. She would just drop Enid a little hint before dinner. The agent's son. No money and socially not quite—quite, he had been asked to fill a gap at the last minute. The fact that he was rather good-looking in a pale languid way struck Lady Charlotte for the first time. Fairly presentable considering his grandfather was a Solicitor, was all she had thought before.

Huge savoury stews, game pie, and brawn and ham and mince pies and cheese; whisky and white wine and port (and water for the young ladies, but Cynthia drank half of Desmond's whisky and a glass of port, looking at him in a quick secret way and saying, "I think I'm cold"; even Lady Charlotte missed this, it was so gently and so speedily done.

Later she stood with him in a wet quiet ride where little white flies and midges were like steam and the late Autumn felt as dark with life as Spring, as they waited there for first sound of the beaters. Desmond was an awfully keen shooter so he only kissed her once a little abstractedly, for it was unthinkable that even one cock should swing across the open ride and tilt and dodge away among the trees unslain by him.

And Cynthia who was always keen for love but particularly after she had had a drink or two, restrained herself a little sullenly, but forgot about some of it soon for she was interested in shooting and in any case would have attended Desmond with fervour if his sport had been fishing for eels or shooting pike in a weir.

Soon there was tremendous popping and banging through the wood. And everywhere some lady smiled and nodded commiseration or congratulation, relieved from making silly speeches by the merciful imposition of silence that attends these woodcock *battues*. At other shoots they had to rack their brains for something to say to their " guns." When they said the wrong thing it was taken for granted. The only things that were not taken for granted were silence, or staying at home or wanting to shoot themselves.

The end of the day came at last. Lady Charlotte

had driven home long ago and the girls were strangely gay and comfortable without her. The last drive was over and the early dusk was running softly through the valley. The smooth river as though its course was on a map curled along below them. The whole evening felt sleek and finished. Good and indifferent dogs were highly commended. The party walked down the soft rides through the hazel groves and glossy cold leaved thickets of rhododendron, the girls gay and chattering.

"This is the wishing well."

"A real wishing well?"

"Yes, The wishing well in Craiga Wood."

"Oh, I must have a wish!"

"And so must I——"

They stooped laughing to the low icy water in the roofed well. They stood in the soft ground where springs broke up and stooped like spotted antelopes among the shining dark leaves. They drank out of their cold pink palms and wiped their wet chins with handkerchiefs. Their eyes were dark—they would never tell their wishes for love and marriage, but lied and said they wished for a new hats or diamond watches.

"A *rich* husband!" Enid said to poor Arthur with daring sauciness. A good thing Lady Charlotte did not hear her, or see her changed eyes when they met his unsmiling answering look, or know the deep-sweet change of his voice in her ears.

Back to the house in the valley then and the ladies scurried upstairs to change into smart clothes for tea. After tea a group played cards, a group leaned round the piano causing havoc among the photographs, singing softly.

Two were missing, Cynthia and Diana. They sat

by the fire in Cynthia's room, Diana talking—thirsty for this new sympathy, a little drunk with the excitement of it, the thrilling discovery of Cynthia. Their moment together was stolen, for they had no business according to the rules of such a party to absent themselves like this. One should not relax in effort; one should be beautifully, endlessly hearty and sparkle meritoriously at all hours. Yet here they sat upon the floor by Cynthia's fire, spread skirts, heads bent talking as the moments raced by. Cynthia saying yes and yes and yes. Her weight leaning on one arm, her strong hand with fingers spread propped on the floor behind her. Diana sat up straight, her voice going on and on, as emotional as wings in the dark beating on, to what end they do not know.

"But, Cynthia, I can't go on living here. I must do something. I must escape. . . . Muriel's all right. She adores Violet, and besides she never thinks anyhow. And Violet, she's sure to marry."

"Yes, and Enid?"

"Oh, I have no patience with Enid. Always some silly, rather disgusting flirtation and then a row with Mother, and tears, and everything forgotten."

"Yes, and you?"

"Oh, I hate young men nearly as much as they hate me. I only want to get away from Garonlea. It's so awful, I know, I've never confessed it to any one before, but I loathe and loathe this place. I never feel well for a moment and there's that awful depression pulling one down all the time like lead."

"Yes."

"You feel it too, Cynthia?"

"Yes, it's overpowering."

"Did you know we aren't supposed to talk about it?"

"How do you mean?"

"The oppression—it belongs to the house. It's a thing we aren't allowed to admit."

"This melancholy?"

"Do you know, it got so bad once, Father got a man here who was supposed to clear up horrid things in houses. But he didn't do much good. We were all sent away."

"How eerie. But the feeling I have is that no one has ever enjoyed themselves here; and if any one ever did anything wrong it's been a sad wrong, not a bit enjoyable."

"Perhaps you and Desmond will change the house. He is so much gayer than the rest of us."

"Oh, Diana, but Desmond and I will be old before we come here to live."

"It is so wicked of me to say but the thought of living here with Father and Mother for the next twenty years seems absolutely unbearable."

"You must live with us a great deal."

"That will be lovely." Her acceptance was casual because this telling was so much more important, all the luxury of first spoken grief was in it. "But Cynthia, if you only knew what it's like, always being watched and ordered about. Everything *known* about one. Even if one writes a letter to another girl—one has nothing of one's own at all. Not even a dog. Father won't let me have a dog of my own."

"My poor darling, I do so understand."

"I can't believe that two days ago I didn't know you."

"Or I you. . . . Should we go down? I'm afraid we should."

Violet was captured too with talk of clothes and bridesmaids' dresses all planned for Violet. She was easily capturable and except for her beauty unimportant. But still Cynthia meant to have her.

"This design won't suit my other big bridesmaid so well—poor Mabel, she'll need a lot of padding—but you will look so lovely in it, I think it's the only one to have. What do you think about the bouquets? . . . Yes, I think you're right there. . . . Yes, I agree about those shoes."

Then with Enid: "What a charmer that dark young man is. And he never takes his eyes off you. Are you interested in him or just amusing yourself?"

How grand it made Enid feel, how worldly and successful, as though she had dozens of beaux.

"Well, I am rather attracted. But I have to be dreadfully careful on account of Mother."

"Yes, I see. But he seems so charming. Why does she disapprove? No money, I suppose. The nice ones never have, do they? Tell me more about him."

It was sympathy that would listen even to Enid's shy rhapsodies about poor Arthur. And unscrupulousness that could sit with Lady Charlotte for an hour and agree and agree:

"I do see how difficult it is. There are very few men here, aren't there? But naturally one must put one's foot down about people of that sort. I don't want to betray a confidence but from something she said to me I think perhaps she is—just a little—— I do hope you will let them come and stay with Desmond and me. There is a charming Major Blake in the regiment. I thought of him the moment I saw Muriel. Yes, a little money and some helpful connections. One isn't a snob but you know I *do* think Family matters a lot—don't you——?

"I've no sisters so it's wonderful to find a ready-made family of them. . . . I expect you remember my mother better than I do. I was only six . . . so it's wonderful to find you. I know you won't mind giving me help and advice and I shall want such a lot, shan't I?"

Diana was discussed too. "I think you're very wise with her. Obviously she's rather difficult. But it's only a stage she's going through, I expect. One does, doesn't one? I was dreadfully tiresome at eighteen but my father was so sweet and wonderful to me."

With Ambrose too she was tremendous. She walked out with him, visiting his young plantations, and paused comfortably at every alder tree they met in the woods. This was not the thistling season or she would have had to pause at every thistle too. She asked him all the questions he could answer about the fishing and the shooting and whether the land would be suitable for bloodstock. For she had a very good plan about getting Desmond out of his dreary regiment and—as she thought it necessary for him to have an occupation which would interest her too—this would be excellent. A stud farm on the other side of the valley. Some really good mares. What a help Father would be about getting hold of them. She saw them standing with their foals in the deep shades of chestnut trees in July. Next year she saw those beautifully bred and well fed and insolent yearlings galloping and wheeling and stopping in the strong railed fields. She saw the prices they made at Doncaster and every one congratulating Desmond on a wonderful sale.

She had seen too a big bright house on the other side of the river. A charming house with every possibility except that it had been inhabited for the last fifteen years by two old aunts of Desmond's, but no doubt a more suitable little house could be found for them when the time came.

Cynthia and Desmond had gone to tea at this house one day. They had eggs for tea in filagree silver egg-cups, and buttered toast and medlar jelly and Sally Lunn teacake, while the Aunts bustled about scolding each other and lavishing flattery and adoration on Cynthia and Desmond.

Aunt Milly had been a beauty and married an exotic Swedish Count who spent her money and shot himself. She was a little mad and enjoyed behaving as though the parlourmaid was three flunkeys in scarlet and gold braid, and generally induced a sensation of red carpets and foreign courts. Her sister Mousie who was little and brisk and loving was the real ruler of the house-hold and queen of the garden wherein was walled their pleasure and chief sport in life, their expectation, adventure and triumph.

Although the month was November, Cynthia was shown the garden. Out through the drawing-room, where old Pieces, small and perfect, were crowded and obscured by bamboo and plush furnishings, and photographs of the late Count Standoof jostled Chelsea figures and deep Worcester saucers patterned in scaled blue and flowers, as rich and thick as sweet-williams.

They trooped out through a long window and across a small sunk lawn bounded by a little nut walk, the Aunts pausing at every step.

"And in the Spring you can't see the grass for the crocus—that bank, my dear, is as blue—as blue now as china with the scyllas."

They talked like so many Irish ladies of their gener-ation with rather plummy brogues which however detracted nothing from the brisk and distinguished pronunciation of each word.

Beyond the lawn and the dark little nut walk was a

37

walled garden entered through a small green wooden
door, and in here they moved from one point of
excitement to another.

"That's my robin."

"No, that's *my* robin, Mousie."

"As if I didn't know my own robin."

"You must see the Christmas roses."

"Well, you can't exactly see them, dear, as I gathered
them for the tea-table—seven lovely blooms, Fancy!
But here's where they grow by the door of the potting
shed."

"And Iris Stiloza under the wall here. The dear
creature, she loves us here and Mrs. Barclay who lives
a mile away can't get her to flower at all. In spite of
her five men in the garden."

"And here's where our prize Auriculas will be a
lovely show in the Spring. Look at all I have now,
and I started with three little plants your father
brought me from a big flower show. Look, Desmond—
You must tell him when you go home."

And so round the empty, tidy garden which to them
was alive with a reality beyond any dim sleeping
promise, and back to the drawing-room where candles
and a lamp that gave as much light as a goldfish in a
bowl were lighting. The Aunts showed Cynthia
their china, about which they were as well informed
as about their flowers, even if they did obscure it with
photographs. And when it was time to say good-bye
they gave her a ravishing Chelsea group. A naughty
musician in a braided coat bending towards a sleeping
lady, thrillingly and for ever suspended in a moment
before a kiss.

They would hear of no refusal which Desmond was
inclined to urge, knowing their china was as much to
them as their garden, and that they would mourn the

piece sadly when the excitements of giving and of Cynthia were gone.

"Ah no, Desmond, you mustn't forbid us such a pleasure."

"Our new niece and all."

"But you'll be careful of it, Child," a sudden anxiety overtook Aunt Mousie, "see, there's not a chip anywhere. And just count her petticoats—she has seven, I declare, and a different pattern on each, stripes, and flowers, and her farthingale spotted with little strawberries."

"I will, indeed I will, and thank you a thousand times."

Cynthia kissed them, making it the most important matter in the world, and set off for her walk home with Desmond, carrying the china figures preciously within her muff.

From half a mile down the road she looked back but she could no longer see the house, only a slight pale tongue of trees advancing from the solid wood at its back, coldly delicately stepping up towards a pale winter field and a pale sky.

She put her hand into Desmond's, her other hand holding the china figures inside her muff, and walked on down towards the river, exquisitely conscious of herself and of him and of a happiness so whole that it seemed impossible it should last. And if it did not, there was no further happiness to look for.

They crossed the river by a twisting cramped stone bridge, its walls niched for the safety of foot people in the coaching days, and in the middle niche Cynthia stood looking down at the river and knowing her reluctance to go back to Garonlea. A quarter of a mile distant the house rose romantically towards its woods, a pale milky mass of stone with the river

39

dark as a moat below it. The strangling sharp river breath of evening caught her and the cold remote river sounds filled her with a curious despair.

"Desmond, Desmond," she said and clung to his hand as if in a dream she saw him lost to her. She set her fur muff with its china burden down on the stone, a last formal little act, absurd beside her need for Desmond and those kisses in the chill and quiet darkness of the evening.

But when he had kissed her and after she knew with that fleeting realisation and sense of power that makes women gay that his love was even beyond her own, there came to her again as they walked back along the smooth avenue towards Garonlea, that sense of oppression that was like dying or weeping in its strength. She pressed nearer to Desmond under the dark tree, but his nearness only made her feel more sad.

VI

THE warm, full light of an August day poured liquidly into the schoolroom filling it with round generosity. At the piano with her back turned to the window, Enid sat playing; turning her music, sometimes she smiled secretly and touched the bosom of her blue dress from which came the faint but thrilling crackle of a letter from Arthur—no longer to her Poor Arthur, but the first and most beloved of men. How, she wondered, had she ever thought differently about him. How had she lived so long only half living, unloved by him. She went on playing her Chopin in the most unconsciously uxorious way, half delighted, half dreaming in the warm room.

Lady Charlotte walking briskly down the flagged path outside the window pursed her lips disagreeably. One knew without looking in that it was Enid being rather embarrassing at the piano. A pity girls didn't understand themselves more clearly. A pity one could not explain anything to them without destroying their innocence. One could only indicate and forbid. And this reminded Lady Charlotte that she must have a serious talk with her Enid. There had been yet another letter to-day from that most undesirable young man. And he was to be home on leave almost immediately. Enid had suggested with nervous indifference that he should be asked to Thursday's tennis and croquet party. To-day was Tuesday, and there she was playing the piano "like that."

"Enid! Enid!—A pity, dear. Too sentimental, don't

you think? I was looking for you to come for a little walk."

Enid blushed cruelly and sprang up from the piano with a violence that set all the photographs rattling. Desmond's wedding group in a silver frame and all the Presentation pictures were still vibrating when with sullen indignation she followed her mother towards the distant kitchen gardens. It was not entirely an auspicious opening for the little homily which Lady Charlotte was about to deliver.

Within the high brick walls all was hot and still and orderly. It was as orderly as a prison yard and Lady Charlotte had as much authority over her children as any prison Governor. She was a dispenser and an arbitrator and behind her was unquestioned power. A habit of obedience overlaid the tumultuous desires and suppressions of her young daughters. There was nothing good about that habit, and had they not been rather stupid girls they would have evaded it with subtler lies and schemings than they were able to concoct.

"——So you see, my dear," said Lady Charlotte after five minutes of uninterrupted speech—"I trust my little Enid and I appeal to her own good sense. Is it quite wise or quite kind to encourage a young man whom neither Mother or Father could ever consider seriously possible? Just write him a perfectly kind, straightforward letter by this afternoon's post saying, let me see, saying, well, perhaps you can think of what you will say. But in any case give him clearly to understand that all this silly letter-writing must cease."

Enid said in a strained, small voice, "But, Oh, Mother, you don't understand——"

"I think I do, dear," Lady Charlotte's interruption came bright and remorselessly. "Mother understands

42

a great deal, you know, about what her little girl is feeling, but she is old enough and wise enough to know what is best."

"But, Oh, Mother, you must have forgotten—don't you remember how you felt about Father?" Enid's voice faltered over this rather bold and embarrassing question.

"My dear child, that has nothing whatever to do with it." And here Lady Charlotte spoke more truly than she knew, for that seemly, unloving and fruitful union had been at no moment familiar with the heat of love. "Nor do I consider that this is a particularly pleasant or suitable matter for you to argue about like this."

Enid's face was crimson. Her breath came interruptedly so that she could not speak clearly. She was helpless to speak in this passion of rage against her mother and the quiet, controlled force of her mother's power. Tears swelled in her throat. Her fingers were stiff as little darts as she fought for the control which she had not. The most wretched and passionate young creature imaginable, she stood in the sunshine between the hot savoury-smelling hedges of lavender, with the high, warm prison walls about—walls against which delicate fruit ripened decorously in the faint shadow of its proper leaves.

"Of course, my dear, if you would rather that I wrote——" Lady Charlotte dropped the suggestion almost casually. It came as the hot garden world was spinning in a bright globe of tears before Enid's eyes and it shocked her into a realisation of more than her own present emotion. Such a letter would wound—shock her Arthur, her love, in a way she could not endure. She gathered her voice and spoke high and uncertainly, almost neighing with nervousness.

43

"No, I'll write," she said. "I see, I see what you mean." There was a moment within her mind of almost maniacal tension when she thought of all she might say. I must marry him. I adore him. You don't know how much I need him. Why shouldn't I marry him ? You could give us enough money, and with his pay we could manage. What right have you—Oh, you have no right—to do this to me!

But to hint of money or to speak even in the shyest, most romantical terms of "sex" (that then unknown and unspoken word) was beyond Enid even in the most desperate circumstances. Her heart might break as she stood there but she could not do it. She heard her mother's voice commending her good sense and obedience and then hoped to escape. But no. She was bidden to hold a basket and snip the rotten paper heads of dead sweet pea. It was not till Lady Charlotte thought those foolish tears and sniffings might attract the notice of her head gardener who came lumbering slumberously down a path towards them that she dismissed her daughter in waiting, saying:

"Go to your room and splash your eyes with cold water, dear. Then lie down for half an hour, you'll soon feel better. Yes, Williams——?"

Although too encompassed and overcome by habit and emotion to tell of her love, Enid possessed too much obstinacy, passion, caution and cowardice to let her love go. It was unthinkable that she should forgo those thrilling embracings, and lonely hours of delighted imaginings, all dependent on Arthur and her relations with him. So, although for a fortnight no more letters came in Arthur's writing, letters still came, with writing curiously disguised, and there were fevered short meetings in woods and fields and other outdoor places when the strength of their emotion

44

delighted and terrified them both, and the mental and physical strain became quite absurdly unfair.

Not unnaturally, Lady Charlotte's suspicions soon fell on these peculiar letters. While her imagination could hardly compass the possibility of Enid's disobedience and deceit in such a matter, at the same time she could not feel quite easy in her mind. Then one unlucky day Poor Arthur, emboldened by the safe arrival of so many ardent letters, was taken with the fancy to write to his love not once but twice. Two letters. It was too much for Lady Charlotte's trust. These two peculiar letters arriving together even lowered some of the surprising conceit which forbade her mind to accept suspicion of disobedience in one of her children. Her pride at once transmuted this lessening of her vanity into a sadness. Lies and deceit besides being wicked were sad, but especially sad and wicked—useless as well—when employed by a more than foolish child against Mother or God. The tremendous urgency of a matter that could have made an Enid rebel in such a way did not occur in any just proportion to Lady Charlotte. A court was set in the library, Ambrose in so serious a matter was an unwilling lay figure of extra authority, and a message was sent for Miss Enid to attend.

Miss Enid, who for the last hour had been in that exquisite state of hot and cold impatience which each day preceded post time, arrived in the library with every nerve twittering. She felt like a tree full of starlings.

"Yes, Mother?"

Lady Charlotte came directly to the point. She handed Enid the two letters, saying, "Your father and I would like to know who your correspondent is."

Ambrose cleared his throat as if about to speak and then said nothing.

45

The summer air in the library was astonishingly still. In all the small, scattered vases of flowers not as much as a feather of asparagus fern stirred. From the walls a few pompous ancestors seemed to puff their dark cheeks, strain their bellies against their waistcoats and lean a little closer to this trouble and tension. The photograph of Violet in her court dress smiled divinely on. The stillness could not last. It was swelled to its extremity.

Enid spoke.

"I shan't tell you," she said.

The room seemed to stagger back from her words and before its force returned or Lady Charlotte could gather herself for a deadly rejoinder, Enid, terrified by her own daring and its possible consequences, had turned and run out through the door and, unseen by sisters or servants, out of the house.

Lady Charlotte's first and truest reaction was shown in her first words:

"What a pity I gave her those letters. It was my duty to read them."

Ambrose, though he suffered slightly from shock, was pleased that the matter of the letters had been taken so briskly out of the hands of proper authority. Proper Authority seemed to him at times so cruel if so necessary a weapon in dealing with the young. He thought for a second with a faint quickening of life of Enid fled from the house sitting now in some dark strip of wood, where lately he had cut the elders, where sorrel flowers were blown on the faintest air, reading her letters, her rescued letters, preposterously thrilled by the written word. . . . A pity that it was all so unsuitable. However, Charlotte must deal with it as she thought best. Charlotte was a wonderful woman.

He said, "I think, my dear, I will attack the thistles in the Long Acre this morning——" He hesitated. "If that is all?"

"Poor child," Lady Charlotte was recovering her poise. "It is most distressing. It seems to me most sad, a sad—sad pity that she should have acted like this. Yes, Ambrose, I will deal with her after luncheon."

But the luncheon gong rang in vain as far as Enid was concerned. The cold lamb and salad and raspberry tart and cheese were eaten to the accompaniment of rather strained and spasmodic conversation and the three girls stole questioning looks at each other and could scarcely wait for luncheon to be over so that they could discuss the matter in the schoolroom. But this design was astutely opposed by Lady Charlotte who took the lot with her to pay a round of calls, sending them up to their rooms separately to put on their smart hats.

The due reward of maternal tact, diplomacy and firmness awaited Lady Charlotte when, more unquiet in her mind than she would admit, she went up to her room on her return from that weary round of calls in the company of three sulking daughters. The reward was almost too wonderfully apt to be true, and through it Lady Charlotte experienced a sensation of placidity and pride at this fresh proof of the eternal rightness of her judgment.

The reward was Enid, showing extravagant signs of repentant hysteria. Lady Charlotte, who derived an unknowing and not very pretty excitement from the emotional outbursts of her children, gave Enid's sobbing apologies every encouragement.

"My child is truly sorry?"

"Yes. I am."

"Then what have you done with those two letters?"

47

"I tore them up. But I never want to write to him again or hear from him, or see him, I promise."

Enid was shaking all over. She had the look of one with a chill on the stomach and a fever on the mind.

"Again? Then you saw him to-day? And alone? Oh, Enid, Enid."

Enid leant her head against the end of the stiff little sofa and sobbed steadily.

"Enid?"

"He's going away to-morrow."

"*Oh!*"

"And I don't want to see him again."

"Can Mother trust you this time?"

"YES," said Enid, almost on a howl.

After this Lady Charlotte had sense enough to bestow a kiss in which forgiveness was suspended if not actually granted, before she sent her to bed and ordered a glass of Marsala to go up on Miss Enid's dinner tray.

looking woefully ashamed of himself, had mounted his bicycle and ridden away. Thank God he was joining his regiment to-morrow and perhaps they need never meet again. Never indeed, perhaps, for in two months' time his regiment was going to India. At this thought a feeling of complete desolation took entire hold of Enid's mind, an unfair hold, and beyond it the first faint stir and twitch of a fear.

There followed a week or two in which she tried to forget, with a concentration that made her all the more conscious of the unclean and awful crime that she had committed. It was terrible really, this frightful feeling of shame and self-hatred which quite demoralised her as a rational being, and she was never too steady at the best of times. And it was not so much the thing she had done (which indeed seemed to her punishment enough for its doing) but the fear of what other people would think should they guess her to have done it. She became very affectionate and emotional towards her sisters, thinking of them as pure and different creatures from herself. This only Diana found embarrassing. Muriel and Violet, calm, cordial creatures, accepted this sudden lavishing of affection without undue surprise—they knew so little about each other, a thing like that neither surprised them or weighed with them at all. They were like children, thinking, "Enid has been naughty to Mother—now she's showing how good she can be to us all." They encouraged her affectionately.

Lady Charlotte, of course, accepted all Enid's eager busyness as a proper showing of remorse for conduct magnanimously forgiven. She inquired no more into the sudden closing of this affair, beyond ascertaining the date on which Arthur rejoined his regiment, and the date on which the regiment departed for India.

But Enid, poor, foolish and most singularly unlucky Enid! The weeks (as they do) went past and while they brought her a sort of unbelieving forgetfulness of what she looked on as the most disgraceful and painful act of her life, they brought her nearer to a horrid doubt. She was indeed most unhappy, for what had it all meant to her? Nothing. Nothing but pain, embarrassment, the death of this romance and then—Nothing. A tremendous blank, a chill sense of entire failure. And now there was this suspicion. A despair that doubted towards hope once for every three times that it sank to terrifying certainty.

The terror that lived in everything then. Terror in the morning, waking early and lying still in bed while the swift, full consciousness of fear returned. She would gaze with eyes of terrible envy over at the bed where Diana lay sleeping, dark and sullen and safe from this unanswerable fear. The dread of the morning came on her then. First, breakfast-time and the question whether she should feel ill or faint. The necessity of eating or incurring her mother's comment. Then the later day and a thousand elaborately-sought reasons for solitude. Solitude in which to exercise her body—Enid's soft body that loathed activity. Now she was all eagerness to ride in these autumn days, lumping along solemnly and studiously hour after hour, clap-flop in her side saddle, short stirrup and close pummels. She could find no reasonable excuse for work of any kind though she longed to dig and cut down trees. But in the afternoons, on the pretext of a walk, she would hurry through the yellowing woods, unseeing, until she came to where there was a three-foot drop off a wet rock face, and from the top of this to the ground she would jump with horrid mechanism time after time and hour upon hour. Resting sometimes, her forehead sweating in her hands, and then again, five steps through the hazel bushes she had broken and twisted back, and spring, land, and repeat, spring, land and repeat. All no good. No good as she knew in her secret terror, running back through the wet woods to teatime and early lamplight and the long horror of the evening. All the time she felt ill and yet too frightened to mind feeling ill. She was as if looking through glass and feeling across a distance.

A week was like seven years and a fortnight a lifetime of entire despair. She never wrote to Arthur, feeling nothing but complete revulsion and a tremendous embarrassment towards him. A dreadful little letter of his more of apology than love was the last she had heard from him. He seemed now to be more outside her life than any person in the world, as day after day she took her silly measures against her certainty of pregnancy and each day her mind in its desperation grew further from normal. She had nothing. She had no hope. What did girls do? What happened to them after this? How soon would people notice what was wrong? She would stand quite still sometimes or lie taut in her bed, her whole conscious being only knowing complete fear. And then an agony of small realities would come to her until she would tremble and sweat. She did not often cry because she did not dare.

Should she write to Arthur? Should she not write? No, she could not write. Perhaps everything would still be all right. She had read in an enthralling book called, *Till The Doctor Comes*, of nerves affecting women to this extent. It was a tremendous piece of knowledge this, and it gave her rather a comforting sense of grandeur to state it to herself. It was only a very momentary comfort, when she was trying to be desperately strong-minded and reasonable. But when three weeks had gone her nerve cracked utterly. She could not endure any more. The horror of life at Garonlea continuing its measured daily course overcame her. She longed so deeply to be borne once more on the even current of its going. Writing notes for her mother. Messages to the tenants. Afternoon calls. Delicious confabulations with her sisters on the endless subject of clothes. She saw that state of life as the only desirable way of being, and the state of being in it yet outside it as she now was, overcame what little sense of proportion Enid had. And about this there was no possible foundation for her having the faintest sense of proportion.

One Sunday she was overwhelmed by such a desperate feeling of faintness and sickness during the Litany that a sense of inevitable disaster caused her to touch Diana who knelt beside her on the arm and sway down the length of the aisle down past the pews so well filled by neighbours in their best clothes. A rustle of taffeta as she passed and bowed elegantly-veiled heads turned to watch her for a half-second before they bent again in devotion and possibly disapproval over their prayer-books. Then the cold air in which the beads of sweat felt like frost on her forehead and a wild dash to the back of the family vault where she was sick amongst the nettles and drew in breaths of mouldy air in sharp relief.

"Don't tell Mother I was sick, Diana, please promise me." There was so much urgency in Enid's voice, such energy, that Diana promised quickly, feeling as she usually did as though she stood outside Enid's weakness and hysteria. But to-day as she followed her half-way up the long wet avenue towards the house, Diana was aware of something fixed in Enid, something that was

54

beyond hysteria, something that she vaguely knew was dangerous. It made her hesitate and obey when Enid begged her in a strained, urgent voice to go back to church and leave her to walk home alone. Ordinarily Diana would have played the sensible bully and ignored this appeal, but to-day she went sturdily back to the family pew, whispering to her mother that Enid was quite all right, the air had done her good, and meeting her sisters' wondering looks and stolen nods without any encouraging air of secrecy.

"Well, but, my dear child," Lady Charlotte said with great moderation and reasonable authority, "surely if Mother says she wishes the Doctor to see you, you must know she has her reasons."

This was later in the day when Enid, fulfilling her Mother's invariable prescription for any ill, was lying down. A long Sunday afternoon under her eiderdown, broken now by this new and immediate terror.

"Oh, Mother, really there is nothing the matter. Please do believe me." The dark head and lovely brow were lifted from their pillow. Those tremendous, imploring eyes dark with tears and fear.

"Surely, dear, Mother is the best judge of how her little girl is looking—and I don't want you to go and stay with Aunt Alice next week if you aren't at your best."

"But I didn't know I was going to stay with Aunt Alice."

"Yes. It's a surprise Aunt Alice and I planned for you," Lady Charlotte lied agreeably for she had not yet written to Aunt Alice on the subject—the idea having only vaguely occurred to her during the sermon that morning.

"Oh, must I go and stay with Aunt Alice? Must I, Mother? I'd rather not."

"Now, my dear, this is pure foolishness——" Lady

55

Charlotte felt the moment had come to be brisk. But Enid felt that the moment had come past which she could not endure. She sat up in bed in that white petticoat bodice run with so much blue ribbon and trimmed with so much lace and she wept in terrible despair.

Lady Charlotte left the room more determined than ever to summon Doctor Maxwell and compel him to agree with her that a little change was what the child needed.

Enid's sobs lessened gradually, for what was the use, what relief was there in tears for her? Presently she got out of bed and leant out of the window, her hot eyes burning in the cold autumn air, her bare arms like glass. She was conscious of the unseen stoop of the valley above and behind the house, the downward pressing of the great woods. Her eyes travelled across the wet levels of the lawn, resting on distant groups of trees with the ease that their planters had designed. She was entirely in despair and the thought of death seemed to her the only easy thought. Leaning there in the window she considered means of suicide with the detachment of a person just beyond reason.

Looking out of her window across the lawn at the bare wet walnut trees with their brown soaked circles of leaves below them, she thought of hanging herself from some easily reachable limb—you just kicked the chair away, but what sort of knot do they tie? She caught herself softly by the throat and knew she could not do this. Then there was the river, a deep, smooth-barrelled autumn flood dropping over its weir, dark alders and willow trees twisted round with the pale rubbish the flood had brought down. Enid thought of the strong water catching her, swirling under her, lifting her skirts and closing over her face, drowning her cries and choking her.

56

In personal extremity the least imaginative people
see things with this awful clarity of vision. Knowing
she could not hang nor drown, Enid's mind turned to
the methods of more domestic if more horrid suicides.
Leaning there at her window, her mind turned over
these matters with less concern and excitement than
had gone two months past to the planning of her new
winter clothes. She thought now of chemists' shops
and arsenic and of the dark toolhouse in the kitchen
garden and weed killer. Again she saw the shelves of a
chemist's shop and a neat little packet of Salts of Lemon
marked in neat red letters, POISON, lying on the
counter. It had been bought for removing stains
from household linen, but she understood it to be a
deadly poison. Shelves. Little dark cupboards with
different deaths on their shelves. Her mind knew it
could find something. In a moment remembrance
would be alive in her mind. She was so afraid, she was
so much afraid to remember that it came to her with
dreadful ease. It was in the cupboard above Diana's
washstand, hanging above the muslin splash-cover,
dotted muslin over pink, frilled against the wall.
Open the little cupboard and on the right-hand side
you will see a dark bottle, POISON—NOT TO BE
TAKEN. A bottle of black hat-dye with which Diana
had intended to transform a summer hat, refraining
on Enid's advice and substituting a less drastic change
in the shape of a wreath of Mr. Liberty's brown silk
poppy heads and assorted grasses.

So in the summer when she was happy, before any
of this had come to pass, she had planned her own
death. She took the filthy black bottle out of the cupboard and read the directions on the label with a mind
numb to their unimportance. She took the cork out
of the bottle and sniffed the thick, rather sweet, smell.

57

There was an oily glimmer in the full neck of the bottle and her throat closed up at the thought of swallowing the stuff. With tremendous speed her mind found a reason for postponing the moment and the terror. "I'd better wait until they are all at dinner," she said to herself, "or some one might come in and save me."

It was now five o'clock. She drank her tea when it was brought to her and passed the interval between tea-time and Diana coming in to change for dinner, emotionally, but not entirely disagreeably, in the composition of a letter to Arthur and one to her mother.

"DEAR ARTHUR,

"When you get this letter I shall be dead. You must not think of me sadly or feel that I blame you in any way, but this seems to me the only Way Out. I feel it is only fair to write and tell you I forgive you utterly and trust you will find Happiness and never let the thought of me haunt your life.

"ENID."

Then she wrote to her mother:

"MY DEAREST MOTHER,

"I have done this to save you all from pain and shame. It seems the only Way Out after having been so foolish and wicked as I have been. Remember I blame no one but myself and forgive me if you can.

"Your loving daughter,

"ENID."

The writing of these two letters uplifted in her an extreme consciousness of sacrifice. Now she did not feel any longer poor frightened Enid but a creature

58

able to rise beyond the torment and horror of her circumstances, able and ready to make the supreme sacrifice of Life itself. She felt keenly the pure and entire nobility of her motive. Perhaps her desperation only matched the rather desperate issues of such a situation at that date. Perhaps, had she taken it, that desperate escape would have caused her less pain and less entire spiritual shame than the life that was to follow.

For, of course, that black cup of poison was never drunk. Even if Diana had not come running back, called by some dire sense of disaster when she was half-way down stairs and badly in the wake of the dinner gong, it is hardly likely that poor Enid who had so little resolution even for the swallowing of senna would have managed more than her first sip of that burning and dreadful cup. She put it down with a little cry of horror and her body was shaken with sobs and convulsive nausea when Diana came in.

The bottle and the cup were both beside Enid and her lips were blackened and her eyes staring wildly. Diana kept her head long enough to seize both cup and bottle. Then quite suddenly she lost all hold of herself and ran screaming downstairs for her mother.

But to Enid was at least left the conviction that had it not been for this interruption she would have carried out her plan to its end in death. This conviction mercifully never left her—not even when the Doctor's verdict that twice the contents of the bottle would scarcely have proved a fatal dose was made known to her. The conviction, as a spiritual support, remained unshaken.

Tremendous inner confusion and upheaval masked by a splendid show of decorum followed the discovery of Enid's frustrated attempt at suicide. The family Doctor, called in to prescribe for nerves, was hardly

allowed even to hint at the real cause of the trouble.
Lady Charlotte was, of course, present during his
slight examination of Enid, who entirely collapsed and
in floods of tears refused to answer any leading
questions. When her mother and the doctor had gone
she lay with her face to the wall, quite still and completely vanquished while Spiller, her mother's maid,
sat and sewed by the window in the clear autumn
afternoon. She was not allowed to be alone for a
moment. She had even been forbidden to lock the
bathroom door. She lay like an animal waiting in a
trap for her mother to come back, as she soon would.
Soon she would. Then she came. The door opening
and her voice saying:

"I will sit with Miss Enid for an hour, Spiller."

She would not look round to see. She could hear
Spiller softly gathering up her sewing and softly
going from the room. She heard her mother sit down
beside her bed and felt her leaning across towards her.
She looked up and saw tears falling, running down that
smooth, pale face, caught on the brown shelf of bosom
—almost rebounding off its firm and matronly
contours. The padded shoulders ringed with braid were
leaning over Enid, and nearer still the pale face, the
puffed grey hair.

"Oh, Enid, Enid, my poor little girl—tell Mother all
about it. Mother will see to everything. It will be all
right. Don't be frightened, my darling."

Ah, the easing sweep of such words. The comfort.
The kisses. The tears. Enid's head was laid upon that
maternal bosom. Her tears soaked into its brown silk.
Her cheek, as in childhood, took the impression of its
knotted braid. She breathed the faintly stuffy smell
that was her mother. She knew complete relief and
presently her tale was told.

VIII

ONE cannot deny that the days of maternal omni-
potence had their moments. There is something to be
said surely for the tremendous ease with which the
young could fling themselves upon the mercy of omni-
potence.

But the end was not in that hour of maternal
consolation and spiritual ease. The end was a long way
further off. Enid indeed, never quite saw the end.

"My darling, tell Mother, it will be all right—I
promise you——" And the tale was told and Enid's
whole life was taken out of Enid's keeping, indeed, she
did not wish to keep or order her life, she had no plan
for it. She had ended her present life as surely as if
she had died and now Lady Charlotte like God, must
give her a new one. Really, there was nothing else to
be done except the things that Lady Charlotte did and
she did them with wrath and speed and efficiency and
throughout showed an unflinching social front.

There was a visit to Arthur's parents during which
nothing but money matters was discussed. Womanly
delicacy alone prevented her from seeing Arthur and
telling him what she thought of his past conduct and
how little she hoped for his future happiness. It was
with real regret that she abandoned this hour to
Ambrose and maintained herself only an icy politeness
towards the culprit. It was lucky for Ambrose that she
never knew how miserably he had faltered even over the
opening lines of his speech, "*You are a Cad, Sir*——"
and had with pitiable weakness abandoned the rest,

saying only, "I should rather not discuss your conduct," and proceeded at once to the matter of settlements. Perhaps, as well. For Arthur, being an emotional young man, was very near tears as it was.

As for Enid, she was not allowed to see Arthur alone for a moment during that unspeakable fortnight which preceded that nightmare day when, heavily veiled in Brussels lace and garlanded stupendously with orange blossom, she wavered sobbing up the aisle on her father's arm, followed by her sisters, shivering in muslin gowns and weeping in sympathy.

During the fortnight between her confession and her wedding she was not quite fully conscious except at mercifully brief intervals when the truth of what was happening to her closed over her body and her spirit, drowning her in an agony of shame and revulsion. But the round of occupation found for every day and every second of every day by her mother left her so exhausted that she felt only numb when left in peace though never alone. She was borne up and onwards on this current of ceaseless things that had to be done, on towards the last dreadful necessity of marriage beyond which she would not look.

Daily she wrote notes of thanks for wedding presents and stood for hours while calico shapes were pinned and unpinned upon her before being sent off to a London dressmaker. The gowns which came daily from different shops were fitted and altered by Spiller— enthusiastic but perhaps not very apt. And Enid, who loved clothes with passion and excitement, would stand in a clearing among new tissue paper and half-unpacked boxes, insentient now to the thrill of lace flounces and the sharp whisper of taffeta, the gleam of gold tissue and the mists of tulle on her shoulders, the feather boas and all the hats.

"Oh, Enid, how happy you must be. All these lovely, lovely clothes, and Arthur is really quite nice," exclaimed Violet one day, from the bottom of her heart.

Enid smiled wanly, taking what refuge she might in the mystery and dignity of an engaged young lady.

Lady Charlotte left nothing undone to make this hurried wedding appear a desired and successful event. She could hardly have turned a braver face towards her world. There were parties at Garonlea for luncheon, tea and dinner, at which Enid appeared pallid (but she was always pale) and with a smile pinned on her face to receive the congratulations of the countryside. These were very hearty, for Lady Charlotte was not too popular among the local mothers and they were not displeased in consequence that the first of her daughters to be married should make a match that the most generous-minded could scarcely look on as brilliant.

Towards Enid, Lady Charlotte maintained a demeanour of rather stern solicitude. There were no more tears, there were no more confidences, but there were no recriminations. Enid made only one flutter towards escape. When she was told that the date of her wedding was fixed she grew pitifully red and said with nervous effort:

"Mother—don't make me marry Arthur."

Lady Charlotte replied, "My dear Enid, you have left very little choice in the matter either to me or yourself." And so the subject was closed. That faint shaft of sarcasm left Enid silent and powerless. The horror of acceptance was less than the horror of explanation. Perhaps it was as well that Lady Charlotte discussed the matter no further for, however clearly Enid had been able to explain her present fear and dislike of Arthur, the situation would in the end have remained unaltered.

There was one moment of which Lady Charlotte felt herself unfairly defrauded by the unfortunate incontinence of her daughter. But it was a moment she was to fulfil gloriously a few years later when, twenty minutes before Violet departed among showers of rice and confetti for her honeymoon her mother told her (with perhaps more delicacy than exactitude) what she might expect in Married Life.——"The poor child was really upset," she told a very intimate friend later, "although I veiled it all as much as possible."

The intimate girl-friend replied with forgivable maternal pride, "When I told Little Mabel I saw to it that she had a good breakfast first, but even so she cried so dreadfully, I thought we should never get her dressed or to church. Such innocence seems very beautiful to me."

"Wonderful," agreed Lady Charlotte, "and especially in a girl of thirty-two."

"Perhaps. Mabel was twenty-nine when she married."

"Really. I thought I remembered her toddling about the winter before Desmond was born."

"Twenty-nine," replied Mabel's mother, solidly refusing to be drawn into the question of dates, "and as innocent as a child."

Only Diana was particularly conscious of Garonlea as a place during this time of bustle and confusion, new clothes and parties, solemn conclaves in the library, Muriel's and Violet's excitement by day and Enid's tears at night.

Garonlea was the house and the reason for this dreadful thing that was being done to Enid, Diana thought quite unfairly. No ugly scandal must touch a Miss French-McGrath of Garonlea. No vaguest

64

clouding of the rich Protestant chastity of that valley. The Family, the Place, the Other Girls, Enid's Good Name, everything but Enid's happiness and Enid's freedom to live. Diana was necessarily a little vague as to what she herself would have done for Enid under the circumstances but she was really bitter over her parents' line of action. Night after night to hear Enid crying—Enid cried very easily, it was true, but not with this hidden insistent despair. Diana would pretend to be asleep for she could not possibly face the scene that must follow if she should show herself entirely aware of the trouble that encompassed Enid. Secretly she was disgusted by the real cause of this trouble. It was a sort of fulfilment of her dislike of Enid's lack of control. It was a slap at her own extreme repression too. She, only, knew what she had found Enid doing. She had not told her sisters, taking in this reserve a sort of defence against the self who had run downstairs that night calling, "Mother, Mother"—and put the whole matter at once into the power of Lady Charlotte and Garonlea. She had not even paused in her acceptance of the real powers in her life. Violet or Muriel who had no resentment towards these things could hardly have acted in a more girlishly conventional way.

Garonlea, the place and the house armoured in all its sham magnificence of Gothic, terrified and overcame her by its inviolability. It was more than usually the placid, ordered house she had always known, with its many chimneys and its stable yard with four fat horses and many sheeted, elegant carriages. It all overcame her. All of it! The slope and rotundity of the valley; the kitchens and many fat, clean servants; the echoing, icy dairies; those passages and pantries and dark little courtyards; the clock tower; the wet, well-kept gardens and dank lawns; and all the paths through the nearer woods swept and raked on Saturdays; the hot greenhouses; the orchid house where the orchids did not thrive too well in that hot breath and unseen dripping of water; the fat voluptuous stable cats; in all these places was that inexact, familiar sadness. You knew it all too well. Nothing lovely, nothing exciting, would ever happen here. The level of sadness and propriety was so secure. There would never be a break or a change in this. No matter what happened to the McGraths who lived at Garonlea —what sadness overtook them, happiness, adventure or heart-break, all was finally subdued to the pattern of Garonlea.

Now it was to be the same. This crisis in Enid's life was hushed and blotted out and drained of its power in the ruthless benignancy of Garonlea and all that Garonlea stood for. It would always be the same, it always had been.

ON the Saturday afternoon before Enid's wedding, Muriel, Violet and Diana fitted their bridesmaid's dresses for the last time. They put them on in their icy bedrooms with Spiller running from one to another with pins and suggestions. Violet wore her dress in that hushed rapture in which new clothes invested her; Muriel shyly, with little tinkling laughs; and Diana looked like a rebellious little policeman in her confection.

They paraded before Lady Charlotte in the library. Lady Charlotte, busy at her writing-table; very busy, affecting extra preoccupation as the chatter she had heard on the stairs hushed at the door and Muriel came in blushing, and Violet followed her, a ravishing and satisfactory daughter, and then Diana, sturdy and irritable, picking up a paper and reading it as she stood there, all to show how little she minded whether her dress was approved or not.

Lady Charlotte gave the dresses a businesslike attention which excluded any possible hint of admiration for the daughters inside them. Towards Diana, indeed, this was no affectation, but she showed herself her maternal impartiality and at the same time gratified her resentment of Diana's dark independence by making Spiller practically refit her dress, altering the position of each pin not once but twenty times.

Diana spent the rest of the afternoon in the woods in a fret of rebellion and disgust. Only one thing stood out for her, Desmond and Cynthia were coming that evening. The thought of seeing Cynthia was like a hot ripple of excitement, making her feel stormy and uncertain. Half of her fury at her mother's quiet bullying had been caused by this unrest. Now, as she walked through the woods, there came to her their undefeated certainty of life to be, the strange strength of woods in winter-time, their arrogant loneliness and sense of power from within. Sitting on a fallen tree that had seeped itself wet and rotten and almost back into the wood's life, Diana gave herself over to the warmth and comforting thought of Cynthia's presence and sympathy. Shyly dramatising their meeting one moment and the next realising ruefully how little able she would be to make any such smart and interesting remarks when it came to the point. A passionate creature, she sat on there in the wood, all her dramas and theories so true to her, there she sat in her green serge, tailor-made, with stuffed shoulders and tight sleeves and a high, starched collar cutting severely into her soft neck. She had a god and waited there alone, for an hour of worship was near.

Cynthia arrived. She was perfect with every one. An enormous pale-blue feather boa foaming round her shoulders—you kissed her within its scented, fluttering depth. Soft wings of hair under a monstrously romantic hat and a carnation flush in her cheeks, when she lifted her veil for kisses.

"I hope you're not quite worn out——" for Lady Charlotte, and a speaking look of sympathy combined with congratulation.

"Dear child." Lady Charlotte felt warmly satisfied with her daughter-in-law.

"Popsie, darling," and she embraced Ambrose, "I have the naughtiest story to tell you—Oh, you'll turn me out of the house." She managed it so that Lady Charlotte missed this and only Ambrose got his kick.

"Popsie darling"—why was it none of his own children could show such confidence in him? He almost blushed with pleasure.

Then Enid was kissed and congratulated so that for a fleeting moment she thought herself almost an enviable creature. And Muriel—"Darling, what an enchanting blouse—you look like the sweetest baby. I must buy one——" Muriel, in her lace and bébé ribbon, fluttered a delighted kiss. And Violet—"My dear, so lovely just to look at you again." For Diana—"Little one," and a closer kiss. But later by the schoolroom fire, between six and seven, they had an hour together.

A great deal of sewing and excitement had gone on in the schoolroom lately with Violet and Muriel stitching at flounces and threading miles of narrow blue and pink ribbons, ribbons that pulled everything into frills. The Empire tops of nainsook nightgowns, the legs of knickers (below the knee), the gentle tops of camisoles and their foaming laces were all run with blue and pink by Violet's clever, languorous fingers and Muriel's, bird-like and eager. In the air of the schoolroom their thrill was faint and vibrant still.

Old photographs of the girls—taken and framed and carefully preserved since childhood's earliest days up to the big moment when three feathers nodded and the photographer did his utmost, sparing no balustrade where a girl might lean with grace, or any joke that might evoke a smile—stood on every available inch of space, and hung upon the walls too, jostling dark-brown reproductions of the works of Mr. Watts, Mr. Burne Jones, and Mr. Rosetti. The carpet was faded and the pale chintz more friendly from careless use than the covers in any other part of the house. The bookshelves were still full of schoolroom stories—

romances of the Upper Fourth, or the Lower Sixth, and adventures of gorgeous young girls and gorgeous young Brigands, Hunters, Trappers, Sailors and Explorers. They were fine books, sometimes three hundred pages in length, yet no matter how intimate the circumstance of adventure, never once on one of the three hundred pages did sex ever show its hideous head. Violet and Muriel read them still, with keen excitement and really felt more moved by them than by some of the novels which Diana and Enid read in privacy and hid behind those virtuous and friendly red and blue backs on the schoolroom shelves.

"Is Enid happy about this, do you think?" Cynthia sat on the white fur hearthrug. Her lovely hat was gone and her hair seemed as pale as white broom, not glancing or golden in the firelight.

"Cynthia, she's miserable. I think it's such a terrible thing that's being done to her. I suppose it's her own fault, but is it? After all, if Mother had been human to her about Arthur. But Enid's always so frightened and then does something wild and uncomfortable for everybody."

"My dear, what *do* you mean?" Cynthia was alive with curiosity.

Diana told her what she meant.

"But Lady Charlotte can't possibly do anything else," Cynthia said soberly. She entirely accepted this fact and her sympathies were directly with Lady Charlotte and her condemnation for Enid. She was very hard. She did not condemn, she dismissed. She who knew love only at its utmost best, seemed isolated by this very fact from sympathy with love that had gone astray. Sordid was the word for it then. Her own heat and passion were justified sacred mysteries, ennobled by marriage and suitability. Without any pretence she despised Enid and sympathised entirely with Lady Charlotte.

"I know it has to go on," Diana sighed. "There's really no way out."

"No. There's no way out."

"I think it's so unfair," Diana spoke bitterly.

"Well—unfair? But weren't they insanely stupid? What did they expect?" This was not really the way a young married woman ought to talk to a young girl, Cynthia knew. But she knew, too, how flattered Diana would be by this assumption of knowledge and equality.

She was right. Diana valued it more highly than she had ever dreamed of valuing a relationship. She fought on to explain her reaction about Enid's marriage.

"You see, it's for us and Garonlea and the family name, it's for everything but for Enid's own sake that they are making her marry him."

"But where would she be if they didn't? And if they had let her marry him three months ago you'd all have been delighted and said how right, wouldn't you?"

"Yes, perhaps. Yes, we would. But now you see, the awful thing is that she knows what it's all like and she doesn't want to. Is it always like that for every one?"

"No, no, you've got it all wrong!" In a flash, Cynthia was the clear, passionate creature that bound others to her. For half an hour she talked to Diana with simplicity and no embarrassment, telling Diana all the things that she would never need to know. It was in times like these, natural crises, that Cynthia was really unafraid and almost great. And it was by such moments that Diana saw her and by such stray acts that she was blinded and lost to the lesser things in Cynthia.

Cynthia was the greatest help through Enid's wedding. Much more beautiful than she had been, she was a live and present example of what marriage should mean. She was most efficient about helping Lady Charlotte without once taking the responsibility for anything. She comforted Ambrose simply by being herself. She walked out with him and sowed the seeds of her stud farm plan most cunningly in his mind; slightly dramatic, impressively businesslike, not too ambitious: "Just a few well-bred mares." And she knew what she was talking about.

She could have been of great help to Enid but Enid shied so desperately away from all near contact that Cynthia simply could not take the trouble, or embarrass herself sufficiently to get through Enid's defences. After all, Enid was neither attractive to her nor useful, so she let her be.

And so Enid was married to Arthur, Arthur who had so lately changed from a lover into a frightened stranger, and was to change from that into a bullying and selfish stranger, but never back into a lover. She went away to India with him and wrote to her sisters by every mail until the following April, when one week there was no letter from Enid, but a cable for Lady Charlotte, and the girls were told how sad it was that Enid's little baby had been born dead. And while they put away the clothes they had made for it very sadly and wrote long letters of sympathy to Enid, Lady Charlotte sighed with relief. It was astonishing how Providence was always on her side. A dead, premature baby was so much simpler to explain to her friends than a hearty live one. And poor little Enid had lots of time before her.

X

In 1910 Lady Charlotte and Ambrose still reigned at
Garonlea. Ambrose was a great deal frailer than in
1900, but he still kept up a systematic if enfeebled war
against the elders and thistles on the place, and never
faltered in devoted deference to his wife, although he
differed from her wordlessly in those sturdier wars
which she waged against that Rebel Queen, that
Cynthia, whom she had loved and trusted so unfortunately.

It would be impossible to say when Lady Charlotte
first realised that Cynthia, while she asked for advice
in the sweetest way, took none but her own counsel.
It was from this discovery that the declaration of war
dated.

There was a grievance when Cynthia's son and
Garonlea's heir was born in a nursing home in London
and not at Garonlea, as Cynthia had agreed it should
be, and as Lady Charlotte had told all her friends it
would be.

A tremendous stand was made over Desmond's
leaving the army, but again Cynthia won, helped by
guile and patience and a fate that ordered his regiment
to an impossible place at the critical moment in the
struggle. Aunt Milly's opportune death helped her
here too, for it was much easier to remove one sad but
unprotesting old lady from the house you wanted
than two sad and perhaps, from mutual support,
determined ones. So Aunt Mousie and all her parcels
of plants (for she took a very small piece of almost

73

everything in the garden with her) went to the nice, suitable little house nearer the village. And there, surprisingly enough, for she was old for changes, she found new life and interest in making another garden and in improving the suitable little house. Cynthia never failed to send her game and fish and *The Illustrated London News*, and was her most regular and popular visitor. In spite of all Lady Charlotte's efforts to enlighten Aunt Mousie on the methods by which Cynthia had caused her removal from the house where she had spent so many years, she remained unshaken in her allegiance.

There were a thousand other matters of variance too. There were the hounds which Desmond produced to hunt a country from which the chase had been too long absent. Through these hounds he became almost entirely lost in a peculiar, myth-like glow of popularity. He was adored where Ambrose was mildly liked and respected, and his house of Rathglass had an importance far beyond that of Garonlea and its shoots. Whoever was young and gay in Ireland came to Rathglass, and strange and rather grand English visitors came too. And were they all explained to Lady Charlotte and was she asked to meet them? Almost never. Worse, there were many parties full of possible husbands at Rathglass to which dear Muriel, growing each year a little thinner and a little more uncertain and eager to please was not bidden. Where now were all Cynthia's plans and promises for the marital advancement of the girls? It had been in no way due to Cynthia that Violet had married Lord Jason Helvick, who had once been thought of as only suitable for little Muriel. But when Violet was twenty-four and still unmarried the outlook seemed gloomy enough to justify the acceptance of the gentle ornithologist's slightly abstracted suit.

And like Aunt Mousie, Violet seemed unexpectedly happy in her life. She liked to bully Jason and enjoyed being just a little grand and married towards Muriel when she had her to stay. Aside from vague yearnings for the dignity of the married state, she had been happy at Garonlea, and now, that state achieved, she found all that she knew how to expect, and was untroubled by any vague yearnings in any direction. Certainly she was the most satisfactory product of Lady Charlotte's upbringing. She was in fact, the exact material that such an upbringing demanded.

But past all such grievances as these, and in a different sort from the small ones (as, for instance, the ridiculous way the children's hair was cut, the absurd clothes they wore and the useless and expensive dentist who attended to their crooked teeth) there was always the constant itch and grievance of Diana's happy escape from Garonlea to Rathglass. More and more of her time was spent there, losing, as Lady Charlotte saw but could not explain, more of herself each year and growing to find all her happiness in being Cynthia's slave and shadow.

It was strange that Lady Charlotte was able to see so clearly exactly what Cynthia's form of friendship for Diana meant. All her own life she had been Queen and had ruled her children sometimes by main force, sometimes by emotional appeal, but always behind these things had been, for her at least, a right and a reason why she should demand obedience. There had been no absence of principle, however tortured the principle. She herself was self-sufficient, she was beyond those whose lives she ordered and commanded. But Cynthia was of their lives and in their lives. She took the utmost from them, far more than she gave. She failed them and charmed them to her again. She leaned upon them and queened it over them at the same time.

Lady Charlotte grew always more bitterly jealous and resentful of her, dreading the time when Cynthia would take her place at Garonlea; watching Ambrose's health with a fostering eye for this as much as for any other reason; making long lists of china and silver and linen that were her own and to go with her out of Garonlea. She took a tremendous interest in all that happened at Rathglass although she pretended to none at all. In reality the children could not have stomach-ache or a kitchen maid be sent away that she did not hear of it and derive satisfaction from the smallest disaster. Of course such major calamities as the July when they lost their yearling in a thunderstorm or the time that Desmond had six hounds poisoned—were a sort of gloomy heaven to her, even if they were offset by a successful sale at Doncaster or by the felicitations of some old friend on her superbly able daughter-in-law. None of her friends really knew of the distrust and jealousy of Cynthia which were like aching and hidden corns in Lady Charlotte's soul.

Only Diana guessed the strength and the depth of that dislike. She could feel it in her mother's changed breathing if Cynthia's name was mentioned, in the faint excitement and desire to quell which held the air close to her. Then if Diana felt in form for argument was the moment for her to produce a remark about some horse at Rathglass, or some exotic rhododendron lately flowered with success, or even some story of what the child Simon had said to, or of, the infant Susan. As surely as it woke in Ambrose a curiosity and pleased interest, quite as surely it called forth a sort of muffled virulence in her mother.

Suppose Diana said, "Simon has a hen sitting on ducks' eggs. He doesn't know how to wait for them to hatch." The reply might be, "I wonder when Cynthia is going to get that child's adenoids attended to. If she leaves them in much longer he'll be quite idiotic."

Ambrose would say, "Idiotic? Brightest little chap I've ever seen. Dearest little chap in the world."

And Lady Charlotte would swallow back her hot feeling against Cynthia, it was like a dark web within her, a fibrous tangle like the roots of plants in too small a pot. It would have been too contrary to her idea of dignity to have said openly to Ambrose or anybody else all that she knew and felt about Cynthia. But on each contact she made with anything done or said by Cynthia she cast a curiously lessening shade. And always when she could in any way compass it she prevented Diana's being at Rathglass.

Although Diana was twenty-eight and Muriel thirty-six, they were still treated much as they had been at eighteen and twenty-six. They were the daughters at home. They had no fixed allowance beyond a small amount to cover such expenses as stockings, gloves and stamps. Their clothes and their journeys were paid for and they were still told what to put on and where to go. True their clothes were bought from the best shops and they travelled first-class but for all that it was an absurdly unfair system, making Muriel so entirely dependent as to be nearly half-wit, and making Diana entirely rebellious, souring and closing her spirit in all directions except towards Cynthia, the one towards whom it flamed unquenchably, towards whom it must for its own sake refuse all disillusion. For there is no romantic creature living who will be entirely thwarted of Romance.

Rathglass in these days was like a court, a small and fashionable court well filled with courtiers and ladies-in-waiting, and of these ladies-in-waiting Diana was the chief. Behind her came perhaps a dozen younger ladies to whom Cynthia sold her horses, sometimes gave her old clothes, often gave confidence and advice and criticism. She told them what to eat and what to drink, what to wear and what to think, and she was never bored by their continual worship. On the contrary she was always thrilled by it and it called out at moments a dramatic feeling of goodness and humanity in her, rather an imitation sensation perhaps and one that never lasted long enough to cause her any serious personal inconvenience. Besides the girls there were their mothers, with whom she was more than wonderful and who only rarely mistrusted her, blaming their own foolish daughters if they could not ride the horses Cynthia sold them or caused them to buy.

Sympathy was the chief thing Cynthia gave to her train. She gave them sympathy for the hardships of home, the tyranny of parents, the shortness of money, the waywardness and backwardness of lovers—sympathy and a tremendously personal interest.

"Don't let him maul you about, child, you're much too nice."

"Never let me see you in that dreadful little hat again. It quite hurts me."

"I wouldn't let everybody ride this horse, but I always say you have quite exceptional hands."

"Blue, dear? *Not* your colour. I think I see you in rather a sweet shade of mauve."

It required a certain amount of concentration never to let one of them go and to make each feel that she was the superlative one, but Cynthia had immense physical strength and was almost never caught at a moment when she felt too tired or too ill to contend with every aspect of life. She was always hard at work or play, and her court had to work and play (but especially work) with her.

There were in those days lots of boys for the girls to play with, for Ireland was full of soldiers who hurried willingly to such houses as Rathglass to shoot and hunt and do reverence to Cynthia, while they flirted with and sometimes married into her Court. It was really rather an exciting and powerful position for Cynthia; not that she was unique in having soldiers eager to go to her house, but she was almost unique in the way she could always secure the ones she wanted and if she planned a match she often succeeded in making it.

No wonder that the mothers of the county were eager for her patronage for their daughters. No wonder that they encouraged them to hunt with Desmond's hounds, however cowardly and indifferent they might feel about the chase, and endured with patience the long tales of Rathglass and its doings which were almost the only subjects the girls could talk about. It was amazing how Cynthia filled Rathglass with a sort of glamour that was real and thrilling, a thing no one could escape, intoxicating to any ardent young Miss of her train.

Another curious thing about the Cynthia worship was the unbridled spirit of imitation which it engendered, so that soon even the moderately independent girls tried to dress like Cynthia, use the same soap and face cream, ride like her, and do their hair for hunting like her. Beyond this, they even tried to talk like her, copying faithfully those little words and phrases which each person has peculiar to themselves.

What did the train talk about then? They talked about Cynthia and her clothes and their own clothes. They talked about her dogs and their dogs. They talked in an ignorant and excited way about hunting and endlessly about how their horses jumped and who was behind who in the chase, and who was closest to Cynthia. Where the hounds were hardly mattered in comparison. They also vied with one another in the collection and life-like delivery of funny stories about the Irish, disasters recounted by cooks and grooms having a special vogue. But no peasant could really open his mouth without saying something they wished to remember and repeat. A good Irish story went with an easy swing in those days even when told by an English soldier.

Nothing is so revealing about people than the aura or influence they put forth round themselves. They cannot be strong enough to maintain it if it is unreal. It is not, unless it breathes of the very life of the person it surrounds. And this romantic excitement which was of the very air round Cynthia was a true reflection of her life at this time. She was tremendously happy and immensely painstaking in her happiness. The first and truest thing about her was that she was still in love with Desmond and needed him with as wild a longing as she had ever done. Perhaps this was the core of her romantic appeal. Then life was not a thing settled and over, it was vividly of the moment. She had Desmond, yet because she loved him so she had him not. This put the keenest edge to her vitality in living. It lent a glow to all effort. It was for her a matchless thing to have in life.

True, Desmond had never shown the remotest desire to stray from his domestic felicity, but this was not quite the point. Not only must there be no faintest thought of straying, but life together must be beyond solid domestic felicity, and here she was a winner—through ceaseless elaborate, tireless effort she never let Romance quite out of sight. In a thousand hard-working ways she gave herself leisure for this glamour, setting her whole life so that Love might be first.

Her house was lovely. It was comfortable in a way not as usual in 1910 as it is to-day. For instance, there were several bathrooms, invariably provided with delicious essences for softening and scenting the water which was always hot. The beds were soft and sheeted with the finest linen. She often changed the decoration of her drawing-room but she never changed her cook, who was an artist with an imagination of divine versatility.

In the garden where Aunt Mousie and Millie had grown their favourite flowers in nooks and patches, she made sweeping and exciting alterations. Round the house where the ghosts of walks and paths had threaded their way for years through dark tangles of Portugal laurel, she cut down and opened out and planted with azaleas and rhododendrons and rings of fuchsia, putting rustic seats in all the places where people might sit down for conversation or for Love. She did not really mind about gardens, and had an unfair way of achieving success with them, using flowers as decorative furniture.

It was in her garden that she made great use of her court. She would give a young man a day's hunting on a young horse she wanted well schooled and the following day he was made to sweat and toil cutting down laurels and rhododendron. It was rather a good plan and really the young men liked it, especially as their girl of the moment was so often sent out at the same time with the bulb planter and a few hundred bulbs. Of course they complained a lot in a comical spirit, for gardening was not then the fashionable excitememt which it is now.

Only Diana, who had never thought of gardening at Garonlea, indeed it would have been rather a silly and embarrassing occupation under the amused patronage of Lady Charlotte and Williams, took to it now with a real passion and delighted excitement. The ideas were generally Cynthia's but their successful fulfilment were more often than not due to Diana's streak of gardening intelligence and hard work. For good sweeping ideas don't make gardens really grow, although they sound so assertive and unanswerable that they seem bound to lead to a "blaze" here and a "mass of colour" somewhere else, the disheartening thing about them is that the blaze is so often the meanest flicker and the mass of colour fails to appear at all.

However, between Cynthia and Diana and the court workers and two able-bodied men, Cynthia's garden five years after she had gone to live at Rathglass, was a lovely and romantic garden and quite as successful as the rest of her life.

On a quiet exciting afternoon in March 1910, Diana
left the Gothic towers and battlements of Garonlea
behind her and set out on her bicycle for Rathglass.
She looked exactly the same as she did in 1900. Everything was still shut up inside her. Her mind still
registered the same protests, except towards Cynthia
she had not expanded in any way. Garonlea still held
her prisoner, depressing her health and her spirits
unnaturally. It cannot be good for a person to live
in a place towards which they have as strong a dislike
as Diana had for Garonlea. But where else should she
live? Was not Garonlea her home. And girls should
love their homes although they must be willing
enough to leave them when Mister Right comes
along.

It was one of those March days when there is the
first strong feeling towards life in the year. The end
of winter, a false end because it does not really end here
at all. For a moment of great sweetness and drama
you see a deep red in the blackness of the hedges and
the distances are like smoke, honey and irises. The
fields that were white as bones and dry as meal, although there is as yet no growth about the grass, look
kinder and more bosomy. Afar off on a turn of the
river a fisher stood, his ghillie squatting, an attentive
blot, on the bank near him. There were hazel catkins,
hanging straight and architectural curtains up the
flights of the woods, nothing whimsey or faëry about
them ; they had on the contrary an exotic and orchid-
like effect in the absence of other flowers.

Diana bicycled through the day feeling separately pleased by all these things in the detached and finished way that people can who have never confused the spring weather with their emotions. Connect such a day with a memory or with a hope about love and it loses its edges and that sharp ring which such days as these have for the very young. Diana had kept this quality of seeing days as days entire and unconfused.

At Rathglass Cynthia's baby, Susan, was asleep in her perambulator in the shelter of the bank where Aunt Millie's scyllas flowed down towards the little lawn and spread in brilliant flat pools across the grass. Cynthia had planted a thousand more each year and ordained that her child should sleep there covered in a blue blanket. She took great trouble and pleasure in an effect of this sort, for it is not easy to buy a blanket the colour of scyllas or to have a child as beautiful and healthy as Susan.

Diana took a look at her niece and a longer look at the scyllas. The house was empty, with cross little terriers lying in patches of sun on the steps and window sills. A red japonica, very correctly pruned and trained against the wall, held its first flowers stiffly on stems thick with buds. There was a rich, trim air about the house. The rooms inside would be warm and full of flowers. There would be a very delicious tea when Cynthia and Desmond came in from hunting.

Diana went off to weed the rock garden. She felt quite silly with delight, the day was so fine and she would see Cynthia soon. Then the purple primroses she had planted caught the afternoon sun and made her wild with pleasure. The shaft of sun went through them as though they were a bottle of port.

At five o'clock Cynthia came back from hunting

84

in their tall motor car. She wore a check tweed coat with a velvet collar over the new fashioned safety habit that Lady Charlotte thought so indecent. She wore a silk hat and a veil and a bunch of violets, now limp and dark and very sweet from being worn all day. She stood in the hall looking at the post with her arm absently through Diana's. The post is always of curious importance when people come in from hunting it brings them back with a sudden turn into the other excitements of life.

"Only a catalogue from Stevenson and a letter from Mrs. Walls to say the chestnut mare ran away for three miles down a road with that tiresome Mabel. I thought she might do that. I'd better take her back and let them have Silver Tip. He won't run far, he's too crippled. I think I'll have a bath now and tea afterwards. Desmond's feeding his hounds, we may as well wait for him. Well, darling, I'm sorry you weren't out to-day. We had a nice little hunt. You'd have enjoyed it. Come and help Cynthia have a bath."

On their way to Cynthia's room they looked in at the nursery where Simon sat with his smooth round young head, olive blond as a linnet's wing, bent over his plate. Susan dragged and tramped her way up and down her railed cot.

Simon who had been lost in the stupor of nursery dullness came to life with a flash.

"Mummy, Mummy, did you kill anything?"

"Not a thing. Those silly old hounds of your silly old Daddy's couldn't catch a fox. They can come down in half an hour, Nurse. Well, Susan, old lady. See you soon, Simon."

"*Yes*," Simon shouted.

Cynthia was rather impersonal about the children. If they had not had decorative value and if they had

85

not excited Desmond so much, she would have had very little to do with them. Perhaps when they were older and started riding they would be more interesting. But Desmond wouldn't let Simon ride yet although he was five and it was quite time he began. Cynthia was always seeing ponies that would be the very thing for him.

Cynthia's bedroom was very good of its period. It was pink and silver with a pale grey carpet and a pink satin eiderdown. Pink and silver brocade curtains and a dressing-table covered in stiff white dotted muslin over pink calico, with all Cynthia's silver-backed hair brushes on it and a vase of pink carnations and on the writing table a silver bowl of violets. It was a nice setting for a gorgeous blonde.

Cynthia undressed in the fine graceful way she did everything, pulling off her rather pointed boots with a nice gesture; laying herself at last in her deep warm bath and only giving Diana her attention when her first exquisite shudder of delighted relaxation was quite over.

"So Mother said I was to drive to Barretstown as she and Muriel were going in for Muriel to see the dentist. And what, I said, do you intend me to do? Wait for an hour at the Ladies' Club looking at the Punches? As a matter of fact I didn't say that. I said (quite quietly) 'As a matter of fact I happen to be going to tea with Cynthia!' She said, how curious it was how I always kept my movements quite secret, and might she ask, if it wasn't probing too far, whether I was doing anything with Cynthia next Thursday as she would like me to go with them to lunch at Summertown. 'Next Thursday' I said, 'let me see. No I think Wednesday's a hunting day next week.' "

Cynthia vaguely knew that it did Diana good to

report these almost fictional and extremely boring conversations with Lady Charlotte. Whether or not she had ever been so calm and balanced as her version of the contest invariably showed her to have been, when she had told the tale she thought she had and this did her great good.

Cynthia was turning a piece of purple soap round in her hands and then slowly lathering her long fat arms. She said, entirely without malice, "What a horrible old woman she is." And then, "I think you're wonderful to stand her at all. Only an angel would."

"What else can I do, child?" It made Diana feel a little grand to call Cynthia "Child." Grand and easy.

Cynthia always wanted to say, "Come and live with us," because Diana would be so useful and work without ceasing and praise without ceasing from morning till night. But she could not. It would have involved too bitter an argument between Desmond and his mother. And she saw the stupidity of that. Diana dimly saw it too, but she was always waiting for Cynthia to say this as people when they have a long hopeless affair feel that the only important words their lover can really say are, "Will you marry me?" Even when they know he can't or know that they would not if he could. Still it seems the touchstone of all meaning.

Cynthia slid down into her bath, saying instead, "I do call that a pretty tweed. Mine that he made is horrid." It did nearly as well but not quite she knew, so she added, "The hat I had made to match doesn't suit me either. I wish I'd thought of giving it to you two days ago. I've sent it away to be altered now and perhaps it will be all right. If not you can have it."

"How sweet of you, dear. But you mustn't——"

87

Diana took off her own hat and fiddled with her dark hair which would have curled if it had not been so heavy. However, it was the date when the great thing to be able to do with your hair was to sit on it and if you could do this, never mind how it looked under a hat it was still a boast and a beauty. So Diana need not be pitied because of her hair at any rate. She could easily sit on it.

Possibly it did lend a certain heavy air of seclusion to the face below, this massed quantity of hair. The face was like a little bird's pale egg, one had to look for whatever it had near beauty. Every woman could not by cunning barbering and painting be compelled to look her best all the time.

XII

CYNTHIA came down to tea in a mauve dress with a good deal of silk fringe attached to it. She put on her rings too: opals and diamonds and emeralds on her warm hands. She made tea in a slow way, heavily as if she had been captured. Desmond was looking at her. He seemed cold and independent in the hot fragrant room full of cyclamens and freesias sending out wreaths of scent, but he was not.

They were all hungry and ate a quantity of very good food without talking much. Desmond could hardly be very interested in Diana. He asked her a few questions about Garonlea, rather as though she were still a little girl. They were not all at familiar. He thought she ought to try and marry, or else be more like Muriel who was at least an apologetic failure. He simply did not try to understand anything about Diana. Although he had a warm and quick imagination where Cynthia was concerned or his children, he could give no more than acceptance to Muriel and Diana. He saw Diana's infatuation for Cynthia with a faintly contemptuous indulgence. He accepted it as he accepted without notice the respectful adoration of all young men for Cynthia. He knew without a shadow of question that she never had a flicker about one of them. He and she were two people apart. Really apart from others. It was a very strong thing of which to be so aware.

He was pleased when the children came down. He and Cynthia played with them, throwing cards into a hat. Cynthia said—"Dash!"—that rather modern slang word when she missed the hat. They both preferred playing with their father which did not annoy Cynthia at all. She would have liked Simon better if he had been sturdy and curly like Susan, instead of being rather long and thin with soft straight hair. And she would have liked Susan better if she had been another boy. Susan was enormous, she burst out of her white dress when she laughed. Cynthia kept buttoning it up again behind but the buttonhole was too large for the button. "Terrible figure the girl has," she said in a detached way.

"Oh, she has a lot of improvement in her," Desmond said.

Simon was looking at a book.

Cynthia said, "I hope that child's not going to turn out clever."

"It's all right. It's only a hunting book."

"Simon, old boy, stop reading and come and tell Mummy what you saw on your walk to-day."

"I was reading."

"Yes, but what did you see?"

"Didn't see anything."

"Oh, nonsense darling. Where did you go?"

"Down the Green Lane."

"And you didn't see a thing?"

"I saw some rabbits," said Simon, driven to rather obvious invention.

"Pity Daddy's hound Grampion wasn't there, he's very good at catching rabbits."

"Daddy's hounds catch *Foxes*" Simon knew this much. And the remark was so successful that he was allowed to go back to his book.

Presently the children were fetched away to bed and Nurse was told about the back of Susan's dress. The curtains were drawn and the scent of freesias redoubled in the room. Desmond finished writing endless lists of which of his hounds had done what, that day and came back to the fire where Cynthia and Diana were drinking sweet brown sherry. One of the pleasant things about Rathglass was that Cynthia was much ahead of her day in providing drink when it was needed between meals. She liked a nice drink herself and saw no reason why her girl friends should not have one too. She managed to invest this drinking with a sort of sacramental quality, for then women did not drink between meals and at meals hardly enough to do them any good.

Here it was warm and there was a feeling of intense life. Diana, just the faintest degree drunk, stared into her sherry glass. There was wine and confidence here and the dearest thing in life. Diana was thrilled by Cynthia's life in the way she had been thrilled when she first read Mr. Kipling. Forbidden moments with *Plain Tales from the Hills*. This could not be better, with *Puck of Pook's Hill*. Complete romantic illusion with the Brushwood Boy. It was touching that for reality of comparison she had to go from the lives of others to reading of the lives of others. She had no criterion of happiness in herself. Seeing Cynthia and Desmond and Susan and Simon it was as if she read a thrilling story about them. Now she had come to the end of a chapter. Now she had to go. She looked round the room putting it once more in her memory—the deep soft reflections in the polished faces of bureaus—the breathing flowers stirring in the warm stillness—the solid turns of heliotrope cushions. Cynthia silent a little tired, her mauve dress agreeing mildly with the cyclamens.

Now she must go.

91

"Wednesday, Thursday, let me see—— I'm going into Barretstown on Friday. I could pick you up. If it's not a frightful day, could you be on the bridge at three? Or I'll pick you up at Garonlea if you'd rather."

"Cynthia, lovely! I'll be on the bridge. Good-bye, darling."

"Good-bye, darling. Thank you for all your work in the garden. Oh, one moment; shall I give you the pattern of that dress Muriel was going to make for Susan? Tell her to make it a bit bigger for the monster."

Then Diana went away, riding her bicycle carefully along in the last of the dusky peach light. The evening was utterly quiet and the turning reaches of the river untruthfully apricot under the sky, and the woods were heavy like smoke against it. The edges of old fields were enfolded mysteriously in the wood's earlier darkness and the hazel catkins in their heights and lines were only little hanging ghosts. The slope of a newly ploughed field looked rich and promising. Presently she passed Aunt Mousie's house and saw a faint twist of smoke from a bonfire in her garden and a brilliant ring of these same blue scyllas low round a tree stump near the gate. So like Aunt Mousie to decorate a tree stump instead of grubbing it up. A light in the window of the post office and another in the shop that sold everything. The pink and white sugar-stick and bunches of bootlaces and packets of starch behind the square small panes of its window front looked quite sophisticated and urban, so illuminated. Diana felt rather gay looking at them. But when she crossed the bridge at the valley foot the feeling of loss came through her like a draught, thin and penetrating. An air blowing from the other side of life. Not the other side really, but her own side to which she belonged. This was her own Life from which she could not escape, and the warmth and beauty at Rathglass the fantasy. This was the real she, this rather soured and flattened creature wheeling her bicycle into its shed, taking off her gloves among the stubby little suits of armour in the hall, and blinking in the lamplit library where her mother and father and Muriel looked up at her from the world where they belonged and she belonged.

"A little late, perhaps, Diana, for bicycling round the countryside?" Lady Charlotte suggested. The atmosphere was full of acid and broken glass. Yes, the other life was the fantasy.

"It was quite light till I crossed the bridge and came through the woods," Diana said in rather a Russian way.

"Curious. It was dark here at least an hour ago."

"Oh, nonsense, Mother."

"Thank you, Diana. Thank you very much indeed. Since when may I ask have I been in the habit of talking nonsense? And since when have I allowed one of my children to speak to me as you have just done?"

Diana looked at her darkly and sat down. "Children?" was all she said.

But they all knew what she meant. Poor little Muriel blushed hotly, she could never get over the shame of her age and single state. Ambrose looked as discomposed as though some one had talked about adultery. Lady Charlotte's grand bosoms shook like jellies with rage. No one in this house ever got even into the shadow of reality, so to say "children?" like that was more daring than it sounds.

"I never expect you to be particularly civil on your

93

return from Rathglass," Lady Charlotte was really trembling with rage. "The place seems to affect you in a most unfortunate way, doesn't it? Doesn't it? Well? Can't you answer? Can't you find an answer?"

The room was too full of hidden grief and helplessness and power misused and grown weak and wicked. No stir in the air of past love or pleasure. No remembrance of indulgence for foolishness. No jokes or good honest quarrels.

Ambrose turned his paper miserably inside out and Muriel bent her head in diligent absorption over the jersey she was knitting for Enid's Ambrose. She minded so much and felt violently towards Diana for this breach through her pretence of contentment, in being the invaluable daughter at home. The pretence which her vanity demanded was only too necessary to her. And all Diana had said was "children?" It was appalling that one word spoken like that should mean so much and cause such devastation of spirit.

Lady Charlotte's demand for an answer was purely rhetorical. She expected neither answer nor retort, and would have been still further incensed if either had been forthcoming. Diana knew this perfectly well and she had neither the wish nor the will to provoke any further trouble. The sense of nausea and desolation that these scenes left in her, while it did not quite stop her provoking them, was still too powerful to allow her to come to any real issue. Now she could feel the twittering apprehension of Ambrose and Muriel, small birds frightened in their cages, saying to her, "Please stop. Oh, please, no more." And her mother's heavy swelling anger. What was the use of protest? One had no power. And what have I done, Diana thought with despairing resentment, what

94

have I done beyond spending an afternoon weeding in a garden and having tea with Cynthia and Desmond? It's too ridiculous, too humiliating, that it should lead to this. It's so cruel and so unfair.

But she knew how much more than this she had done. She knew how tremendous was her escape from this atmosphere and place to Rathglass. In moments like this the contrast was so dramatic that she could not find herself believing in either one place or the other. Two extremes, met, leaving her somewhere suspended in a sort of mental mid-air, from which in a moment she would fall sickeningly into the complete reality of Garonlea.

XIII

Muriel came into her room before dinner.

"I wonder if you would just fasten this for me."

Diana stopped the intricate work of coiling her hair round and round layers of hair inside which was her head. She was fond of Muriel though often exasperated by her and she was strangely jealous of Muriel's pretences about happiness. Muriel could fill her life so full of small importances and interests that unless it was cruelly brought home to her, the indelicate subject of her real unhappiness need hardly ever be faced. Besides through all her rather simple-hearted pretences she arrived at some unexpected points of joy. She maintained her interest in clothes as keenly as ever. She had far more success than Diana ever did with her nephews and nieces, taking immense trouble to find them interesting toys and to tell them interminable serial stories about a family of frogs. Violet and Enid really enjoyed having her to stay with them, she was so flattered by confidence and uncritical in comment, whereas Diana was a defiant and embarrassing kind of sister who obviously knew in a silly, theoretical sort of way, far more than an unmarried girl had any business to know.

Now Diana stooped, peering and hooking her way up from Muriel's narrow hips to her neat bird's neck. Her clothes were always exquisitely fresh and neat, every smallest frill stood out primly separate from its fellows, and there was complete absence of scent about her, only a feeling as fresh and smooth as a little bird's body caught in between your hands.

"Thank you so much, Diana."

"Did you have a bad time with the dentist? I forgot to ask."

"Oh, not too bad. It's never very pleasant, is it?"

"No, horrible."

"How were Simon and Susan?"

"Very well. Cynthia sent you the pattern for Susan's dress."

"Oh splendid! I found a delicious piece of stuff to-day—a sort of duck's egg blue."

"How sweet."

"Unless Cynthia would rather keep her in white."

"I don't suppose she minds."

"Wasn't it a lovely afternoon?"

"Yes, wasn't it. Is that the gong?"

"My dear, do hurry! I suppose I'd better fly down. One of us ought to be in time."

"Oh do wait, Muriel. If you're late too, it won't matter so much."

In her old green velvet gown Diana looked like a greasy little middle-aged toad compared to Muriel. She had none of Muriel's gifts for the smaller pleasures of life. That trivial little chirp of conversation just now had been typical of any sort of communication with Muriel. Diana went heavily and as slowly as she dared behind Muriel's fluttering, skipping rush down stairs. They both paused outside the library door and looked at each other like naughty little girls who are late for prayers again. Diana was ashamed of the reality of moments like this. Having been caught by herself and Muriel feeling like a silly little girl, she felt angry and unbalanced and summoning a sort of sulky dignity she walked unto the library as boldly as she could. And at this time Diana was twenty-eight and Muriel thirty-five years of age.

97

But then neither of them was married or ever likely to achieve that state of bliss and dignity. They had no caste. They were the girls at home.

They escaped reproof—beyond that which was conveyed by Lady Charlotte's dignified, "Well, shall we now go in to dinner?"

In to dinner they went, Lady Charlotte on Ambrose's arm, as she did every night if there was no other lady present. Muriel pattering and Diana striding solidly down the hall behind them.

All the leaves had been taken out of the dining-room table and it was now a very small square in the centre of the room. A long way from the dark indifferent pictures of ancestors swallowed in the gloom of red wallpaper. A long, long way from the screened door, through which the butler and his satellite brought course after course of dull food and cleared it away again.

Ambrose was not unnaturally more interested in his food than in his daughters, who could hardly be said to sparkle as they sat one on his either hand.

"What's this, Coulthwaite, what's this?" he would ask impatiently at every course, for he was too blind to read the menu written on a little china slate propped in front of him, and too forgetful to remember it though it had been read to him once by Muriel and once in a loud, defiant voice by Diana.

"Delicious soup, sir," Coulthwaite would say in a reverent and at the same time indulgent voice. He managed to invest the most horrible food with a quality and rarity it never possessed.

"The fish? Grilled salmon, I think, sir. No, sir. I'm mistaken. A nice piece of cod from Cork."

"Spring chicken, sir."

Ambrose chewing away at a fowl that had seen

many an active springtime would have liked to meet somebody's eye and wink, but there was no such possible person, so he felt slightly cross and grieved instead.

"What sort of day did Desmond have? What's this pudding? What is it, Coulthwaite?"

"Some farinaceous substance, sir." So did Coulthwaite elaborately disguise his master's pet aversion, sago.

"Sago? No. No thank you, take it away."

"Just a spoonful, sir—very nourishing. Would you try a spoonful of black currant jelly with it?" Coulthwaite's posture was such that one imagined he only just refrained from patting Ambrose on the back like a coaxing nurse. His stoop was a respectful benediction.

"What sort of day did you say Desmond had?"

"Good, I think."

"Where did he have a hunt from?"

"Oh, I don't know. They just said they had a nice hunt."

"Dear me, why didn't you ask, child? Well, where was the meet?"

"I forget, Father."

He tried Muriel next.

"And who did you see in Barrettstown?"

"No one, Father."

"No one? It must have been pretty deserted. Diana, remind me to tell Desmond there are badgers using that earth in Gibbetsgrove again."

"Yes, Father."

Two plates of shrivelled apples, their varieties labelled, provided a useful subject of conversation when the labels had been read aloud.

"You can't beat Cox's Orange, *I* always say."

99

"I feel sure these labels are muddled," Lady Charlotte announced with decisive asperity. "This apple is certainly off the unnamed russet near the well."

"Yes, I think so too, Mother," said Muriel quickly.

"What do you think, Diana?"

"I haven't had an apple, Father."

"Try a piece of this. I'm sure I'm right. It's a funny thing if I'm mistaken in the flavour of a Cox's Orange."

"I really don't think I want an apple, Father."

"It is too much to ask you to do as your father suggests for once?"

Diana silently cleared her finger-bowl and its lace mat off her dessert plate. Her hands shook and her mouth quivered ominously. It was all too much. Even being bullied about a beastly piece of apple. Had one no right of one's own? Not even the right to refuse to eat a withered quarter of apple?

Across the table over the six tiny packed vases of snowdrops arranged by Coulthwaite Muriel gazed fearfully at Diana, her eyes wide and startled, her hands clinging together on her knees, so that her knuckle bones felt so enormous she thought she would never unclasp them. For Diana was going to cry. Muriel knew it as surely as anything. Knew it in Diana's silence as she took and slowly cut into little pieces her slice of apple, knew it from her head bent as though it were twice its proper weight, knew it again from her long struggling silence.

Lady Charlotte knew it too. She bent forwards, her great bosoms pressing upwards, a deep clefted shadow between them, as she bent, staring down the table, her eyes bright yet calm with satisfaction.

"Well, Diana, and what's your opinion? Is it a Cox or a Russet?" Her voice was almost persuasive as though she wanted to prolong a delicious moment. She peered eagerly in the light of the candles. She shifted one branched candlestick a little as though greedy for more light on this unhappy moment. Muriel drew in her breath quickly, slightly, sharply.

"Well, speak up, Diana? Surely you have some idea. If you were at Rathglass I dare say you could find your voice, eh Miss? But not much left to say when you come home, have you?"

Diana broke down, sobbing into her hands, tears running through her fingers over her plate and apple and silver knife and fork.

Ambrose looked on pityingly, making little ticking noises with his tongue, and Lady Charlotte sat back in her chair, very quiet, in a sort of horrid repose after ecstasy. Muriel went on choking down small pieces of apple and sipping water.

Presently Lady Charlotte said:

"Muriel and I saw the first primroses to-day. Quite extraordinary late—aren't they?"

Diana's dreadful tears were not, because she chose to ignore them. It was entirely typical of her attitude towards her children.

Ambrose was fumbling with the decanter of port, pushing a glassful towards Diana, saying, "There child, that will do you good. Drink it."

Diana shook her head violently. It was a last indignity that all her just grievances should be looked on as no more than a wave of hysteria to be cured by a glass of port. She could not and would not drink it. The glass stood there half-way between Ambrose's place and hers, and he still nodded and murmured and pushed it towards her a little more.

"Well," Lady Charlotte said presently, "when Diana has finished her port, shall we——?"

Muriel jumped up. There was a white sick look about her.

"——I said *when* Diana has finished her port, Muriel."

Muriel sat down again. They waited. Lady Charlotte sat with her hands folded calmly on the table in front of her. Sometimes she looked down at her hands. Then she would raise her head and look down the table towards Diana. Now her face showed nothing but calm impassioned disapproval, a kind of patient resignation.

Poor Ambrose who had precipitated this new crisis sat like an old mouse now. They waited for more than five minutes. Then Diana put out her hand, picked up the wineglass and drank her port at last, choking it down in three quick gulps. She was not crying any more but she looked rather wild and strained with her face blotched blue and white. It seemed unnatural too to see such a square little person trembling so violently.

Lady Charlotte, still watching, drew her breath regretfully in and out through her teeth.

XIV

It was November 1915 and the War had been going on for a year and four months, when Cynthia heard the news that Desmond had been killed. She had never really expected him not to be killed. She had felt quite fatally certain about it from the first. Before he went to France and during his short leaves she had been infinitely various in love and wildly living in the moment. They had each known quite well and each tried quite futilely to hide from the other the foreknowledge that this was going to happen.

It must all have been very much as it would be if it happened again now. All the patriotism and memorial services and kind letters from brother Officers which comforted mothers and sisters, can't have been of the smallest value to lovers. They had endured those horrible brave good-byes, and those other moments when neither could find the other and the time was all too short, flying along. There were moments when Cynthia could hardly endure Desmond's spending so many of her brief hours with the children. It was because there was no strain with them, they were entirely uninvolved and untouched by all the pain. They were calm.

Can one know about things like this? What can one think of the deaths of silent young men like Desmond who are as much part of the place they live in as any field is such a part.

Cynthia had made herself hard over this so often that when she read the telegram and said, "There's no answer," she only felt a sort of hot dizziness.

She had been rushing about after an early breadfast because it was a hunting morning and she had a quantity of things to see to before she rode on to the meet with the hounds. Her mind was on this level and it seemed stuck there, buzzing like a small overthrown machine. Part of her felt furiously angry. She went from one window to another. She could not keep still. Walking about in this most familiar room was like walking about in a photograph. Everything was only like a picture of a thing. Nothing had solidity or four sides or reason any more. Out of a window she saw Susan going out with that untidy and tiresome Belgian refugee governess to feed her rabbits. She must go; she must certainly go. Quite useless.

Then she remembered what had been annoying her so much, what she was looking for before the wire came—the list of groceries that must be sent off by to-day's post. Then she would be ready. She was still searching for it blindly, turning over envelopes on her writing table, when she heard the Garonlea car coming up the road.

"I can't stand this," Cynthia thought in a crazed panic. She flew out through the hall, collected her gloves and whip and ran as fast as she could in her hunting boots down the kitchen passages and out through the back door.

She was determined to go out hunting. She was not thinking, as she easily and truly might have done: Desmond would have hated nothing more than for people to come to the meet and find no hounds there because of him. She only thought that the important and sensible thing to do was to hunt.

She wondered if Lady Charlotte would hear the hounds clamour as they came out of the kennels and down the slope of field to where her horse and the huntsman's horse stood together in this strange day from nowhere. But the high toppling looking car did not appear. They rode on unmolested. They talked about the business of the day and the hounds.

There was old Tarquin, he should be knocked by rights, but the master was so fond of him perhaps they'd better leave him till he came back.

"No, you may as well put him down," Cynthia said absently.

"Well, he only meant to keep him up to the end of cub-hunting really, didn't he?"

"Yes. That's all."

At a wide cross-roads Lady Charlotte overtook them. Cynthia always remembered what it looked like: the flatness of a pond, the common ducks that splashed shrieking across it away from the hounds. The three dirty cottages that the hounds would have investigated thoroughly for twopence. The wet November leaves in the road and a full ditch of brown watercress. Lady Charlotte sitting in the back of her car like a sad and outraged toad—Lady Charlotte climbing out with her chauffeur's help.

"Cynthia, my poor child, you haven't heard——?"

"Yes, I've heard."

Cynthia seemed a thousand miles away from Lady Charlotte, sitting up there on Desmond's horse among Desmond's hounds. She called out the name of one of them that was looking towards the cottage door, in a harsh important voice.

The whole group, Cynthia, Richit, the bright horses and hounds, was outside Lady Charlotte's power or keeping. The whole drama of sorrowing mother comforting the tearless widow was taken from her by this strong, dangerous creature who said:

"I'm afraid we must be getting on now; we're a bit late," and looked at her watch.

Lady Charlotte was very pale, her fat hands in black gloves made futile extravagant gestures.

"You can't do this, Cynthia," she said. The hounds fussing round her and Cynthia fidgeting to be off made her furious and hasty: "It's indecent—it's an insult. To Desmond and to Garonlea and to us all. Wait. Listen——" She was distracted and angry.

Cynthia looked down as if she would like to spit on Lady Charlotte from all that height above her. She gathered up her reins and nodded to Richit.

"Perhaps I'm the best judge of how Desmond would feel about hunting," she said. They rode away and Lady Charlotte turned blindly round in the road as though she could not see where the car was. The chauffeur put her in and shut the door on her, when he had tucked in the blue fur-lined rug.

The powerful excitement of her meeting with Cynthia was over. It had buoyed her spirit up for hours. Now it was over. Her hate for Cynthia and her bitter sorrow for Desmond were of equal and cruel importance. As she was driven slowly back to Garonlea she mumbled to herself plans of revenge and broken troubled words.

"Oh, Madam," Richit said, he was in tears, "I'm that sorry, Madam. Oh, excuse me, but no one will want to hunt to-day."

"I want to hunt, Richit," Cynthia said. She rode on in front of the hounds that went so gaily and proudly down the road round Richit's horse. The day was still like a very glazed photograph to her. Nothing real had happened in it yet.

Every one at the meet—where the news had travelled in the peculiar way it does—was deeply shocked and embarrassed by Cynthia's appearance with the hounds. This quite truly surprised her when she took it in. She said to a young soldier, whose last day's leave it was:

"I'd have hated for you to be done out of a day's hunting, Johnny," and this soon got about and rehabilitated her in that atmosphere of starred romance. Even Johnny, who was rather nervous and unenthusiastic about the chase, felt hallowed and important.

Saying things like that did not disgust or annoy Cynthia at all. She had hardly said them before she believed them.

It is difficult for people who don't bother about hunting to make any sense of Cynthia's strong desire and insistence on hunting this day and causing such unnecessary distress and embarrassment. There are people, and Cynthia was one of them, to whom the mental and physical excitement of hunting are both a religion and a drug. At the moment of happening, at the moment of danger, at the moment of decision or fear, and in waiting for these moments there is not time to suffer or know about one's other life. The brain is tremendously independent of the body in the excitement of hunting, and that excitement fulfilled, the body is curiously independent of the brain. One is filled with either excitement, fear, or content; never, as it were, in slack water.

Then there was the semi-fanatic "must hunt" attitude which is often stronger in women than in men. Cynthia had it very strongly developed, especially since Desmond had gone to the war, and she had done so much of the work for this pleasure, it was a flame that was part of herself.

It is difficult to understand why people with enough money and leisure to lie in the sun for the winter

107

months should prefer to hunt, but it has to be accepted that thousands of them do, and with less reason and passion for the sport than Cynthia possessed. To-day and during the months that followed she hunted because she loved it as well as for the sake of oblivion. For hours she found this, though not always exhaustion enough to compel sleep because her body was particularly strong.

To-day she did not once feel entirely conscious about Desmond because mercifully they were hunting nearly all day, and when she and Richit rode home with the hounds it was quite late in the November evening. She was tired with that complete weariness which sets in on a long ride home after hunting. Poor Richit was far more aware than Cynthia was of the desolate and dramatic sight they presented when they said Good-night to the small field and rode away together with the hounds. It was the sort of emotional drama that Cynthia either consciously or unconsciously was always staging through her life. Desmond and Cynthia had always been alone and at a distance together. Now she was alone without him and his hounds and his servant were dearer and of more importance to her than any person.

X V

THE evening was quite dark when they got back to the kennels. A quarter of a mile down the road Richit blew his horn in warning of their coming and for some reason the sound of his horn in the darkness seemed to catch sadly at the still air. Sound shapes like wreaths for coffins. All that had been, all that would never be again, made Cynthia bow her head and her cold face nearer her tired horse's neck for comfort. But there was no comfort, no help in being near anything or any one forever again.

From the kennels to the yard she rode by herself, under the beeches, up the narrow back avenue. The air was very soft and full under the trees and there was with her so true a ghost that she half-turned in her saddle to speak, as she had so often done at this place and time before. But she was by herself, whispering alone to prove her love.

Diana was waiting for her when she came in.

Cynthia slipped her arm into Diana's and went over to the table to look at the letters lying under the lamp. "You'll stay with me, won't you?" She was sorting the letters, putting Desmond's in one heap and her own in another and suddenly stopping and putting them all together. "A bath, I think, and then tea—— Come and help me."

On their way they looked into the schoolroom where Mademoiselle sprang off her chair and seized both Cynthia's hands with unspeakable looks of sympathy, and where Susan dropped the toffee she had been busy making into the fire and said with becoming drama:

"Oh, Mummy, it's not true, is it?"

"Who told you, darling?"

"Granny."

Cynthia's first feeling of relief that she did not have to do the telling changed to one of perhaps justifiable annoyance at Lady Charlotte's action. Even now she did not know what to say. There ought to be a formula. She never knew what to say to these children. Was it fair that they should mean so little to her? She stood there tired and stiff in her habit and said nothing except:

"I'm sorry, darling. I am so sorry," behaving quite truly as though Desmond belonged more to Susan than he did to her. Susan looked terribly embarrassed. Inside she was saying:

"Daddy's dead, it's so awful, Daddy's dead," and trying very hard not to laugh. But the desire to laugh was uppermost. It had been all she could do not to laugh when Granny sent for her to the drawing-room this morning and told her. However, she had enjoyed the pomp of the day, with Mademoiselle allowing her to do just as she liked. No French. She needn't ride her pony in the afternoon—indeed no, your grandmother would enrage herself. They did history instead of geography and had a wonderful tea. She was indulged with a gloomy importance that pleased her very much. And now how quickly it all dropped into a deadly dreary reality with Mummy saying, "I'm sorry." A feeling of suffocating desolation like the first morning after the holidays, with Simon really gone back to school. The beginning of a term was as long as forever. Ah, now she saw what it would be. At last she had a comparison. Tears slowly stiffened her throat and the wish to laugh was over.

Cynthia in her room (which had been changed from pink and silver to cream and delphinium blue) sat down by the fire and leant her face against her hand. She looked defeated but very alive behind all the shock and fatigue.

"Do you suppose your mother sent a wire to Simon's school too?" she asked.

Diana said, "Probably," and gave her a drink that she had brought up from the dining-room and went off to turn on the bath.

"I don't think I'll have a bath after all," Cynthia called out suddenly, "I haven't fed the dogs and I forgot to tell them something in the yard." She went off again in a determined sort of way, flying about the house, feeding all the dogs, out to the yard and back to the house again, still in her habit and stiff dirty boots.

Diana stayed with her, trotting about silently. Holding this. Fetching that. She did not talk. She was like a cosy little black bitch, silent and loving. She said nothing about going back to Garonlea and it was not till bedtime that Cynthia said:

"Gracious, darling, are you staying the night? I forgot to tell them to get a room ready for you. Or perhaps you could sleep in my bed and I'll sleep in Desmond's. I would like that, I think, it's lonely enough when he's just away." She was sitting at her writing table getting through a pile of hunt correspondence, claims and stopping cards and lists of meets.

At about eleven o'clock the Garonlea car appeared to take Diana back, but Cynthia sent a message to say that she was staying the night. Cynthia sent the message out in a quite unconcerned way, but she said to Diana, not with displeasure:

"There is sure to be trouble at your staying—isn't there?"

"Hardly. I mean I suppose there is."

Cynthia said, "Well, they may as well face the fact that you're going to live with me now, mayn't they?" She did not even pause in writing out the card she was doing as she said this, glancing from the book of addresses back to the card under her hand.

Diana sitting by the fire felt a constriction of happiness that almost hurt her heart before it let it go in pure delight. Yet when you have what is in your dream, you have no dream. Your dream is in your poor hands to bruise in ignorance and lose in confusion. It is the moment before the moment that really counts whether the matter is love or hanging.

Cynthia would not go to bed till very late. She thought of a million things to do and when they were done and she couldn't think of anything else she turned to Diana her face of shocking despair—despair beyond exhaustion and all spending—and said they had better go to bed.

Diana was so tired that her sadness and her delight had both fallen long ago into a complete gulf of fatigue, but each time in the night that she woke she knew Cynthia had not gone to sleep yet.

Simon in his bed at school did not sleep either for what seemed to him a long time. He kept turning and turning and what he kept thinking was, "Did Daddy tell her I needn't ride Mitty at Christmas. He said to me I needn't. He knew I couldn't stop her. Oh, I wonder if he told her." In his mind he could see Mitty's piggish pony's eye. Mitty knew who was boss. He could feel her now taking all his power away, making him feel sick and despairing, and Mummy's voice saying to some one, "She does rather take charge, but she's such a perfect jumper," he had heard her say that one day when all his secret wish had been to see Mitty boiled up for the hounds and know that he would never have to ride her again. Now all his thoughts about Desmond were focused and centred in this. He knew he would never be able to tell Cynthia that he was afraid to ride Mitty. Had he remembered to tell her? If he had forgotten now there was nobody left to trust. Next day poor Simon looked pinched and miserable, they all thought the little boy was taking it terribly to heart, but really he was thinking more about Mitty than Desmond. And about Desmond only that there was no one you could depend on now.

XVI

Soon Lady Charlotte had everything she could possibly require to complain about. There was no horizon or limit to Cynthia's monstrous and cruel behaviour towards them all and towards what she insisted on calling "Desmond's memory." There was simply nothing she did not do, from making them change the day of Desmond's Memorial Service so that it did not clash with hunting, to selling two of his horses for troopers. Lady Charlotte discussed her actions with the sort of fervour that belonged to religion in the days that religious people burnt one another. And she was always hearing of some new thing that Cynthia had done. How she had locked away or destroyed every possible thing at Rathglass that was his, sold some of his saddlery and given even his clothes away to any of his friends or her friends who could wear them. Always she came back to the question of the Memorial Service and Cynthia's rage that they had chosen a Wednesday. How she had flown into a temper and insisted on the day being changed because she would not go unless they changed it. How all the farmers and horse copers and vagabonds in the countryside came to Rathglass to sympathise and would stay there half the day or half the night drinking. One night Lady Charlotte heard they had played cards. Again she heard of a party. And this was not a month after Desmond's death. She sent for Diana; if Cynthia could entertain the countryside she did not want company. But Diana made excuses and would not come back. Lady Charlotte did a thing she had forsworn. She went to Rathglass once more, grieved and bitter beyond measure, and demanded to see her daughter. They said she was out. Lady Charlotte said she would wait. But she waited in vain, for Cynthia had taken the precaution—knowing Diana's vulnerable nature—of locking her up in the straw-house. Lady Charlotte had to drive away without seeing Diana and next morning there was a letter for Ambrose saying that Diana thought she would stay with Cynthia indefinitely as she was needed far more at Rathglass than at home. It was quite a reasonable letter and quite a reasonable idea. The consequent ferment that boiled up at Garonlea was out of all sane proportion. Lady Charlotte made poor old Ambrose (not herself) quite ill with her emotional and half religious attitude towards the matter. Now she would mourn for a lost child—two lost children, in fact—then she would abuse Cynthia by the hour with a sort of false justice and acute perception of all her worst side.

Ambrose, who was brokenhearted by Desmond's death and devoted to Cynthia and her children, had a miserable time of it. His chief pleasure in life was taken from him—there was nothing he enjoyed more than toddling round Rathglass with Cynthia, seeing all the horses and the hounds, advising her about shrubs, and doing a bit of pruning. He got the same sort of life at Rathglass as any of Cynthia's girl-friends. And she was so sweet to him.

"You must come and look at the Teneriffe colt—you were right about him and I was wrong."

"Look at that shrub you cut back so savagely—isn't it good? And I thought you'd certainly killed it."

She showed genius in treating him as a contemporary. She never compelled him to see his grandchildren or expected him to like them. Simon and Susan liked him. He had not much more to say to them than he had had to say to his own children, but he never put his thumb and forefinger into his waistcoat pocket without bringing out sixpence or a cough lozenge, and now and then—Oh unbearable suspense, will it be now!—what he called half a sovereign and they called ten bob.

Much as he loved Garonlea it was a sort of traditional affection, and his passionate interest in Rathglass was that of an older woman for a young lover. It was as alive as that. It was no wonder that he suffered and failed in health after Desmond's death and this now open hostility with Cynthia. He gave in to Lady Charlotte over it all, of course, he was not strong enough in mind or body for any sort of contest. But he changed his will a little, leaving Diana more money than she would have had in her mother's lifetime. During the winter what sap there was in his life seemed to run out. They said his arteries were hard and his heart was tired. They said he must have no worries. One faintly warm day he pottered out to slash at his young enemies the elders; he had been forbidden to grub, but he found one so lightly rooted in a sandy bank that he believed he could manage to pull it out. They found him dead in the shelter and sun at the edge of his wood. He had defeated the young brute of an elder first and settled himself quite composedly against the bank.

Cynthia's hounds did not hunt till the funeral was over.

Lady Charlotte, who had never given more thought to her own death than was warranted by her assurance in perpetual and respectable comfort hereafter, now began to dread it very much indeed. And her fear and angry resistance were due to her horror of the idea of Cynthia succeeding her as mistress of Garonlea.

The thought upset her so much that she would tremble and feel sick when it was in her mind. She was convinced that Cynthia was only waiting and watching for the moment when she should step in to Garonlea and overset the thoughtful ordered train of its present state, outcome of so many years of economy and right living and dignified hospitality.

She felt that Ambrose had behaved a little traitorously in pulling up that elder and dying without an effort under his bank of celandines. If she had been there she would not have let him die. And she found it hard to forgive his ample provision for Diana. It was a cowardly undermining of a possible time when the poor child might have been brought back to reason and Garonlea again . . . with Cynthia married perhaps . . . if she had had no money or less money, Diana would have been forced back to Garonlea. Lady Charlotte thought that Ambrose should at least have discussed the possibility with her.

The first result of her fear of death was the quite practical plan of going to bed if her finger ached or if she sneezed. This was hard on Muriel who grew thinner and more twittering than ever as she ran in fifty directions to do the bidding of the muffled old queen in her bed upstairs. There were all sorts of wartime economies that required perpetual supervision. Though there was abundance of food of every sort in Ireland, patriotism gave Lady Charlotte a fine opportunity for meanness in many directions. She could hardly trust the cook to divide the tea into small enough packages or to issue bacon and butter to the household in strict enough portions. Lady Charlotte had made a great many discoveries about food which she tried hard to convince every one were particularly delicious. Barley meal and saccharine and margarine. There was no end to the substitutes she found for food which was really there in profusion. Cutting down expenses became her greatest interest, justified by the war and by her determination to save all the Garonlea money she could for Muriel and Enid. She was to live at Garonlea till her death, or till Simon came of age, and somehow Simon's coming of age seemed to her much nearer than her own death. To her, more than to most old women, her age was simply an illusion of other people's. She could not let go in any direction.

Muriel with all her birdlike busyness and sweet temper found the days at Garonlea very long in the year that followed Desmond's death.

Breakfast by herself in the schoolroom—for the dining-room and drawing-room were both shut up. Perhaps an egg, perhaps not, if Lady Charlotte wished to send four dozen into the town instead of three dozen and eleven. Barley scones and margarine and jam. The room was heated by a black stove that burnt sawdust, but the heat it gave out was so negligible as to be almost an insult.

After breakfast Muriel would go briskly into her mother's room which was even colder than the schoolroom, but Lady Charlotte, with a white Shetland shawl over her head and two hot-water bottles early provided by Spiller, did not feel the cold.

"Good-morning, Muriel dear," she mumbled, her lips for a moment near Muriel's pale cheek, and her hands in white mittens rustled the letters that the post had brought her.

"We seem to have a busy day in front of us. Quite a busy day. My tonic, dear, on the wash-hand stand. Yes, Muriel, it must be there. No, not that bottle, It's the new tonic."

"Won't Spiller know where it is, Mother?"

"Perhaps, dear, but Spiller has gone down to her breakfast and I want to take this immediately after food. Perhaps the parcel from the chemist is on my writing table in the library. Just run and see."

Muriel ran obediently down the passages and the long stairs and the hall and back again, shaking her head.

"I have a letter I want you to answer—it's all right, dear, Spiller had left it on my tray. The letter is from Mrs. Barclay saying she would like to come to tea on Thursday. Say I should be delighted. Then there's a letter from Enid——"

"Oh, from Enid?"

"Yes, to say the children have had whooping cough and may she bring them here for a change."

"Oh, Mother, how lovely!"

"Really, Muriel, I hardly think we can manage it. We have a minimum of servants in the house—Enid seems to imagine that there is unlimited food and luxury here. She forgets that we do our little bit to help to win the war too, and I really can't have the heater for the bath-water going any oftener than I do."

"Oh Mother, do let them come. The poor little things, they do love being at Garonlea."

"Trust me to know what is best for us all, Muriel. Then, will you write to Roche and say only seven of the egg boxes were returned and what has he done with the other one. I expect a brisk walk would do you good, so trot off to the church and collect the holders for Sunday's wreaths. And walk, Muriel, remember, don't take your bicycle."

They had luncheon together, a carefully served and rather nasty meal.

Muriel had a little news:

"I met young Byrne and he told me they had a good hunt on Wednesday and Diana had a nasty fall."

"Hurt?"

"No."

"Oh."

"He said poor Simon cried at the meet and Cynthia shouldn't let him ride the pony."

"My dear, should you discuss Cynthia with young Byrne?"

"I didn't. He just told me."

"I pity those poor children. Did he tell you anything else?"

"No. I think that was all."

After luncheon they might drive round the place in a round pony trap, Lady Charlotte delivering herself of admonishments and charity. She was extremely popular with all the people in the place and never spared time or trouble to help them. Then back to tea and the long hours before and after dinner when Muriel knitted socks and read the war news aloud. Till the war is over life will be like this. Muriel looked towards the end of the war as a sort of term. When it was over she did not quite know what she would do, but she hoped vaguely that life might be different.

News from Rathglass came through Spiller as often as not, for she had struck up a friendship with the Belgian Mademoiselle who had somehow survived at Rathglass. Spiller would tell Muriel who listened with thirsty excitement, and sometimes she would impart little pieces of information to her Ladyship, who longed to hear more but was by principle entirely cut off from listening to any servants' gossip. Principle, however, did not go so far as forbidding Muriel to hear all that Spiller could tell.

"Oh, Miss Muriel, there's no margarine in the

servants' hall at Rathglass, heaps of butter and bacon and Mrs. McGrath doesn't keep them to the rations at all—— Mademoiselle says it's terrible the way those children are made to ride their ponies every day and the poor little things would rather be reading books or off on their bicycles some place——

"Miss Diana is forever working, work! she never stops, Mademoiselle says. Exercising the horses and digging in the garden, yes, and feeding the chickens and pigs——

"Oh, Miss Muriel, such a thing Mademoiselle told me to-day. Mrs. McGrath has bought Miss Diana a pair of breeches and boots like one of these Land Girls in the *Daily Mail*. Whatever would her Ladyship say if she could see her? And Mademoiselle says she hardly ever takes them off but goes striding about in them like a real little man. Mademoiselle says she doesn't think it's suitable at all.

"There was a party at Rathglass last night, Miss Muriel. Fancy, Miss Muriel, and poor Mr. Desmond as you might say—— All out hunting all day and back they came without a word to the cook at seven o'clock in the evening. Six hot baths and bacon and eggs and champagne and the gramophone playing till three o'clock in the morning. Three gentlemen home on leave it was, and two young ladies. And at three o'clock, Miss, out they went hunting rats in the stable yard. Yes, and Mrs. McGrath in a beautiful black satin dress with a sequin tunic and the fish-tail skirt and her mink—imagine that gorgeous coat—thrown over that and out with lanterns and torches through all the dirt.

"Well, I did hear a thing to-day you won't believe. Miss Diana has bobbed her hair. Yes, really, and Mademoiselle says she looks ten years younger. She saw her the other night—don't pass this on to her Ladyship for I think she'd have a stroke, if you'll pardon the expression—in a pair of blue silk pyjamas and her bobbed hair and she smoking a cigarette and drinking a glass of whisky (and now Confidence, Miss Muriel), down in the drawing-room in that turn-out talking to a gentleman. Yes, fancy, imagine. And Mademoiselle said it gave her quite a nasty feeling and at the same time she said Miss Diana looked charming you wouldn't believe.

"And they had poor Miss Mousie McGrath up there one night and giving her champagne and a kidney omelette till she went on quite undignified. Yes, Mrs. McGrath wouldn't let her go home though it's only down the road as you know, but a bed had to be made up for her and a fire lit at ten o'clock at night in the blue bedroom, and Mrs. McGrath put her to bed herself in one of her own pink satin nighties and a pink Shetland coat. Really, I think it would have looked sort of pagan on an old lady.

"And hunting, out hunting every day in the week and Bob Richit up in the drawing-room after tea and a big glass of whisky for him and talking of the hounds and the horses and the gallops they had till near dinner-time. It's a wonder, Mademoiselle says, sometimes she wouldn't keep him to dinner. She'd sooner talk to him than to any of the nice young gentlemen she has there."

Spiller's accounts of life at Rathglass were fairly truthful. All the events she described with delighted horror and popping eyes really happened but with less forced and pagan joviality than appeared in her accounts of them. At the same time, Life at Rathglass was—in times of sore tragedy as well as in times of happiness—a terrifying contrast to life at Garonlea.

XVII

SINCE the news had come of Desmond's being killed, Cynthia was like a person wilfully keeping herself half-anæsthetised. The final reality of being without him was more than she could endure. Also she was a person to whom the present was of more importance than what had been or what would be. If she could fill the present moment so that she need not look before or behind it, she found that she had some ease and quietness of mind. Hunting she thought was best, but what really made her nearest to forgetting was her perpetual and indefeatable success with the men. It did not do her any good but it seemed to stop that gap of loneliness sometimes and it gave her a sort of wild satisfaction too, continually denying the love she could have had. In every possible way that was hard and stupid and unkind she laid Desmond's ghost, but though she denied him, trying to save herself from pain, he was still so real to her that she dared not think of him. She was really afraid, in the strength of so much sorrow and took any means to put it away from her and exhaust her mind and body towards forgetting.

Life without Desmond seemed to kill all the warmth and kindness that had ever been in her and leave bare a sharp wolfish twist that looked for satisfaction in different ways. And found it. She found it in exciting men about her and denying them love. In their need and unhappiness she felt less alone. She found it in a sort of wild, hard gaiety, in those silly parties that Mademoiselle chronicled so faithfully to Spiller. The only people to whom she was a little kind were her dogs and Diana. The amazing change in Diana amused her a lot and gave her a fine sense of victory over Lady Charlotte. She was never so pleased as when, having persuaded Diana to ride astride, as her side-saddle was always giving sore backs, they met Lady Charlotte and Muriel driving the fat pony in a narrow lane-way and Lady Charlotte was compelled to maintain a glacial front for nearly five minutes while Cynthia pretended that she could not persuade her horse to pass them. This gave her the keenest amusement and even Diana was able to treat it with a sort of swagger. Such was her escape from the old power at Garonlea, that she did not even feel sympathy for what this meeting meant to Muriel—Muriel beating the fat pony, blushing up to her eyes and bravely saying, "Hullo, Cynthia! Hullo, Diana!" Diana had nearly forgotten what this must cost in effort at the time and in trouble with Lady Charlotte afterwards.

At this time Diana was deeply and entirely happy. She was not tiresome or jealous in her love for Cynthia. She worked for her because she liked working—not in romantical slavishness. All the new excitements like cutting off your hair, and drinking whisky, and dancing to the gramophone (she and Cynthia would practise tricky steps together because dancing well really did matter in those late war years and after) satisfied her desire for pleasure and excitement, even her sense of being entirely daring was gratified by the constant thought of that tremendous shadow of displeasure across the valley.

The knowledge that she was at last of real value to Cynthia made her in a dim, unknowing way, grateful for Cynthia's wild and driven unhappiness. Although she could not by any standards measure the loss, she knew something of the despair that Cynthia fled in that grasping at life which so scandalised Garonlea.

She saw another thing too, which she put out of her mind because it frightened her a little, and that was the vein of cruelty in Cynthia's relationship with Simon and Susan. She called it by other names. She called it "Cynthia making Simon hardy." She said Cynthia was so wonderful about not spoiling Susan. How good it was for Susan to be teased. How well Simon rode and how Cynthia would stand no nonsense from him about not liking his pony, or from Susan either for that matter. Let them fall off, she always said, that's the way they'll learn that falls don't hurt. They must shoot and fish, too, and swim and play tennis seriously in between. Cynthia was wonderful about Simon's holidays. She never let anything interfere with hunting, or riding in children's classes at shows, or with his shooting or fishing or any other manly sport in which she wished him to be proficient.

Those were all the things Diana said to herself and to other people about Cynthia's ways with Simon and Susan, but secretly she could not help a creeping feeling of dismay and fear at the hardness of Cynthia's ways with them.

Since those first holidays after Desmond was killed, when Cynthia had passed over Simon's one small, desperate effort about not riding Mitty, he had gone uncomplainingly through the sort of martyrdom nervous children endure through being compelled to ride their ponies and take part in the chase.

Susan, who rode much better than Simon, disliked hunting very much, too. It frightened and bored her, even if she didn't have quite so many distressing falls. When they were older and discussed the matter without shame or passion, they could not think why they had not admitted their fears to one another at the time. Those evenings in the schoolroom, with the meet at

Drumbeg next day, and the first fence away from the first covert a stone wall with a dirty, unseating drop the other side, were quick with unadmitted cowardice.

"Ah, Simon, do stop playing scales; I can't read my book beyond it," Susan was a pale, overgrown creature with long arms sleeved in blue velvet. She had a face short and blunt as a cat's, but more irregular, and shy, stripey eyes.

Simon stopped obligingly. His heart was not in his playing to-night. He came to lean over her shoulders at the scarred table.

"You aren't reading, Sue, you were at that page an hour ago." He moved away to the window, an elegant, soft-looking creature. He opened it, leaning out: "Still freezing. Hope it won't be too hard to hunt to-morrow."

"Hope not. But it won't be."

"No, it's only the second night's frost." Simon's voice was thin with pretence. He dropped the curtains and went fiddling over to the fireplace.

"If it *was* too hard to hunt to-morrow," he said, "what would you like best to do?"

Susan said, "I'd rather hunt than anything."

"Yes, of course. But if we really couldn't."

The pupils of Susan's eyes expanded as they always did when she imagined hard, growing so dark that they spread across her eyes. "I'd like to stay in bed with my favourite book and my favourite rabbit and have all my favourite food for breakfast and lunch and tea."

The contrast between such a day of luxurious idleness and the stern reality of the real to-morrow was almost too much for Simon.

He said, angrily, "I hope that beastly Dolly Bryant isn't out. She always shoves me out of my place, and Mummy always seems to know when I'm last in the hunt."

"Last in the hunt"—that shameful position.

"How does she always *know*?" They were a little closer to truth and sympathy.

"And she never even noticed when I jumped that high wall on Wednesday behind Richit."

"I saw you. I jumped it on your left as a matter of fact." Susan was a little grand. Then she caught Simon's eye and added, because obviously he knew, "where some one had knocked a big hole."

Susan was a shade more sophisticated about being cowardly than Simon. Once, in her very young days— before Cynthia had given up persuasion and taken to more arbitrary methods—she had wasted about twenty minutes trying to induce her daughter and her daughter's refusing pony to leap some bushes in a gap.

"Look at that little boy, wouldn't you like to ride like him?" She indicated a tough child who drove his pony at the place with commendable bravery and energy. Brave and enviable as he was, he unhappily flew off as his pony leaped, landing first in the next field, much more hurt and frightened than he was allowed to admit. "You see," Susan said, "that's the sort of thing that puts me off huntin'. . . ."

Of course, it was not too hard to hunt next day. Just not too hard. They went into Cynthia's room after breakfast to have their stocks tied. Their noses were pink, like half-ripe strawberries in their pale faces, their stomachs were queasy and unsteady and not much breakfast in them.

"Turn round, darling. Safety-pin? Safety-pin? Turn round. Too tight? Never mind, it's very tidy. Now, Simon. It's time you learnt to do this for yourself. You might try on Friday."

"Am I hunting on Friday?" Simon's face was

faintly greener. He hooked at the tight white-linen stock with his finger.

"Yes. Now I've let the cat out. It was to be a surprise. I've got Myer's grey cob for you again. It's too far to send you on your pony."

"Yes, that's what I thought," said Simon faintly, "thank you so much."

"Poor old Sue will have to go round with the terriers."

"Or couldn't I ride the grey cob half the day?" Susan asked, impelled by the force of despair that she knew was loosed in Simon. He was the only person who hated Myer's hot, grey cob (such a perfect jumper if a bit quick on her fences), more than Susan did.

"Well, we'll see, darling. Run along now and let me get dressed. And Simon, for heaven's sake fasten your leggings inside your shin bone, not round the back of your calf."

Nearly an hour with nothing to do. An hour feeling quite cold with rather a nasty taste in your mouth before Mummy came down, Mummy in her iron-dark habit and little tight pieces of hair like flat, netted coins under her velvet cap, Mummy, so brisk and hard and clean, picking up her white gloves and her whip in the hall and shouting for the terrier that was to go in the car with them, on the edge of being very angry if he had gone off rabbiting. But Diana had shut him in the car already so that he couldn't. They were glad she was hunting to-day. She often helped them when they were in trouble and never told if she saw them avoiding a stone wall.

Simon ran stiffly across the gravel, its little knobs frozen solid under his feet, to open the gate with a wreath of ice on its bars and latch. Crouching under rugs in the back of the car, he shared with Susan a period that was blank except for that slight, persistent feeling of nausea. They did not seem to belong to their own bodies at all. Those large hands in string gloves were strange people's hands. They crossed and uncrossed their legs clumsily as if they were wooden legs. They did not mind about each other, each feeling that the other was better equipped to contend with its beastly pony. Susan remembered that Simon's Dinty was much easier to sit on over a wall than her own Mouse, and Simon remembered that Susan was much more adhesive over any sort of fly fence than he was. Each felt that the other was less to be pitied and each longed for the end of the day with an ardour which saw such ease a month away at least.

They looked at the fields and the fences as they drove past them, seeing them with a curious relationship of fear and not fear. The fields were not ordinary fields as in summer. They were places you had to get out of, that you were inexorably carried over. That field now, with green plover waddling and pecking about on its dark, sheep-bitten turf, was a dreadfully unkind field with wire on its nice round banks and as cruel a coped stone wall across one side of it as a frightened child could face. The very young may be sick with cowardice about a fence but they are more afraid not to jump it than to jump it. Then there might be a wide, benignant field with an open gateway in a corner, or a ditch full of dark, cold water, but only a low bank in front of it, very heartening. It was not soft falls in water or bog that they dreaded, but that shameful, hurting, falling off and the moments before you fell, their agony seeming to endure in interminable uncertainty before you went with a sort of sob and the ground hit you from behind, strangely like a house falling on you, not you falling on a house.

XVIII

THEIR ponies were waiting for them at the meet,
looking hard and unkind and as if they would gallop
and jump for weeks without ever being tired or ceasing
to pull like bulls in those round, smooth snaffles.
Cynthia liked the children's ponies to look fit and well
and they were beautifully turned out, their tails pulled
to the last hair and their manes plaited up. They eyed
them for a moment with hatred from the car before
they remembered to get out and fuss over them as
keen, sporting children should.

"Thought we should hardly hunt coming along,"
Richit said; his face looked red and blue and his eyes
were as pale as a jackdaw's under the peak of his cap.
You could not possibly have mistaken his mouth for
the mouth of any one but a hunt servant.

"We'll hunt, all right," Cynthia stood among the
hounds, digging the heel of her boot into the grass at
the side of the road. The smell of the hounds' bodies
was like a warm, low cloud round her. She looked at
the list of hounds Richit handed to her, and when he
said, "The north side of the fences will be like iron,"
she did not answer.

"I shall have some lame hounds coming home
to-night," he said after a pause. The fact was, Ricket
was scarcely more anxious to hunt to-day than Simon
and Susan. He hated his first horse and his piles had
been giving him great trouble. Cynthia guessed about
the horse though not about the piles. She went over
to the public-house at the cross-roads and sent him out

a double Irish whisky. She and Diana drank a glass of disgusting port each.

"Isn't it filthy?" Cynthia said, "I suppose I'd better have another."

The children visited a field before they mounted their ponies. When they had mounted they wished to visit the field again but hardly liked to. Their insides were most unsteady. Presently they were on the steep hill beside the covert, with that wall standing up very dark and tall at the foot of the hill. The covert was on the left-hand side of the field. They could only see the very edge of it. There was a steep lime-kiln, grey and white like the sky, and the gorse bushes appeared to be navy blue in the cold morning.

Their ponies fiddled about and tore up mouthfuls of grass and smelt other horses and screamed and then stood very still, galvanically still, with their heads up, gazing into the blue gorse bushes as though longing to see a fox. Once a hound came out under Dinty's tail and was kicked. Cynthia was on the other side of the covert and they hoped she did not hear the hideous noise it made. They were supposed to know most of the hounds and they wished they could remember the name of this one but they could not. A man they did not know said, "Do the brute good." He looked at them in a friendly way and said, "She's all right, you hardly touched her. Cowardly, that's what it is. So am I." They found this rather an embarrassing remark so did not answer. They did not have time really, for at that moment one of those hunting cries that are supposed to excite hounds and horsemen to a perfect fever pitch of determination and endeavour split the cold air with awful clarity. They gathered up their reins and looked wildly down towards the stone wall.

Their ponies stood like trembling rocks about to hurl themselves over the edge of an abyss.

Then the strange man who had been looking up over the hill did a queer thing. He turned his horse round and galloped up the hill and away from the wall. Simon and Susan looked at each other. He couldn't be right. Foxes always went away at the bottom of the covert. The field always had to jump the wall. But still they whirled their ponies round and hurried up the hill after the cowardly stranger. He had jumped a little bank into the top corner of the covert, squeezed through a gorse-filled corner and was out in the field beyond. Screaming to each other to get out of the way, they followed.

From your point of view when you are a child a hunt is purely a question of obstacles and not falling off. It is a good hunt for Irish children if there are no high walls or pieces of timber and if you are not quite last in the chase, and if straggling hounds keep out of your pony's way, because in any case you can't stop. Susan and Simon were often borne along for miles with tears pouring down their faces, and every one said how much they were enjoying themselves and how well they went for children of their age. And it is the best sort of hunt of all if it is short, with a long dig at the end of it.

To-day that breathless moment of cowardice which had given them determination enough to turn their backs on the stone wall and follow the strange man out through the top of the covert put them in a surprisingly commanding position in the hunt, for any one who knew the covert was of the same opinion as they had been, as to the direction in which foxes usually left it. So for some time the stranger, the first whip and Simon and Susan clattered over the country in rather grand isolation with the hounds.

Susan's pony elected to follow the whip's horse. As his horse changed feet on the bank, Mouse took off and sprang up behind him, arriving in the next field just a second late and just mathematically without a collision. The first whip was an unimaginative boy and riding a steady and solid cob, he did not mind. But it was another matter for the stranger, who was riding a sketchy sort of hireling, to have Simon just missing the small of his back at every leap. He tried saying "Not so close, old man," and "Take a pull, child," but he soon saw that no admonition could have any effect, and at each mistake his hireling made he resigned himself to the worst and was mildly surprised not to find himself on the ground with Simon and his pony and the hireling all on top of him together.

In the end it was one of those tiresome, straggling hounds that caused all the trouble. The stranger, with misplaced consideration and *ésprit de chasse*, waited for it to scramble over a fence and the delay was just too long for Simon's pony. As the now rather weary hireling leaped, so did he, they met with a tremendous impact, and the hireling, without making the smallest effort to reassemble its forces, fell out on its head in the next field. Simon's pony staggered, recovered himself and jumped off the bank straight on top of the stranger whom he kicked in the head and clouted in the ribs before he galloped away in pursuit of Susan and the first whip. Simon, in the confusion and shock, had fallen off.

Cynthia was very upset about it, although pleased that Simon had been going well enough to knock David Colebrook off a fence and jump on him, and concuss him. She knew the stranger better by reputation than her children did or they might have faced the wall in preference to pursuing him through the country. As it was, he was put into the car and driven back to Rathglass and put to bed. The blinds were pulled down and his ribs were strapped up by the doctor.

The children were told they could only stay out to see one more covert drawn as they would have to hack home. Owing to their carelessness in nearly killing a famous jockey and spoiling Mummy's day's hunting, there was not room for them in the car. They expressed suitable regret and then how they prayed that the covert might be blank. They prayed like anything. Their prayer was granted, too, because Richit was not very keen about hunting to-day either, so when they got back to a road they were able to turn their ponies' heads towards home almost before two o'clock.

That was a good day. They stopped at the pub and bought themselves ginger biscuits and ginger beer. Their insides felt marvellously steady and glowing again. It was not like one of those dreadful evenings when a court of inquiry was held over their failures as they sat in the back of the car and Cynthia shot questions at them over her shoulder all the way home: questions that always caught you out somewhere.

"Susan, how did your pony jump the wall near Lara Wood?"

Susan cast wildly about in her mind, how had she been so fortunate as to escape that wall?

"Jumped it well," she might say.

"And didn't shift you?"

"No, I was all right."

"And how did you get on, Simon?"

Simon said cautiously, "I almost fell off."

"But not quite? I see. Well, I think it's a pity you

should tell lies as well as being little funks because I saw you both going through a gate."

They felt ill with shame.

Then there were the Bad Scenting Days when she got behind them and beat their ponies over places when they stopped, knowing exactly where their riders' hearts were. And whether they fell off or did not fall off, they were certain they were going to, which is acute mental strain and agony.

Cynthia, too, minded dreadfully that they should be such pale, uncourageous children, these children of hers and Desmond's. Why could they not love hunting and dogs and ratting and badger digging and their ponies, as all right-minded children should, instead of having to be compelled and encouraged to take their parts in these sports and pleasures? The moments when she said to people, "Simon? I think he's digging out a rat in the wood," or, "schooling his pony for the gymkhana," or "looking for a snipe's nest," and all the time knew that he was playing the piano in the school-room or drawing one of those hideous, left-handed pictures, so unspeakably like Susan or Diana or the sewing-maid were really bitter moments for her. She did not love her children but she was determined not to be ashamed of them. You had to feel ashamed and embarrassed if your children did not take keenly to blood-sports, so they must be forced into them. It was right. It was only fair to them. You could not bring a boy up properly unless he rode and fished and shot. What sort of boy was he? What sort of friends would he have? Besides, Simon was to live at Garonlea.

Since Garonlea was so soon to belong to her son the place had achieved an importance with Cynthia that it had never held when she lived so delightedly with Desmond. But now she had lost Desmond with the most cruel completeness. She had lost the feel of his hands. She had lost the look of his hands. She could not imagine the sound of his voice now. She had lost him entirely. Without him the importances that had only been the outside of her life before became all the life she had. She took them eagerly, driving herself into hunting and parties and hard work of all sorts. She was striving to forget. Soon she had forgotten. She could not realise the danger of that thirsty, unsatisfied part of her life that had been so entirely complete when she lived with Desmond. She did not know that she gave herself a masochistic pleasure in her treatment of his children. She denied this satisfaction to herself and had a thousand reasons which entirely justified her in her own mind. It never occurred to her that she bullied them and drilled and ordered and tortured the miserable little lives out of them because she wanted to, because she did not love them, and needed to hurt them. If the whole importance and love of her own life had not been swept from her she would at any rate have left them alone. But as it was she could not leave them alone. They were another importance to fill that blank and solitary place.

Of course, what Cynthia needed was love with another man. But although men's praise was the breath of life to her and she lived for their admiration and their company, she had a romantically obstinate pose at this time about her faithfulness to Desmond's memory, Desmond whom she had struggled to tear out of her consciousness. Desmond, whom she missed in the flesh so atrociously that she must exhaust herself before she could sleep because of the hopeless cruelty of nights and days and weeks and months without him. She had that extra mould as of glass upon her which people who are physically very brave and very strong too, can maintain uncracked. It shivers very soon with weaker people, leaving them soft and vulnerable. It did not shiver and splinter away with Cynthia, and because of it no one ever got really close to her. Inside her glass mould she would act and go on and produce effects which came through distorted and dramatised and with very little relation to the pain or the savagery that was behind it all.

Another thing was that no one ever criticised Cynthia. This left her more alone. She could not even tell when she was being absurd. The criticism from Garonlea of course, was so far-fetched and fantastic that it cancelled out on itself. Beyond that she was one of those popular Queens who don't meet the people who think them either comic or boring or vulgar or selfish, or any of the things which they so often are.

XIX

THE day that she drove David Colebrook, very shaken
and definitely concussed, back to Rathglass, Cynthia
took into her house a man who never really minded
what he said to people and was not inclined to be
impressed by anything he had heard of them. He had
knocked about the racecourses of England for some
years before the war, and he knew how to look after
himself. He was quick and hardy and very tidy in his
ways of riding a race, or telling a story, or wearing his
clothes, or making love. Being rude to people was rather
a pose with him really. But it was a pose he got away
with nine times out of ten. Especially with women.

On the way back to Rathglass he came out of his
trance of stupidity to say to Cynthia:

"Pretty silly thing. Child riding a pony like that.
Couldn't begin to stop him. How would we like hunt-
ing out of control all day? Tiring. Frightening.
Terrible thing. So dam' silly."

Cynthia felt furious before she remembered that he was
concussed and didn't know what he was talking about.

But ten days later when he was still at Rathglass
(he had a genius for prolonging a visit if a house suited
him, he could compel even people who loathed him to
beg him to stay on indefinitely) and made almost the
same observations to her, as together they watched
Simon's and Susan's unwilling progression round the
riding school, she felt really cross. If he hadn't once
ridden the winner of the National she could have
turned on him and told him he didn't know what he
was talking about.

138

"Their ponies are nappy little—oh wells, too. Some one ought to set about them. God, what beasts children's ponies are."

"Those are two of the *most* beautifully bred ponies you could find and perfect jumpers," Cynthia said coldly, and then added in a voice thin with disappointment, "It's the children that want hotting up, I'm afraid."

"Hotting up? Poor little—oh wells. Their ponies want civilising and a diet of hay and turnips. Civilising. Look here—you and I will clout them round those little fences and give them who began it."

"We're too heavy for them," Cynthia began. She felt for once a hot little wave of fright go through her. She really hated jumping the smallest fence astride, and on one of those refusing little eels of ponies and with Susan and Simon looking on, it was preposterous. But before she could possibly think of an excuse that did not include saying, I'm afraid of falling off one of these perfect ponies over that tiny wall, the children had been called up and dismounted, Susan's leathers were lengthened and she was sitting perilously with apparently nothing in front of her and a great deal of strong and obstinate pony behind her. And there was David bouncing about, feeling for Simon's stirrups.

"You make these children ride far too long," he said, and beat the perfect pony heavily. It thought once about refusing the wall, but thought again very quickly and leaped with zeal.

Cynthia's pony, coming after him, refused, whirling round and stopping just as it so often did with poor Susan. Flame in her cheeks, fire in her eye, but misgiving in her heart, Cynthia presented it at the wall once more.

"I'll beat him for you, Mummy," Susan said. She felt quite stilled by the rapture of this moment. This

139

exquisite reversal. It was almost embarrassing. She danced about between the wings of the fence, screaming at her pony and her mother and Simon. David, who had gone on to jump the little bush fence on Simon's pony came back to look gravely at the drama in progress. After a full minute he gave Simon his pony to hold and dealt with Cynthia's solemnly and very severely from behind. It sprang over the wall at last and scuttled away towards the next fence. Cynthia, who had lost an iron and only got back into her saddle by a miracle, found him not too easy to stop. At the bushes there was another refusal and the children panted up and threw small sods and stones and shrieked advice, emboldened by their mother's second failure.

"You see," David said to her an hour later as they walked into lunch, "you do put them throught it, poor little devils."

"You should be ashamed of yourself, shouldn't you?" Cynthia gave the matter one of those flashing twists—"putting me to shame like that before my children. Don't you think it was a bit hard?"

Instead of turning to her that look of penitence and romantic promise that any admission of failure from her surely called for, he said:

"You did look pretty silly. Frightened too, weren't you? Do you good. Makes rather a change, doesn't it? You aren't often frightened, are you?"

Cynthia said No. This tough young man was no older than she was, but he made her feel years younger. She was afraid that he could fool her and frighten her if he felt like it, and rather respected him in consequence.

But he spent most of his time with the children, giving them arbitrary riding instruction, getting twisted snaffles, quite sharp ones, sewn into their ponies' bridles, shortening their leathers and making them lift their little bottoms out of their saddles when they jumped walls. Silencing all Cynthia's protests and giving the children brief, illuminating counsels— "You set about that pony—give him a good job in the mouth as soon as you get up on him. Let him know who's boss—all right—Children never like their ponies till they can bully them."

He went away quite suddenly one day after he had stayed for more than a fortnight. Diana fell over his suitcases in the hall and told Cynthia who only said that with such a tough and peculiar man they ought to count the spoons before he left. She was feeling rather sour about him since the evening before when he had told her that he thought she drank too much and also he thought her stud groom was robbing her —two absolutely shattering insults. But when he had gone she had to work very hard to keep herself free again from her terror of loneliness. Anyhow, it gave her something to do getting the children out of all the bad riding habits he had got them into.

A person who succeeded in hurting Cynthia and getting under her vanity got dangerously near her and found her both vulnerable and uncertain as vain and lonely people so often are. David had gone through her glamour and seen three true things, not very pretty things either. A woman who is unkind to her children, who is robbed by her stud groom, and who drinks is cruel and foolish and weak. If one of Cynthia's idolising train had suddenly turned and told her these things about herself the impression caused would not have been half so deep or so secret. It was in the shock of her failure to impress a stranger that remembrance lived. She did not see David again for about three years—when he came to stay with them at Garonlea.

XX

LADY CHARLOTTE was dying. Her blood pressure had got the better of her at last and two slight strokes had left her "a dream of what she was, a breath, a bubble," a poor, silly, muttering old woman with the last tyrannies possible to her taken out of her power. She hated her nurse because she could not bully her and tried hard to cling to Spiller's meeker ministrations. But the nurse was firmly established by Enid and Violet after the second stroke and before their mother recovered enough of herself to frighten Muriel into dismissing her. Lady Charlotte whined about it a great deal. She had turned into a very old woman now—pathetic, repetitive, futile. All the narrow dignity of her mind had become as the mind of a not very attractive child. Her wits were soft and blunt. She was greedy as a pig about her food. Sometimes there was a flash of her as she used to be, an order given to Muriel about the garden or house or a quick, backward thrust of memory over some quite complicated matter to do with Garonlea. A word to Enid that sent her spinning back to the helpless, dependent creature she had been twenty years before, a commendation for dear little Muriel that made her feel indeed but the solemn little elder sister of four pinafored children and one fat little boy in a sailor suit.

Violet and Enid took it in turns to stay with Muriel at Garonlea during the last year of their mother's life and in a way it was a vague, comfortable year for Muriel at Garonlea. It was fun to have the sisters there and fuss after their comfort as they hardly did over hers when she stayed with them. There was a feeling of luxury and power about ordering more fires to be lit and better food to be cooked and more frequent bottles to come up from the cellar. The soothing lies she told her mother about all these things spun a thread of independence through her—her confabulations with doctors and agents and gardeners strengthened the thread and gave her for the first time in her life the thrill of being a person held accountable, a real person with responsibility.

Enid and Violet were very sensible about leaving things to Muriel and not fussing her in any way. They encouraged her to see Cynthia and Diana too, and this secret coming together with Rathglass and especially with Simon and Susan gave her tremendous pleasure. Cynthia was very sweet and gentle with Muriel. Indeed, she had never been anything but charming to her, for she would rather have even Muriel like her than dislike her.

Lady Charlotte did not ask to see Diana, only saying, "bad! naughty!" as one speaks of a dog, when she was asked whether she would care to see her. For her part, Diana had not the smallest wish to see her mother, or the woolly, futile shadow of her mother. She did not feel the least emotional about her. But what did upset her so that she could scarcely think of it was the day, uncertain but possibly near, when she must go back to Garonlea.

Cynthia was forty now. Desmond had been killed when she was thirty-six. Four years she had been without him. She had quenched that terrible reality of a ghost that was no ghost. Never again could she let herself suffer as on that first evening when she had heard he was killed, when he had ridden home with her,

143

nearer than in any perishable fever of love. She had done with that time. She had got away from the closeness of his love that had pursued her and filled life full again. Hunting, children, friends, parties, gaiety—sordid, silly, rather desperate gaiety—and tremendous, unassailable popularity. Her position was so strong that she could have done what she pleased in that countryside with rich or poor. And now, very soon now, she was to leave Rathglass and go to live at Garonlea. She looked forward to this as another battle and occupation and excitement for her life. She had not forgotten her dislike and mistrust of Garonlea, but she had such tremendous self-confidence at this time, she felt the oldest, most inward character of a place must yield to her if she should determine to change it. And she was defiant about changing Garonlea.

"Cynthia, couldn't you let the place?" Thus Diana, emboldened by a glass or two of sherry. It was lunchtime in September. The sweet, hot sun was in the room. Her eyes were occupied fondly with a great rich decoration of autumn flowers that she had put on the table. Flowers she had grown, gold and purple flowers and spotted lilies—a bowl for Ceres. She was thinking of the red-walled dining-room at Garonlea and of Mr. Price-Harcourt-Price, a picture that a grandmother had brought to Garonlea, and all the other pictures, cold and dark and not very valuable. She looked out at the little hot lawn. The scyllas got half-way across it now in the spring. Just a few down the bank in Aunt Milly's and Mousie's day, a short spring flood when Susan was a baby, and now a lake bluer than any possible water. "Surely you could find a tenant, darling, couldn't you? I mean, with the fishing and shooting and everything."

144

Cynthia looked sadly down at her ringed hands. In this as in everything, she was going to take her own way without abandoning the pretence that it was at the cost of enormous self-sacrifice.

"Tenants? Fishing tenants? Shooting tenants? My dear, what sort of state do you think the house would be in when Simon is twenty-one? You won't mind so very much coming back with me? I couldn't get on without you possibly. I've forgotten how to. And darling, we're going to change it so enormously. Pull the whole place up by the roots—cut its hair, paint its face, give it some royal parties. Oh, we'll lay the ghosts."

"I don't think you'll ever change it," Diana said, quite gravely. "You'll only break your heart trying to."

Cynthia made a face.

"Reassuring little thing, aren't you? Never mind, I'm going to try, and you've got to help me. We've got—how long? Six years till Simon comes of age. It will be a gay house then—you'll see, darling. And when Simon marries some lovely girl, you and I will come back here and end our days like Milly and Mousie. You'll survive me and Simon's children will whirl you into a cottage like we did to Aunt Mousie."

Cynthia laughed her rich, sweet laugh at the absurdity of it all. She was blind to the possibility of failure or age or sickness, except in the most practical ways. She looked after her skin seriously and kept her body fit and hardy. She would easily have passed for a woman in her very early thirties, except perhaps for a slight heaviness and thickening through shoulders and hips. Her clothes, that she used to wear with a breath of strange romance, seemed rather more full of solid flesh and bones now. But she was lovely to look at still, her skin had neither withered nor coarsened

145

and her hair, cut short now, was almost as good a colour as it had been. And yet there was a look about hair and skin and body of being skilfully and indomitably preserved, a certain rigidity. She was a beautiful woman still, people said. They forgot what they missed. It was that look as of a young river, which had lasted excitingly past her earlier youth and her children's births and had passed from her when Desmond was killed.

Susan had this changing, fluid look. It was the only thing about her that was at all like Cynthia. She was thirteen now, sulking in a fashionable girls' school in the south of England. Simon, who had surprisingly good taste for his age, was almost the only person who admired her. Cynthia certainly did not admire either her daughter or her son at this time. She found their pale-green skins and pale lips most unattractive. Susan's hair was inclined to curl, that was the best that could be said for it since it had turned from its promise of Cynthia's true gold to become almost the same olive colour as Simon's. With their faces she could not be very pleased, but their figures were all right. Long thin creatures like the thin green shadows of a poplar's bones in winter. Another thing that pleased her and proved her right about them was that they both now rode well, no longer apparently afraid, they no longer fell off or made their ponies refuse at stone walls, no longer were sick on hunting mornings. So while they managed to conceal their distressing and disgraceful boredom with the chase they did her due credit among her hunting friends. She was very much afraid that a terrible day would come when they would admit and even boast of it. So far as she knew, even to Diana they had not done this, but she feared that they would.

146

Simon and Susan told Diana a great many things and she confided in them too, although her confidences were distressingly abridged by her terrific loyalty to Cynthia. Not so much loyalty as faith, for loyalty implies criticism and Diana's love had never taught her to look closer and see what her idol's feet were made of.

Even now, in this matter of Garonlea, though she felt miserably unhappy and apprehensive, she accepted all Cynthia's reasons for leaving Rathglass and did not see as far as that chief tremendous motive of vanity which impelled Cynthia to scour the very shadow of Lady Charlotte as swiftly and brutally as possible from the place where she had imposed her will for so long.

With all her schemes and plans for Garonlea, Cynthia could scarcely wait for the news of Lady Charlotte's death to come across the river. The drawing-room at Rathglass was stuffed with parcels of patterns for hangings and chintzes, Cynthia was itching to get estimates for this and that—Light and central-heating, painting and new baths.

When at last an incoherent note came one afternoon from Muriel saying that Mother was almost dead and anyhow, quite unconscious, so would they come over as it could not possibly upset her now, the doctor said, and it would be better for the family to be together. Diana felt that Rathglass was now so permeated with the idea of Garonlea that she would be glad to see Garonlea as a real and separate place again. She could not bear its importance so flowing into Rathglass.

It was the most perfect and quiet autumn afternoon when they walked over to Garonlea. A clearness like a bell's note hung in the air, if days and sounds were interchangeable. In the village, children's voices were loud but rather sweet, Hydrangeas were shockingly blue in a cottage garden and apples on branches as definite as a story that starts about an apple tree. Birds were flying so low over the river that their dark wings seemed to pierce its smoothness. The shadows under the arches of the bridge were black still. In the change succeeding summer the distances were unfamiliar, heights and levels seemed altered. You saw a little valley within the summer valley, the form behind the accomplished shape of things was briefly obvious.

Then they came to the Garonlea woods across the river and the day's clearness and purpose were submerged and lost for Diana in that stillness and pressure of woods; in the dark reek of laurel and rhododendron that masked the very stems of trees, leaving visible only confused worlds of branches, blocking all distance, shoving close round the turn of paths and walls, eager with rankness and strong life.

Cynthia's chin was up and her eyes shining as she walked through the woods. She gave herself no room or time for melancholy.

"You see, darling, we'll clear out all this smelly jungle of laurel."

"Woodcock like laurel," Diana said dispiritedly.

"They don't want all this laurel. Anyhow I shall clear and cut some marvellous vistas."

"You will have to mine the valley before you find a vista."

"I can see you're going to be an enormous help to me." Cynthia had good reason to feel irritated with Diana who was fiddling along behind her in an absentminded, dreary sort of way and making disheartening answers in a manner very different from her usual thrilled acquiescence and delighted elaboration of any plan of Cynthia's.

148

So far it was not dramatic, this return, although it had every element of drama: the old queen dying: a change at hand: a house waiting its new possessor— a house muffled in history and tradition about to be stripped and woken. The coming of a new, angry Queen through the woods did not change their calm by a breath, her strength would use itself against Garonlea in hopeless endeavour. The house was calm too, confident and unafraid. The afternoon sun lay on its face as if on the face of a cat, a yellow, fat cat, tigerishly striped and blotched—autumn creepers gorgeous on the walls, red geraniums splendid in the sun. They walked in at the door, the sunlight mounting the four steps and lying flatter than water on the floor, and beyond in the dark hall a smell of coolness, a reserved and musty breath. Their feet made a faint clatter on the diamonded, tiled floor, coldly noisy between rugs. Pots of blue lilies were scattered here and there among the little suits of armour. ("How typical! Why not a group?" Cynthia's mind busily criticised.) There was a basket of earthy-looking green apples on a table outside the library door with a pair of worn gloves curling from the use of hands beside it. In the room the dead ashes of a wood fire lay white and cold in the sun.

There was nobody in the library, only the dead fire and the empty hush. Diana stood on the white curly fur hearthrug, apart, Cynthia was no help or protector to her now. Again there swept through her that feeling which in four years she had almost forgotten, the feeling of helplessness and depression that she would never escape at Garonlea. Standing there in the library she felt as if her life had already begun to drain and trickle away from her. She was changing already, slipping back into the real she, that sad creature of

149

vague despairs and rebellions. The room was so unchanged since she had last been in it that a night and its dreams only might have lain between her going and her return. There is something strange about a room staying just the same after a long time even if one has been happy there. As it was, Diana shivered under the memory, brought so much too near, of Lady Charlotte sitting in that chair exactly at the writing-table, a vase of flowers in its usual place at her elbow, the silver inkpot and paper-knife shining still. The three different sorts of notepaper still there: one grey-blue with *Garonlea, Co. Westcommon, Ireland,* stamped on it in black letters. The thin white sheets on which letters to shops were written and the address here:

From: The Lady Charlotte French-McGrath, written on top. And the small thin sheets with no stamping that she used economically for notes and lists and writing to the children, they were all there with their appropriate envelopes in the divisions of that upright, red leather box. The chairs and sofas were in their same positions to an inch, and on all the tables the same small vases held the same tremulous little bouquets, dim and unoffending but a little sickening in their silliness and failure to decorate. Everything in life seemed less real to Diana than the terrible sameness of this room. Already Garonlea was soaking up what power there was in her. She felt herself slipping into its dangerous strength, absorbed and half-languorous in the melancholy.

Cynthia was outside its power still. She had no hag-ridden memories, and the rebellion she brought was fresh and strong within her. Her mouth was taut and her eyes half-closed now as she took in the room, concentrating on its possibilities. Calculating, condemning, wondering, changing her plans every moment and fully conscious of her power.

To-day she was wearing a soft yellow silk shirt and a short brown skirt. She leant back in a corner of the sofa facing the room. Her strong thin legs were crossed. Her short hair was bright with health and her thick smooth skin faintly brown. She looked invincible and eager and amazingly young. Diana, who thought that her own terror of Garonlea had been understood by Cynthia and a real thing to her, saw that she had been wrong. Cynthia did not understand anything about fear connected with Garonlea. It had only been her wonderful faculty for sympathy, that gift of hers, that had misled Diana. But Diana was not quite right. Once Cynthia had been afraid of Garonlea.

Presently Muriel came in, a little figure, stooping together and crying sadly, with Enid on one side of her, looking rather quiet, and Violet, fat and lovely and rather flustered because she was not crying too, on the other.

They sat Muriel down in the middle of a sofa and then sat down themselves one on each side of her, patting at her and nodding and grimacing the tidings of death towards Cynthia and Diana. None of them could say, "Mother is dead"—"Gone"—"Passed away." Why are there no simple words for poor, ordinary people to use about death. They can only translate their emotion into words that pervert all meaning and embarrass any natural truth.

Presently Muriel looked up with streaming eyes.

"Darling Mother's gone," she said.

Diana said "Yes." It sounded unloving and aggressive but it didn't mean anything.

Cynthia said, "My poor darling, you must be completely worn out——" When had she said that before? In this room? Yes, to Lady Charlotte at the time of Enid's wedding. She was bending towards Muriel, her eyes full of sympathy, her arm lying with ease along the back of the sofa. Diana could remember that other time as she could remember so much about Cynthia . . . the scented, plumey pale blue feather boa, the ease of speech and movement. And now she was bending towards Muriel, her eyes full of sympathy, her voice deep with kindness, her arm lying with the ease and strength of flowing water along the back of the sofa.

"How long have you been here, Cynthia?" Enid asked in a whispering voice as though she were in church.

"How long, Diana?"

"About half an hour."

"About half an hour, Enid."

Muriel looked up with changed eyes although they still swam in tears.

"Then you *were* here when she—Passed."

"I expect so, darling."

"Oh, I'm so glad. It makes a difference, doesn't it? Anyhow, we were all together at Garonlea. She would have liked that if she had been herself."

"Yes, of course she would." Violet and Enid echoed politely and Cynthia and Diana bowed their heads in agreement. Though all four of them knew perfectly well that had Lady Charlotte indeed been herself it was the last situation she would have countenanced. But this immediate romantical and wilful falsifying of a dead person's character was both natural and comforting to Muriel and less embarrassing to the rest than candour can be at such a moment.

Presently Diana said, "Cynthia shall we go? I have

152

fourteen clocking hens to let out and—and—a lot of things."

Immediately she saw that Cynthia was displeased, and when Muriel said, "Oh, do wait and have some tea and after tea nurse says we can see dearest Mother again," she looked pleased and even a trifle eager. She wanted to renew her mind with the upstairs possibilities of Garonlea. Neither was she at all averse to the idea of taking a look at her old adversary lying dead. It was not a girlish desire, but Cynthia was a strangely natural person and had no recoil about death or birth.

Diana was afraid of all three. She got very red and then, very white and said, "Oh, Muriel, I think I won't. I don't think I need, need I, Cynthia?"

"Of course not darling. Much, much better not. You go quietly back to Rathglass now and I'll be after you in no time at all."

Diana stood about for another minute, uncertain whether to kiss her sisters and go, or go without kissing her sisters. It seemed rather extravagant to kiss them and yet just to walk off seemed as if it was going to be rather a bald departure. Cynthia helped her again, saying:

"Tell them I won't be back to tea," and on this note Diana smiled at the three on the sofa uncertainly and left the room.

"You did understand my encouraging her to go," Cynthia said with soft persistence. "Anything like *that* upsets her for days. I've known her not to sleep after a badger dig."

They all nodded mournfully but agreeably at Cynthia, accepting without question some obscure link between a badger dig and their mother's deathbed.

XXI

In fact, Cynthia was curiously but absolutely right.
It was death that made Diana shrink and shrivel as
though spiritually she had eaten a lemon and caught
a chill on her liver both in the same hour. She hurried
back to Rathglass now, and when she was over the
bridge took deep breaths of village air and avenue air
and sweet, airy breath of flowers in the hall. She stood
breathing it in, waiting for Rathglass to heal the
sickness of Garonlea within her. Very slowly she felt
the first thin returning tide drawn through her, drawn
by her as though she was the moon. It came slowly,
curling, falling small waves where sands had dried and
bleached again in the sun that afternoon. Sands that
had been covered for four years.

She waited another long moment, almost as though
in an embrace, as if this was a lover's breath, this warm,
light smell of flowers and air. Then she went out of
the hall and down the tidy warren of passage towards
the kitchen. The warren was blowing with the delicious
and intoxicating fragrance of boiling jam—such a
warm and reassuring odour. Smiling quite certainly
now, Diana opened the kitchen door to ask for tea and
to say Cynthia would not be back.

In the kitchen the cook (a handsome, fat young
woman with flat blocks of hair, to whom Diana was
greatly attached) had just whipped a bowl of cream
and was stirring strong coffee into its smooth thickness.
Blackberries were boiling on the fire for jelly and
bottles of whole, rosy-looking plums stirred about in

154

a black pot of bubbling water. There was a heap of
tiny plum stalks lying in the sun on the window sill.

"I'm sorry to hear of her Ladyship's death, Miss,"
the cook said. Diana was pleased with the detached
way she spoke—as though her Ladyship had died in
Arabia years before. And no doubt she was burning
with curiosity over what changes this might mean for
them all. Diana went down the steps to the scullery—
a cold, clean place, smelling of earth—to see if her
chickens' food was ready. The kitchen-maid was
startled to see her, a toothless, stupid creature, older
than that kind and able cook, she jumped inside her
blue-flowered pinafore. She was pathetically full of
another idea, cabbage and small green marrows in her
hands. Oh, yes, the chickens' food was ready, Miss,
and she was so sorry to hear of the trouble at Garonlea.
Again the Trouble at Garonlea had been disconnected
from the life at Rathglass. Comforted and set to rights,
Diana went back to the drawing-room to drink her
little brown pot of Indian tea which she liked so much
better than Cynthia's scented Chinese brew. Strong
kitchen tea with a little cream lining its surface was
what Diana liked, and that sweet cook had made a
delicious Sally Lunn tea-cake, thinking they would
return in need of her restoring art. Presently she went
out to her chickens, and at about seven o'clock, since
Cynthia had not yet come back, she took one of those
aimless but enthralling evening wanders in the garden.

It was one of those curious autumn evenings, perfect
completion of the clear day. There was a light and
particular distance and glamour over near and
familiar things, as though one looked down through
sea-water into the end of the day.

In her garden, Diana came to a group of lilies in
half a drift of faint smoke, a mist of pale apples on

boughs behind them. Their thrilling animal scent was blown across the faint smell of burning weeds.

She thought that forever this scent would haunt her with her present fear and unhappiness. How deeply exciting she should have found it. She could see it all so sharply, but she could not feel it. She could not be quite conscious, she was looking through glass or through water. She went out of her garden and would not go back because she was so afraid.

XXII

"Susan, my darling, you look perfectly beautiful. Really. Show me the back. If I wasn't Simon I'd choose you to-night. Do you think any of Mamma's smart friends will choose me and my gardenia and my smart coat?"

"Oh, Simon, don't be chosen or I shall have no one to talk to or dance with."

"Nonsense! Our kind mother has asked several suitable young natives of both sexes so that we shan't feel out of it with her ' crowd.'"

"I'm such an awkward age for her," Susan said despairingly, "an old fifteen. But if she would have her dance in the Easter holidays——"

"We mustn't blame her for that. She wanted the magnolia and the cherry trees."

"And she's so lucky—she's not even going to have a smart hail-storm."

"To-night might be July."

"Without any thunder."

"I like your green ribbons. Couldn't you put a bow in your hair? And why must girls of your age have

156

their dresses just down to the calves of their legs? Such a trying length. You should wear either a long dress or a Grecian tunic."

"Or a leopard-skin, or a crinoline. I follow you, Simon." It was so like Simon to say just how lovely you looked and then little by little undeceive you, calling attention first to the fact that your hair was wrong and would be improved by a ribbon. And then to the unalterable ugliness of the misse's party frock. Better get it all over. "You haven't admired my new shoes either. Enormously long, pale-gold canoes," she held one foot that seemed a yard long, out to him. "Pretty, aren't they?"

Simon sighed. "If only I could dress you. Of course, I shall be able to by the time it matters." They both brightened up at this.

Simon leaned over Susan's shoulder, peering into the mirror. He flicked nothing off the shoulder of his coat and drew his gardenia farther down in his buttonhole. At eighteen he wore his clothes with exactitude, elegance and forgetfulness. Except when he was looking over Susan's shoulder and admiring himself quite naturally in a glass. He was charming and distinguished-looking, with soft, near-fair hair and dark, apostolic eyes. But the lower part of his face did something (besides not having a beard) that spoilt him for being an apostle and tried to make him look like a monkey, but gave that up too. It was an attractive failure. He had lovely though rather faint hands, no hips and long narrow feet like Susan's. He was just but only just as tall as she was.

Susan's face was still a blunt cat's or frog's face with a hole in her chin and a full mouth and green and brown striped eyes, and another untidy sort of hole near one cheek-bone. She was not fat any more, but

157

much too tall and thin, although she avoided the Alice Wondering look by just a little, perhaps because she moved well, slithering through her life with hips almost as narrow as Simon's. Her preposterous white dress had a V-neck and elbow-length sleeves and she wore a locket of turquoises and pearls in a short, stout chain round her neck and about three silver-wire bangles. She liked a nice bit of jewellery, whatever Simon might say.

Her bedroom at Garonlea was not the one Enid and Diana had shared, nor was it the one still known as Muriel's and Violet's. No. Among greater changes at Garonlea, the children's rooms were established where Lady Charlotte had only lodged her favourite visitors. Their windows looked up a strange new distance of the valley where trees had been cut down and a new prospect discovered. A little unmellowed and afraid of itself still, this newly opened world, now vain of its daring, now shrinking and refusing to show off—like a new, unsoftened girl needing only time to be sure and beautiful.

Inside and outside, Garonlea had been pulled about and improved (undeniably improved) and changed beyond all knowing. What Cynthia with her enormous grasp of affairs had done in two years in the way of cajoling trustees and carrying out her plans and spending money when she needed to was astounding and beyond belief. And all done in the name of her love and enthusiasm for Simon. "How wonderful you are," her friends said to her, "really, you are wonderful." "Oh, no, it's been such fun for me," Cynthia would say, "though I can't deny it is a bit of a contest at times, and rather exhausting, you know. But then, so tremendously worth while. And I *think* Simon's going to like it."

And by a strange chance, Simon adored Garonlea. He loved it beyond any fret at his mother's hackings and hewings and paintings and trimmings, in which he had no smallest voice nor opinion. In fact he approved them nearly all. He drew Susan with him in his strange quests and discoveries at Garonlea. They had both begun to be happy, discovering happiness by thrilling moments and disconnected pieces as children who have been battered about in early days do chance delightedly on the possibilities of life—suddenly finding themselves unafraid to touch, unafraid to refuse, unafraid to laugh or find the jokes they should respect quite unamusing. Simon had, of course, gone further in this than Susan but he brought her with him because he was so fond of her. She seemed to him the dearest creature in the world, and as he liked all beautiful things he was glad that she was going to be so exciting to look at, or so he believed.

To-night he was determined that she should enjoy herself and with him. Susan danced very well, she was the person he enjoyed dancing with beyond any one. In 1922 dancing was an enthusiasm that one had to share with one person to get the utmost enjoyment. Simon had taught Susan to dance all those complicated steps which really seemed to matter then. So to see her wasting herself, dribbling about the room with one of the young natives asked by Cynthia to be her partners, when together they could make those intricate fluid, delightful patterns to music annoyed him unspeakably.

They leant out of Susan's window when she was dressed, watching cars come twisting towards the house down the opened length of valley, watching women in fur coats with scarfs on their heads bundle out and if they spoke, their voices were sharp with

the uncertain sort of anguish that precedes parties.
Their men put their cars on the far side of the gravel
sweep and came strolling towards the house, white
scarfs round their necks and black trousers under
their overcoats. They did not mind whether they
were a success or not and took the party easily.
Cynthia's dances would certainly be good and the
drink wonderful.

Below the house there was a white vaporous mass of
cherry blossom, it seemed less solid than the mist
from the river, coming like a crowd in blue cloaks
through the stems of the trees and lying heavy already
on the lowest terrace. Where Susan and Simon were
leaning out they looked down into green fish-like
leaves and slumberous white flesh cups of one of the
magnolias for which Garonlea was famous. They
grew against the house in great white crosses and
there was one on the lawn that looked as strange and
dramatic as though it was the tree of the knowledge
of good and evil.

They drew their heads in at last from the disturbed
quietness of the evening and decided it was time to go
downstairs. Susan looked sadly round her room before
she left it. That sensation of horror which preceded
social intercourse for her made her wish that she was
now bundling into bed in her blue and white striped
pyjamas and finding her hot water bottle with her
feet and placing it on her stomach before she opened
one of the books that Simon gave her to read—books
which so soon sent her to sleep among all the delightful
pictures of horses and hunting by "Snaffles" and Mr.
Lionel Edwards which Cynthia generously gave her
children on Christmas and Birthdays. The cream
walls of Susan's room were covered with them, and
the bright expensive chintz of the curtains and chair

covers was gay with an ever repeated pattern of red
coats and pied hounds and bright bay horses. Over the
mantelpiece was a sort of shrine in which were hung
a fox's mask mounted on a wooden shield, two foxes'
brushes and below them a row of three little dark
pads, all trophies of prowess in the chase which Cynthia
desired her daughter to preserve and respect. There
was a blue carpet on the floor and a blue eiderdown on
the bed and a delicious woollen dressing-gown as
soft as a little fawn folded on the back of a really good
arm-chair. The room was entirely Cynthia's doing.
She could not possibly have taken more pains to make
her children comfortable. It was a constant echo of
the time when Susan's blue blanket had been chosen to
match the scyllas. That part of Cynthia had entirely
survived.

Simon waited for Susan and followed her in a
reserved and dignified way down the wide staircase to
the hall where Cynthia was saying "how do you do"
to her guests. He realised at once that they had made
one of their idle, stupid mistakes again. They should
have gone down the back stairs. Now they were
bound to make an entrance.

Cynthia saw to it that they did. She could not have
been more pleased than she was to watch their unhurry-
ing descent. Pride in Simon had only come to her as
self assurance had come to him. She had taste enough
to find Susan's reserve and possibility of beauty enthralling.

It was not a pretty staircase at Garonlea but it had
width and a wide pallid air, with its new extravagant
pale carpet and the white creamy walls from which
red paper and Victorian progenitors had been stripped
and only one enormous decorative but boring portrait
of Laetita, wife of a Desmond French-McGrath,

remained. Her creamy lace and silks and ribbons and fluffy dog were moony and romantic with no further contrast than the milky walls. She was not at all like either Simon or Susan as they came past her and turned the wide corner to the last short flight.

"Hullo, Angels!" Cynthia said. "You know Captain and Mrs. Church, don't you? Simon, darling, you must see that Sue goes to bed at one and don't let her get drunk. The champagne cup has rather a kick in it I'm afraid."

They let this pass though they knew that she knew that Susan never and Simon hardly ever drank anything except water. It was the same thing as jokes about their smoking which they were not very fond of either. Simon put her in good humour in the most able manner in the world by saying, "Can I have a dance with you, Mamma, some time? Or are you quite full up?" She looked so young to-night in her spring and summer lilac gown. She would want to show off with him in his new coat. He would have to, so he might as well do it properly.

"Yes, Simon, of course we must. Before supper. No. 9. Find yourselves programmes, children. You aren't to dance together all night."

"Oh, I can remember No. 9," Simon said and moved away. He was glad it was to be before supper as he was just beginning to realise that he disliked being near his mother when she was slightly drunk. Before supper was best. He must have one with Diana too.

"Simon, I shall be so busy with these cocktails," she said. Cynthia had made her mistress of the cocktail bar and she felt pretty grand about it though a little confused and inclined to mix them far stronger than necessary. Also the smell was beginning to make her feel odd already. She was looking rather smart in a sort of white linen bar-man's coat which happily concealed most of the old black satin gown which had been Cynthia's and altered to fit, though hardly to become, Diana's solid little figure. But the coat had a real flick about it and looked gay with her trim dark head, its short bird's-wings of grey brushed back across her ears.

"One-third Cointreau, one-third gin," Diana muttered distractedly, too taken up with the responsibility of her post to spare even one wondering or incredulous thought to the improbability of a cocktail bar in the library of Garonlea.

"Don't make them too strong," Simon advised, "or every one will be drunk and drunks are often sick and I do hate that in my house. Good-bye for now, I see you won't dance with me or listen to me. Where's my Sue? Will you dance with me my darling?"

So far only the band was in the drawing-room. It was vast and empty and at one end Diana had excelled herself in the construction of an exotic and impossible tree decoration of cherry blossom and white lilac. It had taken a long time and much skill and patience and wire but its large and beautiful result against the water-green walls was most exciting. The windows were open to the floor, so that you could walk out towards the other cherry trees and the nearer brooms and lilacs and thousands of tulips, with which Cynthia and Diana had planted the terraces. The room was as full as a full tide with the river and the garden and the evening. A strange dark man in the expensive band smiled at Susan and Simon and although it was the wrong tune for the start of a jolly evening he played such a waltz for them as he thought the night and the lovely empty room and their empty youth demanded. Susan and Simon waited a moment half shivering with

163

pleasure before they went swirling and sweeping round the pale cavernous room full of music. Together they danced with such skill and accord, it was as if each danced alone.

In the hall some one was saying to Cynthia, "Look, Cynthia—we've brought an extra man. You do know him? Do you mind? He would come."

Cynthia was saying, "How lovely to see you again," to David Colebrook. He spoke to her as if he had not come with his party at all but straight from Africa to be in time for her dance. He looked brown like all sea-voyagers do and spoke in a detached way and looked at her as though she was not a Queen by right, but he was not sure that he would not think she was, which would settle the matter beyond argument.

Cynthia in all her two years' tremendous business over Garonlea, her terrific occupation with every sort of affair—from distemper at the kennels to the discovery and successful proving of a Gainsborough at Garonlea—had not quite escaped her memory of David. He was the preposterous young man who had made her feel absurd and afraid and told her she drank too much and would ruin her skin. He had gone out of her life for three years. Now here he was back again, and once more she was conscious of that chill air of reality, a draught between her skin and herself, that she had known before when he had roused her interest and hurt her rudely and gone away for three years. Now here he was back again on a night when she only looked for praise and comfort and worship. She would not let herself be upset or disturbed by him . . . and yet she wanted him to know that she had not spoilt her skin by drinking; that every one thought her the most wonderful woman in Ireland; that this lovely house and party were all of her making and that she was desirable and inviolable beyond any man's wish. She wanted him to know all these things because he had once dared to hurt her.

Before David had said anything to her, for he always took his time before he spoke, some other arrival called for her attention, and when she thought about him again he had gone. He had not even waited to ask her to dance with him. Perhaps rather a crude way of keeping her attention on him. But this was not his reason, he had gone because he wanted a drink and he wanted to think. He did want to dance with Cynthia. He wanted to spend as much of the night as possible dancing with her. He would leave her alone till the right moment and then he would step in. He was an extraordinary good judge of pace both in love and racing. He never lost a race through winning it three times over and his methods with the women he wanted were rather the same.

Three years ago he had been excited by Cynthia right enough, but at the time he had been feeling too sick and muddled from his concussion to pursue the matter with any venom and since he had left her his life had been pretty busy. He had not thought of her in any way that really counted or disturbed him. It was hardly likely that he should. But now and then he felt alive about her and remembered what a brute she was to those children, and how he had got home on her by finding out that somebody robbed her, because obviously she was eaten up by a sense of her own infallibility. And then the thing about drinking—that got under her guard too, past all that beauty and dignity she was so fond of parading. Yes, what a brute she was to those children—and how grand she looked on a nice quality horse, as though she had been blown straight into the middle of her saddle.

Beautiful horse-woman she was . . . and how shamelessly she worked that little black-haired companion.
Diana had had a birthday when David was at Rathglass
and he had been enormously tickled and impressed by
the fact that Cynthia had given her a new spade for
a birthday present. . . . Not that she was afraid of
work herself. Then that morning when she had
worked a cross cut saw with him, and with a simple
skill that made endurance look silly. A ravishing
creature and as strong as a bull. . . . He was glad he
had put her through it that time over the children's
ponies. David was an experienced and successful
jockey before he became a successful trainer so he
knew what it was like to feel extremely frightened
quite often. For him it had been a question of his
livelihood, but he never quite saw why children should
endure the same sort of thing over a matter that was
presumably for their pleasure.

David watched Diana for a little making her cocktails; then he said, "You'll have everybody blind in
half an hour. Some of them are swaying now." He
stepped in behind the bar without being asked and set
her going on different lines, doing it all in an impressively expert and absent minded sort of way. Diana
did not resent his interference at all. She felt rather
flattered by his presence, and in her slightly fuddled
state she liked having some one she could rely on.

He stayed with her for nearly an hour, until the
party had really become itself and arrived. A lovely
supper-party it became with every one enjoying themselves a little wildly. There was a great deal of dancing
to the expensive band, which persuaded people to come
in from their love-making under the smoky ghosts
of the cherry trees; to abandon the elaborate refuges
Cynthia had constructed all over the house; to set

166

down their drinks, wet glasses on any table, and extinguish their cigarettes in the nearest and easiest way —in fact to do practically everything except spit on the floor.

There was not one corner of Garonlea where you could escape from the excessive success of this party. It swept and surged through the house and through all the people who took part in it. There was a gorgeous untramelled atmosphere about it, in which the least forward girl found success and the more forward became daring and romantic to a degree. The heartiest people blew hunting horns (rather an unnecessary trumpeting perhaps in the spring of the year) and slid down the banisters and threw méringues and rolls of bread at their jolly girl friends.

And such a night for love as it was. Not only under the cherry trees couples were embraced, but wherever you went it seemed to Simon impossible to find a place where you could talk to Susan with dignified composure. He thought it was not right for a young girl to see so much love-making. He had the forethought before he took her up to bed at one o'clock to go and see if her room was empty. It was not.

"I am terribly sorry to interrupt your talk," Simon said in clear childlike tones, "but Mamma has sent Susan to bed and this is her room."

He watched their embarrassment without any pity —this unyouthfulness belonged with his previous momentary affectation of childishness. There was harboured within Simon some of the chill horror that his grandmother might have felt at this debauch and rioting. Cynthia would never lay the last ghost at Garonlea because its lodging was in the blood and bones of the McGraths themselves.

Simon locked Susan into her room and put the key

in his pocket. "I will let you out when Mother's last guest has gone," he said.

"And if they set the house on fire I can climb down the magnolia, I suppose," Susan answered. But she was really pleased to be protected from the savage horde below.

Simon, rather detached from human intercourse now that Susan was safe in bed, met David, also alone, in the hall. A dance was in progress. There was not so much love lying about.

David said, "Remember trying to kill me?"

Simon did. He said, "You forgave me that a long time ago."

David thought, "Like his mother. Just the same twist. But he's too sober for her party. Nice boy." He said, "Enjoying yourself to-night?"

"Not more than I enjoyed jumping stone walls when you saw me last."

David grinned and yawned. "You should. It's a wonderful party and you're the right age for it."

"No, I'm too young. Are you liking it?"

"No, I'm too old. But I hope it's going to improve soon."

Simon said, blushing rather, because he knew David was rather a grand man, "Are you in Ireland for long? Do come and stay with us if you can manage it."

"Yes, I'd like to. I could come over to-morrow."

"Yes, do that."

XXIII

CYNTHIA came out of the dining-room. She had
nobody with her. She had just been seeing how things
like food and drink were going on and given a few
brisk simple orders to the servants, who were rushing
and muddling about as the best servants do at dances.
Obviously she thought the hall was empty when she
came in. She stood a moment collecting her forces
and then went over to a mirror and taking a little
comb out of her bag began combing out the short
soft bits of her hair.

David and Simon stood very still because they
were both taken aback by the reflection of her face
which they could see in the mirror. It looked grotesque
with her pretty hair and her strong assured shoulders
coming out of her beautiful lilac and pink dress.
Perhaps it was only the reflection, Simon thought,
distorted in the glass. But was that haunted, hungry
face the real Cynthia, or the smooth bare back and
strong pretty neck and quick able hands more like
her? After all the figure was real, the face was only
seen in a glass. Still in the looking-glass they saw her
shut her eyes quickly, shutting them hard as a child
shuts out sight and sticking her fingers in her ears
venomously, as though all the tumult and music and
hunting cries and lover's whisperings of this party
were hateful and terrible to her. Then she opened her
eyes again, frightened. And with a sort of wrench,
a tremendous effort and change, her face in the glass
became smooth, as bland as her back, almost, and she
picked up her bag and went running very light on

her feet, as though she chased the party spirit itself down the hall towards the drawing-room door from which the music came so sweetly.

Before she reached the door David had caught her up. Simon saw them dancing together that dance and the next and after that he did not see them. He felt a little sorry that he had asked David to stay. He expected to see him entranced and captured and cast away just as he had seen others in the same process. Sometimes he had thought that they were hurt unfairly and unnecessarily. He did not really mind about them. His mother's beaux seemed to him peculiarly boring and unperceiving people, well deserving of their courted distress. But about David he felt rather differently. Already he had some understanding of the cowardly and highly strung children he and Susan had been and he thought a man who had done so much and so quickly to help such children must be a very nice person. With a nice twist of humour too, Simon thought, a little wave of remembrance breaking in his mind as he remembered Cynthia, angry and insecure, on Sue's pony. No. That sort of person should have more sense than to hurt himself unnecessarily. He should certainly have more sense.

When David had caught Cynthia up at the door of the drawing-room he had not said anything. Nothing like, "Marvellous party!" or, "May I dance with you?" Or anything at all. He put his arm round her and moved in among the dancers without a word. When they had danced once round the room, still in silence, Cynthia said:

"I hope you've been enjoying yourself to-night."

David said, "Not much really. I only came here to see you and I've been wondering why you look so damned unhappy."

Cynthia, the Life and Soul. Cynthia, the Wonderful Hostess. The Successful Mother. The adored Chatelaine of Garonlea, accused of the indecency of looking unhappy. Unhappiness was an attribute of failure. It was even uncivil to one's guests to dare to look unhappy at such a party. One should never let go for a moment. Besides she was not unhappy. Only once during the long carnival of the evening, ghosts had nibbled at her heart, for a brief distracted moment at the height of this brilliant success. She had been delighted with this night of her making, with ordinary parts of it:—such as the delicious food and drink, the flowers, the band. And with stranger parts of it:—her destruction and recreation of Garonlea, Simon's poise and finish, Susan's soon to be beauty; all, her house, her creatures, her doing, praising her. Then for a moment all this splendid sense had died in her, died like wine in a drunken woman, leaving her alone, sick with her hopeless necessity for love that she had never taken. She truly believed it was only Desmond that she needed with such anguish then. Desmond to sleep with and wake with and eat with and talk to. All the rest of her life was a dangerous shell of pretence, a thin shell against her ear full of screaming whispers.

Now she said, "My dear man what a cracked thing to say." This was in her worst sort of smooth voice.

He said, "I'm the only person who is ever likely to talk sense to you or tell you the truth ever."

"How touchingly interesting."

"Yes. But stop defending yourself. No use."

"I find you rather boring—— Or should a hostess say that?"

"Perhaps. To a guest she hasn't invited. Or if he says outrageous things. Perhaps then she may."

171

"No. But I don't like your saying I'm unhappy," Cynthia wanted to touch this subject again. It gave her a strange vibration of fear and pleasure. Like touching a spiral wire within herself.

"But I often think I know when people are unhappy —or frightened. I'm often wrong."

"Frightened too? What am I frightened of?"

"Of your age. Of Love. Of losing your success. Perhaps of your children. They were of you once. It will soon be your turn."

"*Oh*," Cynthia said. "What a ridiculous, boring conversation we are having. Please shall we stop?"

"Are you bored? You know you're not." David stopped dancing when they got to an open window. He put his arm through Cynthia's and took her out with him on to the warm flagged path that ran outside the windows above the terrace.

"Of your age," he repeated to the night, "or Love. I don't know which frightens you most."

"No," Cynthia said, "I think I'm most afraid of losing my nerve."

"Your riding nerve? But you don't have to jump fences if you don't want to. You can ride about on a nice mannerly horse and talk to your friends. That's my ideal of an enjoyable day's hunting."

"You don't understand about it."

"Yes I do. I've been frightened all my life. I was delighted when I couldn't compete with my weight any more. But do we have to talk about horses all the time?"

"What about then?"

"About unhappiness. Or love, perhaps."

They were walking down the first flight of terrace steps, Cynthia silenced, her head bowed and the silvered panniers of her dress making a hushed rasping

172

sound on the stone beside her and behind her. The cherry trees were blanched unearthly wreaths of flowers below and the sliding and turning of the river beyond and below again was insistently audible in the night. Cynthia went to the left at the foot of the steps, walking on in silence between a thousand faint uncoloured globes of tulips. They stopped and leant together in an outward curve of the Gothic balustrading, brooms and lilacs, their flowers drenched and bloomed with mist directly beneath them, and then the river, always turning and repeating its sounds in darkness.

Cynthia felt terrifyingly languid and capturable. Of Love and Unhappiness, he had said. Of love and unhappiness, she heard it in the river sounds and walled and muffled in the mist it assailed her spirit. From her garden and from her lighted house above, again the desolation renewed its power. She was alone. She had been too long alone—too long peerless, unquestioned and lonely. She would have liked to put her face in her hands and lean weeping for comfort on the high half curved rail of stone. But she did not cry. It was too foreign to her. She did not say a word more. And when David spoke again it was to ask her a question about the fishing. Then they had a long conversation about Blood-stock. After that they went back to the house and danced again, more sympathetically than the first time. Then they ate oysters and drank stout and champagne, and David had a long brisk conversation about racing with a man he put right several times. But in spite of talking to other people he kept Cynthia with him. She was oddly in his power already. The tears she had not shed and the wine made her feel a little shaky and emotional. She did not want to go. She wanted to be near for her own sake. She liked eating and drinking and being with him. It was important. It was something close to contentment. Something uncertain and alive with excitement.

Soon everybody was going home and coming to say good-bye.

"Good-bye, Cynthia, thank you for the most lovely party——" a woman in a black satin gown, very short in front and short behind and long at each side and dashing black gloves turning the elbow.

"Good-bye, Mrs. McGrath. I have enjoyed myself so terribly . . ." a rather tousled young girl with one of those new permanent waves that made hair look like hay.

"Cynthia, you surpassed yourself. Bless you, my dear . . ." an old admirer slightly romanticised by wine.

"Good-bye . . . Wonderful . . . adored it . . . darling. The most marvellous show . . . thank you . . . thank you . . . thank you . . ." As always the incense of praise and gratitude enraptured Cynthia. She accepted thanks and good-byes, wonderful to every one and particularly sweet to the people who did not matter.

David came by himself. Instead of good-bye, he said, "Simon has asked me to stay. I think I'll come to-morrow."

"Yes, come to-morrow," He went away and after that she was a shade more absent over the last leave takings.

She was left with Simon and Diana, both rather pale and cross. The thought of their exhaustion did not come within an inch of her consciousness.

"It's been a tremendous success, hasn't it? You were marvellous in your bar, dear, and you looked so original

174

in your little white coat. That was a good idea of mine."

Simon yawned. "It was my idea," he said. "I thought it would be a good plan to cover up those rattling old sequins. Diana, you must buy yourself a new dress if Mamma is going to give many of these routs. And Mamma, you must buy Sue something more becoming."

"Oh, her dress is quite suitable. It's *such* a mistake to dress up girls of her age. Anyhow I must go to bed. Simon, you asked that Colebrook man to stay?"

"I don't suppose he'll remember to come. He seems casual."

"Well, perhaps he won't. Good-night, child. You looked extremely smart."

"So did you, Mamma."

"Good-night, Diana, darling. Just have a tiny look to see there aren't any burning cigarettes about. Will you be so sweet?   "

"Go to bed, darling. You're exhausted. Of course I will."

"I *am*, just a tiny bit."

Cynthia went upstairs holding her panniers in either hand. She was not at all tired really. But her heart was full of a sort of murmuring excitement. Alone in her room she sighed and trembled and looked at herself contentedly in the glass before she undressed and got into bed. She slept till very late the next morning—a lovely sleep, all the time she felt as if she was sliding between glass and water.

XXIV

SIMON and Diana sent the servants to bed and prowled about wearily, tipping ash trays into buckets of water, recoiling from the more nauseating horror and disorder, putting out lights and closing shutters against the approaching light. In the drawing-room Diana's tree was slack and wilted, an anæmic drooping object. No one could have imagined how lovely it had been.

Diana went about her work with quickness and ability and a sort of grim satisfaction in the disgusting shambles to which the house had been reduced by the party. She did everything that really needed to be done but not a thing more. She spoke to Simon quite sharply when he loitered behind her, examining with disgusted fury the mark a cigarette had made, laid down on the corner of the marble mantelpiece in the drawing-room. There were lots of little things like that as well as grosser signs of misbehaviour on the part of his mother's guests.

They heated some soup in the kitchen and drank it sitting on the kitchen table before they went up to bed. Simon looked about him with satisfaction.

"It seems to be the only room in the house that hasn't been sullied and polluted," he said grandly. "I'm going to bed in a minute and I'm not going to let myself wake up for four days. Then perhaps my house will have been put right for me, though I don't think I shall ever feel the same towards it after the hideous sights we've seen."

Diana glanced round the kitchen restlessly. She didn't care if Garonlea had been set on fire and burnt to the ground. She rather wished it had.

"What about the man you asked to stay? What is he to do while you are sleeping off your mother's party?"

"I don't expect to see much of him in any case," Simon looked at Diana, his eyebrows raised into flying lines of unspoken comment. "Mother seems interested." He got off the kitchen table and went over to the door, waiting with his hand on the light switch till Diana joined him. Then he switched out the light, bringing down a mammoth and startling gloom. The tables and dressers and armoury of copper pots and pans, the two shining ranges, were all obliterated entirely for a moment before their outlines took pallid form and unearthly depth of shadow in the green sickly light of very early day.

When Diana took off her crisp white coat it was rather as if she took off a mask. All the swagger and impudent line was gone, leaving behind a tired but tough small body in a black dress too big for it in the wrong places. The slack dress came off in a moment and after that a great many underclothes. She got herself into the red silk pyjamas Susan and Simon had given her, slapped the expensive cream ordained (but not paid for) by Cynthia on to her face and tumbled into bed.

In the same way that Cynthia had planned Susan's bedroom to be a constant reminder of the chase she ought to love, she had in Diana's gone back constantly on the garden motif which seemed to her so suitable and right in connection with Diana. So there was a delightful chintz covered in rich polyanthus primroses, pictures of well cultivated and luxuriant gardens on the walls, a shelf full of gardening books and a great, great many photographs of Cynthia to remind Diana to whose garden her energies should be directed. Her room was her father's old dressing-room and opened off Cynthia's, who had taken great pleasure in making Lady Charlotte's bedroom into one of which the most expensive French tart might have felt proud.

It was a pity that all these changes at Garonlea altered it so little for Diana. To her Garonlea was more itself than it had been before Cynthia had torn down its red wall papers and hurled the unwanted ancestors into attics with their faces to the wall, accompanied forlornly by the Dresden china black boys (life size) in white socks, who had so long been torch bearers in the drawing-room, and other objects upon which she looked with contempt and nausea. In all these locked, out-of-the-way places—bedrooms still waiting their sack and purification, cupboards where old hoards of rubbish had not yet been dealt with mercilessly, trunks of photographs fading in their expensive silver frames, corners of the shrubberies and old airless parts of woods—the spirit and power of Garonlea still lived with a tenfold strength. It was as if it stored and reserved its power for a future day. Quite literally the breath of such places, the strong camphor-filled breath, on the still laden air of an outdoor place thick with old childish memories filled Diana with hatred and a tremendous consciousness of things as they had been at Garonlea all her life till now.

For Cynthia's sake she had fought hard to get rid of this feeling. But it was too strong and could not be defeated. She had given her mind faithfully to the garden, but she had no quiet and secret thrill out of it, such as she had at Rathglass. She only saw its successes and failures. She did not feel them. She knew Cynthia felt her blankness if not hostility to-

178

wards all her plans and changes and she tried to get round this and deny it by giving Cynthia almost all her income to spend on Garonlea or on what she pleased. Cynthia spent it cheerfully and gave Diana her old clothes and new tools on birthdays and Christmas. Not only spades but pruning scissors and forks and trowls and iron dibbles and many other useful things. She also gave her a horse whenever she wanted to go out hunting, which was not often now, as she felt so rotten and without energy living at Garonlea.

She had, besides her unwavering faith in Cynthia and her love for her, two other chief excitements and pleasures in life. Real pleasures about which she schemed and planned and in which her enjoyment was satisfactory, not a shadow and illusive enjoyment bolstered up by the imagination. One of these was her friendship with Simon and Susan which was a deep and true pleasure, and the other was her unfailing care of the garden at Rathglass. Her feeling about this was comparable to the thrill of making a wild garden when one is a child—an adventure and an excitement beyond any satisfaction to be got from the garden plot supervised and interfered with by kind Authority. Or the thrill of making a house in a wood that nobody knows of yet.

Of course Cynthia knew about Diana's pre-occupations with Rathglass but she had quite enough sense not to bully her on the subject. So long as she gave enough time and work to the garden at Garonlea, Cynthia managed not to grudge her her adoring ventures at Rathglass. Materially it was not at all a bad thing that the garden should still enjoy Diana's skill and imagination. Some day, sometime, no doubt they would return to live in the closed house again. But the date seemed as vague to Cynthia and as far off as Lady Charlotte's death had ever seemed to her.

Meanwhile Rathglass was let to fishing tenants in the fishing season and it was always Diana's chief dread that permanent tenants would be found, tenants (insufferable thought) with ideas of their own about gardens. So far, however, it was only the rich, middle-aged fishing men in their lovat tweeds and mackintosh deer-stalker's hats who came in March and left in May and those of them who understood about gardening gave her praise and interesting comment. She had nothing against the fishers. She would go over to Rathglass in the long spring days when the fishers were in possession, days like those when Desmond and Cynthia lived there and Susan lay in her perambulator among the scyllas, and work in the garden, making plans with the man whom she was busily turning into a useful gardener, weeding in her rockery, her hands busy among the little Iris, their roots like little hairless paws pressing faintly warm against her hands. And when she went back to Garonlea the colour of the purple primroses would get into the dark and into her sleep for that night.

Some day Diana knew they would return to Rathglass to live, and Simon promised often that he would never, not even to make room for Susan and her husband and their children, turn her out before she was carried out in a coffin. Diana indeed believed him and was more than grateful, for she knew that she would never be as brave and gracious as Aunt Mousie had been when she was moved to the suitable little house.

XXV

THE day after the dance Cynthia was in a stormy
mood. She soon lost the peace in which she woke
from her long lovely sleep and started rushing about,
scolding and apologising to people by turn.

Simon and Susan stayed in bed, Susan because
Simon had forgotten to unlock her door and she was
not going to have him woken to produce the key, or
scolded for her incarceration. So she feigned exhaustion and stoically denied the pangs of hunger.
Simon slept and woke determined not to sicken himself
further by looking at the house by daylight.

Diana alone was caught and buffeted about in the
whirlwind of Cynthia's unease. She could not account
for it entirely by attributing it all to the after effects
of the party. These did not explain her hasty questioning looks when a note or a telegram was brought to
her. The uncertainty and changeableness of her
orders were foreign to her too, and so was the show
of temper when they were misunderstood or carried
out wrongly. She was really much too sure and
capable a creature not to foresee and allow for
the stupidity of others. But to-day this seemed
to enrage her. Nothing could be done quickly
enough. Nothing could be done just as she wanted
it done. She cursed and stormed over the ignorant
savagery of the friends she had asked to her house
when she discovered things like the mark on the
mantelpiece burned by a cigarette. She sent Diana
flying about to arrange flowers when a room was

restored to its normal state and then criticised her decorations and pulled bits out and stuck them in again and then said, "Oh, Darling, you must forgive me, I've absolutely ruined it. What made me touch it I can't think. Put it right again, Diana, and I won't do it any more."

And Diana, who had been almost angry, was mollified and touched by the sacrifice of such an apology, and strengthened by a drink and an anchovy on a biscuit at exactly the moment in the morning when she felt she could do no more. That was typical of Cynthia too.

It was a deliciously hot spring day—a day when there is both sun and crispness in the air and flowers look young and well groomed and dewy, not swooning and languid. And beech leaves are light as spring muslins, not black-green shrouds as later they come to be. Cynthia took her drink and Diana's out into the sun and they sat on the hot low stone balustrade of the flagged walk round the house, looking down into the hot, widened cups of tulips on the terrace below. Cynthia grew more placid as she drank and sat in the sun, wine and sun always had the most superb effect on her, and began talking to Diana about the beauty and success of the party, and stopped raging at the awful things people had done. She was wearing a soft, blue wool skirt and jersey, with a black belt low down on the hips, and her bosom strapped in as flat as possible as was then the fashion. Her hair was short and sleek and really gold in the sunshine. She did not look the least exhausted by the party—only as though she had caught a slight fever.

They were sitting there, the benefit of wine still attaining its full capture of the moment, when David came round the corner of the house and down the long path towards them. Then Diana, with that flash towards conclusion that comes sometimes when a person is slightly in wine, was able to answer with a single reply all the questions about Cynthia that had puzzled her during the morning. Cynthia was in love.

Diana's immediate reaction to this was a feeling of excited pleasure. Her thought sprang at once from Love to Marriage. No man could love Cynthia without wanting to marry her. Diana's head swam a little. What with wine and the thought of Cynthia happy and the possibility of leaving Garonlea, she looked on David as a saviour and a saviour who would have her unqualified support. She had nothing to go on in her excited acceptance of Cynthia in love beyond her own interpreting of Cynthia's look and silence. But in this she trusted herself entirely and looked for no further confirmation. Though she had not seen this heavy, half-sheltering half-asking look about Cynthia since Desmond's death she could never fail to know it again. It struck her with a blinder force because she had not seen it. "Mother seems interested," Simon had said to her the night before. He had only meant that Cynthia was out to capture another man as they had seen her yoke so many to her. But Diana knew that this time Cynthia herself was in the weighted nets of love, caught and struggling.

It was only the truth that Diana saw so promptly and clearly. Cynthia was really captured. She was in the power of an imaginative man who was also quite as hard and in a way unscrupulous as any one need be. David wanted Cynthia. He wanted her as his mistress and he was going to have her entirely on his own terms. He knew she was not clever and she had no defence once one got inside her pose and vanity.

She was really one of those super conventional women on whom an almost brutal hardness has a wonderful effect. David sat beside her in the sun for an hour drinking her wine and talking about his wife and child. By lunch time Cynthia for the first time was consciously out to wreck a marriage if she could. Heretofore all the marriages that she had upset had been upset without any further plan and left at that. Now there was a purpose and drive about what she was doing. She was determined to do this. But she was unsure. She was afraid. She was in a panic that he knew she was in love. She was afraid he did not realise it. She was at once in delight and in despair. Desmond, Garonlea, everything was passing for her at this moment. Realities became only a background for her present self and her present want.

And things went against Cynthia. For instance, the weather. Even the dreadful month of May betrayed its inconstancy by a week of days as lovely as this morning; days as gay and shining and delicate as the tulips and cherry blossom; days when each child you met on the road in the mornings carried a bunch of bright flowers to school; days when the thorn trees leant their branches every hour more heavily towards the fields under their thick loads of flowers; days when creatures like foals and young ducks and hound puppies flourished exorbitantly.

During these days Diana saw that Cynthia became more troubled and enamoured. It worried her a good deal as she knew now about David's wife, and goodness knows why she had not known before. She felt sorry about it, but she hardly supposed it would stop him if he really wanted to marry Cynthia. As he surely must. Surely he must. She could not understand his indefiniteness, his lack of response to Cynthia. By responding to Cynthia, Diana meant that people should lay themselves humbly down and wait for her to walk on them.

All through those bright days when they worked and played—sometimes fishing, sometimes fiddling about at the kennels, getting puppies out to their walks or standing looking and listening and seeing an incinerator being built, or watching the mares and foals, or starting early for a distant race meeting, David seemed quite as much inclined for Diana's company or the children's as for Cynthia's. He was quite easy whether he was with her or not, although now and then he kept her with him for no special purpose when he happened to know that she thought of being somewhere else. And now and then he would make tortuous little speeches that excited her while they did not give him away so much as an inch.

"I don't know why I stay here, I'm only upsetting myself." . . . "What do you think? Do you think one ought to take what one wants always and disregard the consequences?" . . . "You're the most disturbing person; I wonder where Simon is." . . . "I was right about Simon and Susan, they'll soon bully you. What you need is a tough and virile man to help you keep them down." . . . "Shall you mind if I tell you something? Are you sure you won't mind? Promise me? Well then—I think that Sesame colt you admire so much has weak, nasty-looking hocks. I wouldn't like them at all, and I'm not sure but I *think* there's a curb forming on one of them already."

That sort of thing went on for days till Cynthia's sleep left her and she looked old and cross and hollow eyed and all this energy and imagination that she had expended on Garonlea brought her in no return now. The results were there but they were no good to her at

185

all. She thought no more about the house or the garden, she felt exasperated and repulsed by it all at this time. All the things that mattered to her— the hounds and the horses and food and clothes and the friends she had taught to make ceaseless demands on her, even their endless feeding of her love of power and her love of praise lost grip and hold on her consciousness. These things were much less support to her now than they had been when Desmond was killed. Then she had absolutely required their importance. Now all importance seemed to hinge for her on the one fact of her success or failure over this love. In the space of a week there had been nearly as great a revolution in her life as Desmond's death had caused. She had loved Desmond really and tremendously. It was the only way she knew how to love. She had not kept in practice all these years. She had not grown less selfishly romantic. She demanded exactly the same return for her love as she had known before. In fact she demanded to have twice what only one woman in fifty has once in her lifetime. She looked for the same thing. She thought it must be the same. She had no possible sense of proportion. She looked for disappointment.

One evening after dinner they went down to the river to fish for trout. Susan and Simon went off, a rod between them and determined not to exert themselves over much or near their mother. David and Cynthia shared the other rod. First one would fish a rise and then they would move to another spot and the other would have a turn. They hardly spoke, the evening was so quiet after a hot day. The turns of the river were olive under the willows. The mountains, high beyond the last bend of water and the woods, were nearer rose than blue and more smoke than opal. The sky had the bloom of fruit. There was lilac somewhere. The river air was washed distantly in its pale billowing scent. The plumey sweet stuff leant far out from the edge of a dark wood towards the bluebells, their first hyacinth colour spread sparsely down the slope of the field, more like smoke than flowers. In an hour's time the lilac was as white as a sea-bird and the bluebells a cold dead grey. The river was black where it had been olive and had caught the first quarter of the moon between three dark stones. The mountains had changed to the exact blue of speedwells and the chill in the air was like a constant infinitely small shudder.

Cynthia said, "We'll go back now. It's too late to fish any more." She turned back along the river bank. She said bleakly, "I wonder what those beastly trout were feeding on? Nothing we've got."

"Wouldn't it be much quicker to go back through the wood?" David said.

"No, it wouldn't. It would be far longer."

"Would it matter?" He took the rod away from her and put his arm through hers. In the wood they kissed. Cynthia kissed like one who has found what matters to her most after long waiting. It was sleep after an endless night of waking. It was as needful as that. But it was like drinking wine to quench thirst. It was lovely but it did not quench her thirst.

David found what he had known he would find if he waited just long enough. But beyond this acceptance of things and himself as they actually were he was moved more deeply than he had thought. He knew that now he would hate to hurt Cynthia. There was something disarmingly and inescapably young, unquestioning and untricked about her loving and kissing. In love she was that same divine and unspoilt

187

creature she had been years past at Garonlea before she was married to Desmond.

They slept together that night. Cynthia was deeply delighted and comforted. She said so to herself the next day, accepting the fact of things as they were with entire simplicity. "It was such a comfort to me," she said tranquilly to David, and this was no more nor less than the truth.

He did not say anything. He was not very good at talking about love, either past, present or future, as Cynthia soon knew. But especially about future love he was disquietingly vague. Cynthia was so captured by him, so held by her newly recovered delight and peace, that for the present she would not risk provoking any discussion that might, in its unsatisfactory outcome, prejudice her happiness. She was shy too, with that shyness which especially hampers vain people. Like most of them, Cynthia could not ask for anything she wanted directly, a refusal would have shamed and angered her too much. And now that she had put herself where she had no terms to make, she was forced to keep up with herself a new pretence that the position was of her own choosing—that the present was enough—that she preferred to leave the future undefined—that (most useful of all phrases) *the whole thing was too difficult*. His wife was a Catholic. One could not expect her to do anything rational. Soon it became that for the children's sake Cynthia would rather Leave It Alone.

All these things were said to Diana and at the end of a month when David went away Diana knew quite clearly that there would be no divorce, no new marriage, no leaving Garonlea. None of the things she had thought David would mean had come true. Not one. And yet she could not dislike him or grudge him his victory. She liked him personally so very much. He was almost dear to her with his unaffected thoughtfulness and interest in her and in Simon and Susan. She could not have believed that she would ever like and respect Cynthia's lover so much. She helped Cynthia to bolster up all her pretences about the impossibility of anything further being done. She underpinned her vanity and extreme happiness, a loyal loving prop to both. It was wonderful to see Cynthia so happy, good-tempered and laughing and with something of her old rich gentleness coming back. She was more free and careless too, less distressed about detail. Daily the change became plainer to Diana.

XXVI

ALL that month while Cynthia and David were making love there seemed to be a kind of skin of ease and tranquility over Garonlea. Even Diana felt momentarily at peace with the house and the place. As one lovely day followed another she was aware of a sort of blooming on the new Garonlea of Cynthia's making. The rooms became more real to her and less as if they had only been painted on top of the rooms she had always known. It was as though Cynthia had really and at last won the house over from all its dismal hauntings. Those pale light rooms opening out of one another, full of bright chintzes and spring flowers—tulips and broom and irises and rhododendron—were at last really different from the old rooms with their shut and curtained doors, red carpets and rich dark covers and tiny scattered vases of flowers.

Outside too, the new rides and clearings and vistas in the woods had covered their scars. The woods no longer screamed to one another and to those who looked at them, "My wound that was a T is now an H." The battle was over. The dead were gone and the living throve without them.

In the garden Diana felt a surprised thrill of pleasure in the success of the schemes she had worked out with so little enthusiasm.

The children seemed well affected by the weather and the air of happiness too. They were particularly gay. Delighted to find Cynthia daily in better humour, they pursued their own concerns without having to fight any of those exhausting battles against her plans for their entertainment or improvement. They were very sweet to Diana, keeping her with them all day if they went racing so that she might not feel a bore to Cynthia and would not have to wander about by herself in the crowd, which she hated doing although she tried to be very independent about it at the times when Cynthia did not want her. It always surprised Diana that Susan should appear quite unmoved when loosed off by herself at race-meetings and point-to-points. She would wander about all day talking to all sorts of queer people on the cheap stand (for Cynthia often economised and let the young race cheaply), or on the course and watch a race with a sort of airy and actual observance and intelligence for her age. She seemed quite detached from any feeling of loneliness or self-consciousness in a crowd. She and Simon had been taken racing by Cynthia since they were quite small. Susan had always enjoyed it the better of the two, though she would not have been acutely distressed if she had heard she was not to see a racecourse or a racehorse again. Simon's lack of intelligent interest was maddening to Cynthia. Even now when she felt so happy she complained about it once to David.

"And do you know," she finished, "he doesn't even *quite* always read the racing news."

David was sitting in her room while she changed for dinner.

"No," he said. Obviously he was not attending. Cynthia was doing things to her face in the heavy-handed, absent-minded way that produced such a good result. She paused, stared into the glass, and did something slowly.

"You don't think it's important?" she said when she had done it. "I do think it's rather terrible in a boy of his age."

"Well, Cynthia, why should you bother him about it if he's not interested. Leave him alone."

"Isn't that rather what you said about their riding when they were small? And you see I didn't pay any attention to your advice and now they ride so well."

He saw her completely satisfied smile in the glass, a beautifully shaped smile, sure and a little smug. He remembered the reflection he and Simon had seen on the night of the dance. It was because of that he was here now with his back to the magnolia flowers and the soft evening outside the window. At the moment Simon really meant nothing. He was looking at Cynthia's long, bending, fat arms and strong, clean shoulders and thinking inevitably of the magnolia outside the window.

"He'll be all right in time," he said. "Everything takes time. Boys, horses, yes, everything."

Cynthia made a little face to herself in the glass.

"You're a genius for taking a good long view," she said.

"Did you say I was a genius for taking a long view?"

191

"Oh, it doesn't matter."

"Kiss?"

"Yes. . . . Now it really doesn't matter."

"All the same I know what you mean."

"You do talk so much. I wish I didn't mind about you so much, David. Isn't it a mistake?"

"You shouldn't criticise your lover. You should simply be delighted by him."

"Yes, I am, I think. But don't talk so much."

"Well, I think it's a good plan to keep talking at times . . . especially when it's nearly dinner-time."

"Well then, darling, please go away and talk to yourself severely."

Diana heard their low laughter. It took the air very quietly and seemed to be a necessary part of the gentle glory of the evening. It was not insultingly secret and apart as lovers' laughter can be, but like a smooth stone thrown in a smooth pool, its rings widened out and out through all the changes in the house. It was part of a whole. Perhaps Garonlea had never seen a love affair like this one. If so, the place took to it most indulgently.

On the day David said to Diana, "I'm going away on Wednesday, I think."

She said, meaning a great deal more, "And when are you coming back?"

"I don't know quite when, but I hope very often."

She knew she would get nothing more definite out of him, and in a way she was content with this. There was no reason why she should be easy in her mind about the continuance of his relationship with Cynthia but she did feel confident about it. She saw Cynthia's happiness and she believed in Cynthia's hold over any person she wanted. In this way she thought Cynthia entirely indefeatable. But here she was not a very sound judge for she had only seen Cynthia holding one man, Desmond, who had loved her romantically and singly and quite as deeply as she had loved him. And at that time she had been at the very summit of her beauty and her mind had been imaginatively sharpened to keep her love. Sharp and yet elastic because she was younger then and the years of complete power and success had not become as solid behind her as they were now. Of all the other men who had courted Cynthia she had not wanted one for loving, or if she had, her vanity had denied her flesh such indulgence. They were no more than courtiers and escorts. She had charmed them and hurt them and sent them about their business and whistled them back again when she wanted them without shame and without difficulty, because she did not mind about them.

But with David it was disastrously different. She was defeated and delighted and undone and uncertain and for all these reasons she loved him the more. But because he had defeated her and seen through her vanity and got closer to her than any one else, he had himself destroyed that very quality of strength and hardness and intense unhappiness in her which had first made him love her. For him it was now no more. Instead of this woman he now had one the more intensely dependent and wildly loving because she had depended so little and loved not at all for so long. It was like capturing a wild-looking white town on a Spanish hill and then discovering that its streets and houses are as only a Surrey garden-city, orderly and complete with every modern convenience and houses with "Tivoli" and "Mon Repos" written on their gates.

There were hours when he found her enthralling and his domination of her pleased his vanity and he felt fond and indulgent and romantical, times such as that when he watched her using her beautiful heavy arms sitting in front of her glass and thought of the thick skin of magnolia flowers so near in the evening. And there were times when his complete power over her exhausted and disgusted him. She was so much a less interesting person than he had thought. And still he could not escape from that feeling he had had on the first evening when he kissed her in the wood, the feelin; of her unquestioning confidence in love that had disarmed him of any unkindness towards her. He had known that he could never expect from her that hard and reasonable acceptance of a love affair as a passing thing which is the simplest expression of disillusion and absence of real success with so many women.

It was quite true that he could not without the exercise of tremendous and cruel determination have compelled his wife to divorce him, and he did not even distantly contemplate taking so unkind a step. He was fond of his wife and it was Cynthia's happiness which must take its chance. But he would not hurt her more than he could help in either her love or her vanity. And most skilfully he did not. He parried the openings she gave him for any suggestion of permanency in their love with assurances so vague that a girl of sixteen could not have founded any hopes on them. And though sometimes he implied that Cynthia was the one unattainable never really to be and so never really to end romance of his life—he must leave her for a little. He must go back to England and his horses and his catholic, unrelenting wife and his one tough, ugly, witty little daughter. But soon he would be at Garonlea again. Soon and often, with her at Garonlea. It must not be too long before he came there again. He would have to see her soon. He was only happy with her and at Garonlea.

His touch was very light. He was so uncaring and apart, for all his understanding and kindness, that even such promises became unduly weighty. Cynthia could not reasonably keep him now. And it was beyond her vanity to look for any definite assurance. But on the night before he went she whispered for hours with desperate earnestness that need for his love to continue . . . Do you? . . . Are you sure? . . . Will you always? . . . Do you still? . . . Ah, I know you don't. . . . There were tears and kisses and love and deep sleep and a feeling of safety and peace that lasted for some time after he had gone away.

Soon after that, Simon and Susan went back to the schools they found such a heart-breaking bore, and Diana and Cynthia were left alone again at Garonlea. Cynthia had more plans than usual for that summer. She had asked the people she most wished to impress with her skill and success over Garonlea to come and stay in a continuous succession. She had what she thought an undefeatable show-horse which she had intended to take round to all the agricultural shows during June and July, ending up with a big triumph in the Dublin Horse Show in August and the right size cheque from the right sort of American going home to Garonlea instead of him. Cynthia liked riding in shows and did the thing to perfection.

Then a month after David had gone, Cynthia was abandoning all her plans, all the summer procedure that she had set herself. Putting off the visitors (except poor little Muriel whom Diana could look after), finding some one to ride her horse in all the shows except Dublin. Yes, she neglected Garonlea, her proper setting, and went to England for the summer and bought herself a great many lovely clothes and was a trifle importunate altogether.

Tactically, of course, it was a great mistake. If she had really wanted David and desired no more than to keep his love, she should, as it were, have remained in her tower. At Garonlea, which was both Cynthia's tower and kingdom, her love with David had taken its course with ease and dignity, or as possible a measure of both as the circumstances allowed. The worst small, furtive aspects of an intrigue at least were absent, the time and the place had been so right for love.

But that summer in England, things went very differently. A word from him had brought her flying over for a Newmarket meeting. They stayed together in a friend's house. He had wanted to see her then badly and it was all a great success, worthy of any attendant difficulties. David had never seriously considered Cynthia as necessary to him, but he kept on rediscovering that he minded about her far more than he wanted to. Her loveliness and eagerness and success with other men and her full, perfect and sufficient love for him made her very dear and exciting. If after that week she had gone back to Garonlea she would have gone near to reversing the question of power in the matter.

XXVII

But Cynthia did not go back to Garonlea. It was beyond her power and beyond her imagination to do so. She spent the summer in England, staying with her friends and seeing David in their houses and in London. She saw far too much of him under the circumstances which were bound to be wearing and difficult. He was quite enough in love with her to make tremendous efforts to see her and be with her, and the necessity for secrecy and intrigue soon more than equalled the good he got of the affair. Soon there was not so much secrecy as intrigue about it. All this can be very exhausting. It came to be so with Cynthia who, used to being supreme and unquestioned, was inclined to conduct the affair like an importunate ostrich. She must see David. It was necessary to her happiness and well-being, and if she chose to presume that none of her friends suspected the truth, that was enough. She did not allow herself to think they suspected it.

It was enough for Cynthia but it was not really much help to David when his wife's closest friends took him aside and tried to have long confidential talks on the subject. He realised, if Cynthia did not choose to do so, that people, especially people who liked his wife, were no longer very keen to fall in with plans for parties that included Cynthia as well as him. In fact, they became remarkably cool and tiresome about it in a short time.

But would Cynthia see the growing difficulties? No. She continued to make plans for weeks ahead, writing down dates in an inexorable way in her little engagement book. She looked happier and more beautiful than she had done for years. If tiresome people would not have her to stay on the week-ends when she wanted to be with them, she found other less tiresome people who were only too enchanted to have her and David together. She failed to see that the matter was becoming obvious to others and trying to him. She just went on and on, ignoring the fact that the affair could be recognisable to any one.

The situation became every day more difficult for David to contend with. Since the situation was difficult, Cynthia became difficult too. He still wanted her, but he wanted her far more to go back to Garonlea and wait there for him to pay those frequent romantic visits he had promised her and himself. That English summer, with his wife being rather sour and silent and dignified, and his friends obviously disapproving (though, of course, Old Boy, it was none of their business), was as unfortunate a way of spoiling the possibility of happiness as Cynthia could have found and insisted on. She was asking for failure, while she presumed with imperturbable confidence that this love could develop and endure as that earlier love in her life had done.

At last she went back to Garonlea. She had a most sensible reason for returning—she wished to ride her show-horse conscientiously during the fortnight before Dublin. Even in love she did not leave an important matter like the sale of a horse to chance. She told David it was almost more than she could endure, saying good-bye to him but, after all, it was only for a fortnight. He would be over for the show and would come down to Garonlea afterwards. Another plan was made and David agreed to it. He would have consented to anything then, he so wished Cynthia would leave him in peace to recover some of his wife's esteem and attend to his business of training horses to which she had been a grave interruption.

He did not come over for the show. Cynthia was amazed and angry when she understood that he was not coming. But when she told Diana she elaborated all his excuses . . . Goodwood . . . his wife . . . he couldn't leave the horses . . . to such an extent that she believed in their reality, and only Diana was left in grieved surprise and doubtfulness.

XXVIII

THAT Horse Show was like every other Dublin Horse Show. In spite of the fact that Ireland was at the time in a state of war, the Horse Show was its usual and inevitable social and business success. Triumph and failure, disappointment and disaster rode round the rings together. The whole of Ireland went there. Women in grey flannel coats and skirts, awful hats and brown suede shoes sat on shooting-sticks round the rings and gossiped and criticised and sometimes admired. There were busy men in clean breeches and boots without a moment to spare for anybody, and less-busy men in suits and bowler hats who had lots of time for a drink with anybody who would pay for it.

There were Indian princes in Jodhpurs that zipped down their legs in all directions, and every one in Ireland with a horse to sell coveted their acquaintance. Old ladies coming to see the flower show jostled foreign soldiers in Belgian blue and French grey who would ride their horses in the jumping during the afternoons. It was comic to hear horse-talk going on in French: "Mais c'est effrayant." A gunner, now, would take longer to say just why he didn't like the dreadful horse.

There were lovely girls from England who made the lovely girls in Ireland look nothing. They wore their clothes so much less briskly and painted their faces with less-advanced skill and determination.

And last and first and all the time and through everything there were horses. Horses in the rings. Rows and rows and dormitories of stabled horses.

Horses led and ridden about wherever anybody wanted to sit and have a chat. Never in any place are horses so thick upon the ground as in Ballsbridge at this time. After all, in apology for the nuisance they are, it has to be remembered that this is the greatest horse-show and fair in the world, and it is natural that the warm, close smell of horses and their bedding and their leather accoutrements and their attendants should be heavier on the air than July dust under chestnut trees, as one pushes through the turnstiles to the show grounds, each morning a little later and a little more wearily as the week goes on.

Cynthia was there at ten every morning. Brisk and beautiful in her blue habit, no matter how late she had been to bed the night before, and she was rampantly gay that week, drinking too much, dancing too much, making (had she not been Cynthia) rather a fool of herself for her age. Still, she was able to attend most effectively to business. Her horse won in his class, won the light-weight Challenge Cup, was runner-up for the championship of the show and sold for a vast sum to a fabulously wealthy American. She accomplished what she had set out to do with her usual royal success and among the applause of her friends.

Diana should have felt happier about her than she did. Everything had gone wonderfully and there were six different men at every moment eager to fill the blank left by David's absence. Diana should certainly have felt happier than she did on the Thursday of the show when she and Cynthia made their way down through the main hall, full of rather jaded-looking stall-holders, towards the place where her horse was stabled.

Diana felt absolutely exhausted and sickened by the show, by the sight and sound and smell of horses and the people who got their living or their pleasure by them. She felt if another man with trim legs and tired eyes and a bowler hat tilted over one of them stopped Cynthia and asked her to come and have a drink with him, she would like to scream and spend the day in a stall of Irish lace or nosing round the already wilting flower show, abandoning her job as chief attendant and messenger to any one who chose to take it on.

Cynthia was slightly strung up. The championship was judged in the jumping enclosure where horses could really gallop and her horse took a bit of a hold. She exasperated Diana by stopping to converse with every acquaintance she met on her way to the stabling. Telling each one of them in exactly the same words how cowardly one of the judges was and how she was sure he wouldn't begin to give her horse a ride, and at the same time how a child could ride the horse. All round, the air was crossed and recrossed by broken currents of horse-talk; people standing still, people walking on, people with nothing to do and people in a hurry, all talking and murmuring about horses, their soundness and unsoundness, their slowness and their speed, their stupidity and skill in leaping, their strange inability to walk, trot or gallop as those who bought them wished them to do, and those who sold them upheld they could do.

Cynthia fiddled about her beautiful horse for a long time, talking to her groom and telling every one who passed how cowardly the judge was. She sent Diana on two messages and then went away to keep a drinking appointment with her prospective buyer. Diana had twenty minutes before the class went in. She hustled along to the flower show. Here, too, she was only conscious of the horrid oppressiveness of a crowd. Her mind refused to act about the flowers. She forgot

what it was she had come to see and inquire about. She was jostled and overcome by people who wandered about, ignorant and unseeing, filling in time with no intention of buying or growing any of the flowers exhibited. Their talk, too, made one hate everything connected with a garden, just as the talk outside reduced horses to a plane of extreme boredom——
"My dear, don't you call that wonderful——"
"*There's* an unusual colour——" "Now, Elsie, we must make a plan. You see, I can come by the Dalkey train if you can meet me. Shan't have a lot of luggage—just a light valise——" "Yes, dear. Pretty, isn't it? But I like the salmon pink best myself——" "Darling, those *quilled* puce carnations—*too* lovely——" "Now, how much money have I got left? Well, how much did you spend at this stall? I don't know. Well, ask him. Ask the man who took the order. Oh, could you tell me—I'm so sorry to interrupt you—but could you tell me how much I've spent——?"

Diana nearly ran out of the flower show, hurrying past the monster rainbow bouquets of carnations and the pale, airy-blue spires of delphiniums, past the rich artificiality of gladioli and dahlia, back into the horse-world again. She was very nearly late to put a hairpin straight in the back of Cynthia's veil and watch her mount, superbly confident after her drink, and ride insolently, easy and magnificent as she always was, through the crowded rings towards the jumping enclosure.

The judging of the championship went on for hours and hours and hours. Diana sat in the stewards' box with her chin in her hand and Cynthia's handbag and her own on her knee and watched in the mist of invincible ignorance about horses' shapes (through which she had never actually learnt to see the good from the bad or the better from the best) the long-drawn-out process of Cynthia's defeat. She saw the noble throng of horses moving at different speeds round the great oval of green grass. Rather a lovely sight, really, especially when they laid themselves down to gallop in the shining, morning sunlight. She saw Cynthia among others called in to the centre of the ring while many horses looking to her much the same, sadly left the gay picture. Then followed a period when the judges judged and judged and horses had their saddles buttoned and unbuttoned and their riders ran them up and down and then fidgeted with them ceaselessly in case they went lame standing still.

All Ireland and England that could squeeze into the stewards' box had done so. They sat in tiers and gossiped loudly about the horses and people whose sisters or lovers or cousins were bound to be sitting within twenty yards of them at the most.

"Oh, I think he's a horrible horse. Look—he can't begin to gallop——"

"Forty, my dear, and she looks twenty-five. One can't get away from it, and I hope quite suddenly she'll just break up."

"Johnny gave Madeline's horse a shocking ride. Didn't you think so——?"

"Yes, I've had mine out, but oh, it was horrible. I shall never forget going into that nursing home. . . . Waving good-bye to every one, I was almost in tears; I thought they were all *too* unkind to want to cut poor little Lila up. . . ."

"Now I call that a lovely horse. I've always liked him. I wanted to buy him as a two-year-old and Robin wouldn't let me. He's never been allowed to forget it either. . . ."

"You know I'm not a prude but last night really I

204

thought every one went too far. I don't know what the man on the lift thought about us. . . ."

The Judges were riding the horses now. Diana could not make out which was the cowardly one. She was so hopelessly vague about the people Cynthia knew well and she could not tell by just looking. They both seemed to her to be decorative and able horsemen. And in this she happened to be quite right. Cynthia's horse did not begin to frighten either of the judges. He was beaten because they liked another horse more and, as usual, many people found reason to say they were ignorant and prejudiced and cowardly, and without a doubt about it, Cynthia's horse should have been Champion of the Show.

Cynthia was disappointed but not actively vindictive. She had done more than well and her horse was even better-sold than she had hoped, and her hopes had been high enough, for she never under-rated herself or her possessions. She went back to Garonlea with an enormous cheque in her handbag. And at her heart a less-hungry, wildly-defensive feeling about David than she had known was there and denied to herself when he failed to come. In fact, she felt rather bold and grand and entirely less vulnerable about him because during the week she was in Dublin without him and surrounded by other men she had discovered in her escape from consciousness about David a new spark of attraction and interest towards some one else. She did not tell herself this because she did not know or guess what it might mean. She had no idea of the emancipation that had come on her since her affair with David. But she felt a stronger current of emotional excitement than she had ever experienced with vague, inactual worshippers that had succeeded Desmond and preceded David in her life.

It was tragic if perhaps natural that Cynthia should come off her spiritual pedestal, cease being an unattainable ideal and become a vulnerable creature almost at the last possible moment in her life. With ignorance and vanity and unhappiness she had cloistered herself in the years after Desmond's death, when she could have had one man or all men. Now, in ignorance and vanity she made her belated experiments in love. She was not too old if she had had a more elastic temperament. But she was too set. She had been a queen too long, and the tide of her beauty was too surely on the turn. All her power over men and women was of a very temporal kind, lying in her looks and her bravery, and their admiration for these qualities. Not in any nearness of spirit or sympathy she might have with them. In a way her popularity was a cheap and public sort of thing. Although she had bought it dear enough and kept it so long, it was as unstable and as dependent on her own constant effort and upkeep as such popularity by its very quality must be.

XXIX

JUST as the Horse Show marks a division in the
Irish year between definite summer with horses at
grass and hunting forgotten and indefinite winter
with cub-hunting in view, and fat, preposterous horses
to be caught up and timidly mounted, so this particular
Horse Show became for Diana an ever-fixed mark of
that time when Cynthia's whole life slipped on its
decline. She dated it all from that first week in August
and their return to Garonlea with Gerald Turnbull,
the huge and charming American with the gentlest
voice and the most unquenchable thirst for whisky.

On and off he spent most of that winter at Garonlea.
He was enraptured by the place and the house and the
hounds and the horses and he enjoyed making love to
Cynthia, and when she was faintly drunk she did not
talk quite so much nonsense about love spoiling their
friendship as she did when she was sober.

David had come back to Garonlea in the autumn to
find Gerald almost installed and rather inclined to
show him round. Cynthia was on the defensive. She
had Gerald there on purpose. She produced his best
American jokes and showed him off to David. But it
was no use. David was only amused and politely
ironical. Cynthia had to explain without being asked
that Gerald was not her lover and David said it was
perhaps a pity, a fine, able man like that who knew so
many good stories too. In a week Cynthia was
recaptured, a bird between his hands, and Gerald had
been sent off. But in another week David went away

and did not come back. Gerald came back with new stories to tell in his soft voice and all the money possible to spend on any horse Cynthia told him to buy, or on any present he imagined or she let him know that she wanted. David had given her little pieces of jade and some handbags. Gerald gave her diamonds and a mink coat. She could have married Gerald that winter if she had wanted to. But she did not want to. She felt idle about it and she was always hoping that one day she would win David back to her. She could not know that in each day that passed she grew further away from that hurt, secret creature she had been when for a little while he had found her enthralling.

That winter Garonlea was strangely isolated from the near world of neighbours. A civil war was going on in Ireland, much to the inconvenience of social life. Cynthia and her hounds were so popular that there was never any difficulty over the hunting, but motor-cars were commandeered and bridges were blown up and houses burnt and people frightened away from the country. It was not possible to be out late at night, where neighbours dined, there they must sleep. Because of this, no one thought very much of the matter if Gerald, whose hotel was five miles from Garonlea, should lie five nights out of the seven at Garonlea. The crested wave of Cynthia's popularity had not yet turned. She did so many things that people admired. She swam her horse and her hounds and her unwilling hunt-servants across rivers when the bridges on her road to meets were broken. She compelled her rich American to buy horses from her friends. She had poker parties and lovely food and drink and soft beds for her guests to lie on when they had lost or won. An occasional decorous evening's bridge and a deliciously-planned dinner for the elders of the country kept them still just where she wanted them to be. She was wonderful, they all said. She went her way with all classes. There was a meal and a drink for any man on the run at Garonlea. Or a bed and no questions asked. Yet miraculously the house escaped burning by either party. Cynthia was really loved by the Irish people who knew her. She would have done anything for any of them or they for her. She was closer to them than to her own kind and class. Most of the people on and near the place knew, deplored and accepted her relations with the big American. They looked on her as too great to marry a man they truly thought common and beneath her, and the women would shed tears and spend prayers for the frailty of one not of their blood or religion.

Cynthia felt closed in and contented that winter at Garonlea. In January the thought of David was infrequent and not important. She had never had a better season's hunting. She liked having Gerald for a lover. All through, her spirit and her body felt of a smooth, thick texture. She was possessed of a happiness that seemed inviolable. The atmosphere of Garonlea grew ever richer and easier. It had become a house in which the true luxurious spirit prevailed. It seemed as though Cynthia's victory over Garonlea and the unhappiness bred there was complete. She had sustained without bitterness the loss of her first lover, and established her happiness with another. She had imposed her will of the flesh and the spirit on Garonlea, refusing the draw of its dark, established currents. Not to be happy was unnatural to her. It seemed as though now the years when she had gone struggling and hungry, held by Desmond's memory and her own faithful vanity, were over.

XXX

NEARLY dinner-time at Garonlea. A quarter-past eight on a January evening. Cynthia was down first, trailing about the library in a black velvet dress, one of the soon to be inevitable chokers of big, blatantly-pink pearls casting fat shadows behind their circle on her long, full neck. She poured herself out a cocktail (Gerald's special) and sat down away from the huge fire. The whole room was as equally and agreeably warm as a bath full of water. Groups of freesias and papery sweet narcissi and violets set about in silver mugs gave out their delicious evasive chorus of scent. A little terrier rose from its dark nest and leaped neatly as a robin on to her knee, turning its eyes away from the shaded light behind her. Cynthia looked at it distantly for a moment, enjoying the exquisitely clear and delicate lines of its throat and neck, the bird's poise of its head, before her hand went down to it in a slow, adequate caress. She was tired after a long, successful day's hunting, but not so tired that she did not look with pleasure towards the lovely meal about to be. How right she had been to order champagne to-night. A necessity. Really the only thing.

Gerald came in. Enormous. Heavy shouldered. Carefully shaved. Most attractive in that red velvet dinner-jacket made by a London tailor. He took her glass and filled it again for himself, giving her a new drink. It was a simple sort of love token, rather like a mink coat, but cheaper. Then he lit a cigarette and stood with his back to the fire.

"Cynthia, you're looking very wonderful," he said.

She sent him a look as slow as her hand had been going out to the dog.

"And didn't that horse of mine carry you well to-day?" He swallowed half his drink. "You may say it gives me a big, big thrill to see you ride to hounds on my horse. Certainly it gives me a thrill. You bet your life. I want to see you ride that horse in Leicestershire."

"He's a tiresome sort of horse," Cynthia said thoughtfully. "But once you find out how to ride him he's grand. He's not your horse though, Gerald. For one thing, he's not really up to your weight."

"I bought that horse, Honey, before I ever met you. And listen, sweetie, if you like him and get on with him—he's yours."

"What's yours now, Mamma?" Simon asked. He had just come in. He tipped up the cocktail-shaker and looked from the teaspoonful of drink in the bottom of a glass towards his mother and her friend in a way they could hardly mistake, before he poured himself out a glass of sherry. His grace and assurance had improved. Gerald thought he had a disgusting amount of confidence for a boy of twenty and often told Cynthia that what he wanted was a Father's Authority.

Cynthia said, "Simon, how mean of us. We've drunk it all. Gerald is very sweetly trying to give me Klondyke, the horse I rode to-day."

"I named that horse Klondyke, you see, because I hoped he'd prove a gold mine." Gerald adored naming his horses and explaining about their names.

"Yes, I thought that must be why you named him Klondyke. Yes. I expect he will be a gold mine to Mamma."

"I haven't accepted him yet," Cynthia said with dignity and without humour.

"You don't want to make me mad now, do you?" Gerald gazed across at her, Simon forgotten. "You know I'm crazy for you to have this horse."

"*Pas devant les domestiques*," Simon murmured to himself in governess French.

Cynthia respected Simon now, though she found him trying. She caught back her answering look and said, "We'll think about it, I promise. You must ride him again and if you really don't like him——"

Gerald, who, until Cynthia had told him so, had been unaware of how much he disliked his best horse, nodded wordless agreement.

"How tiresome and late Sue is," Cynthia had no more to drink. "And Diana. I'm so hungry."

"The one good thing about hunting is its lovely hunger, don't you think so?" Simon asked Gerald.

"I'm afraid I hunt for love of the sport itself," Gerald replied ponderously, and could not understand why Simon looked so delighted. He did not know Simon and Susan's game of drawing Bromides from him.

Sue came in. She was wearing a fancy dress that Simon had copied for her from the portrait of an ancestress, green brocade and a quilling of net drawn on a string round the square neck opening. She had a little white dog, a silky, plumey little cur that she sometimes led about on a ribbon if she wished to annoy her mother.

"So sorry, Mamma."

Only Gerald sympathised with Cynthia when she said she wished her children would call her "Cynthia" or "Mum." "Mamma," so ethereal and unhearty and old.

"But Diana's behind me. Oo Simon—just one sip."

"I don't approve of young girls drinking. They don't require any artificial stimulation. But here you are, sweetie, just the same. Why? Gee, because I love you, honey."

Another thing was to talk as they fancied Gerald might do in emotional and unbridled moments.

"Delicious sherry wine to keep out winter chills. Thank you, Simon."

Diana came in then in a spruce little dinner-jacket. Her short hair brushed and pomaded and charmingly grey at the temples. Her finger-nails trimmed squarely. A blue lapis signet-ring. Simon helped her to dress like that. Although he was at Cambridge now, he scarcely realised the implications. In 1922 a great many people did not. Especially in Ireland. He and Susan had bought the lapis ring out of their small allowances and great love for Diana. There was never any difficulty about the names they called her. She had always been Diana and other fond little names.

"Diana, we're rather hungry, darling. But you did have some lunch to-day and we didn't."

"Cynthia, I'm sorry——"

"Don't you be sorry, Diana. You're ten minutes late because you've fed the puppy that has distemper, listened to Mamma boasting in her bath, sewn on a button for me, made Sue almost be in time and after all that you look as tricky as possible."

"If Simon would stop talking we might go in to dinner," Cynthia gave Diana a private smile. Simon was used to do all the forgiving.

Gerald, who enjoyed food and wine and love with elaborate greed, opened the door, courtesy elaborate too. The one thing Simon really liked about Gerald was his knowingness about food and wine.

Strengthened by the soup of extreme strength and clarity they began to talk about their day's hunting again. There was a conscientious preamble about the

213

chase itself: what the Fox had done; what that young hound Tarquin had done; why old Venus must die. But soon they got to the really enthralling part: what Cynthia had done; what Gerald had done; what other cowardly or mistaken people had left undone.

"I did a remarkably foolish thing to-day——" This was Cynthia helping herself to sole in a sauce that just tipped the edge of heaven. "Really one would think a woman of my age would have had more sense."

Gerald said, "You scare me blue, the fences you take on."

"But four foot of timber and into a tarred road. I should have had more sense, shouldn't I?"

"You certainly should."

"But I was in a hurry and I was riding your horse, Gerald. I knew it would be all right."

"Where was it?" Gerald couldn't bear to think he had missed leaping an ugly fence. He would almost go back at the end of a hunt to jump anything he thought might cause his horse to fall or himself to suffer injury. It was a fetish with him.

"Leaving Gramore. I got a little left on the far side of the covert and they ran like smoke for almost a mile."

"Oh, yes, honey. I remember missing that blue back in front of me away from the covert side. Then you appeared from nowhere and none of us could catch you."

"That horse flies."

"I wouldn't mount you on a slow horse, now, would I? Would I be likely to? But, honest, Cynthia, I'd like to have seen his answer to a question I asked my old Foxglove, going away from the covert. The field were all going for a gate——"

("That's US," Susan and Simon said.)

"I dare say you were correct," he accepted their admission kindly. "Well, that was when I pulled out left-handed down a steep slope all rock and bog and bushes and asked old Foxglove to jump the highest stone-faced bank I've seen in this country. Yes Sir, and the ground sloping into it—— Cocked his old ears——"

("I knew he would," Susan and Simon said.)

"—and went right bang to the top. That was his answer."

Susan and Simon took their blessed refuge in an exchange of looks, silent communion from the further side of laughter. Then they devoted their attention to woodcock, a Spanish sauce and a surprising salad. The wine had drifted their fatigue afar. They remembered their day's hunting with enjoyment made clear and understandable at this short space of time from its happening.

Simon and Susan felt very emotional about hunting. There was nothing hearty or commonplace in their enjoyment of it. They refused to talk about it like Gerald or a *Tatler* correspondent. It keyed them up on a higher pitch of imaginative excitement. They loved the emotion they had from it; the thrill going beyond fear. They did not deny the fear. Now they endured it, saying to each other, We have so much imagination, that's why our stomachs are so queasy——

People who like hunting in this way are not tremendously sane but they do experience a great emotion, the contrast between danger and safety, heat and cold, a hungry stomach and a full one, thirst and its assuaging are some of the things that take this sharp importance.

For Simon and Susan there had been to-day a wait beside a dark grove of hazels associated with all the fears of their childhood. The wall that had loomed so

cruelly was not a bogey now. But the feeling of inevitable danger was still there as the hounds, heavy, bright
creatures, filled the grove with their voices. Filling the
little aisle between the close-grown spears of hazel and
the secret fox-ways, pale bents and brambles where
little foxes ran and turned and ran on again with the
noise of their hunting. The air of the day was lost in
the clamour. Hot or cold or fair or stormy, it was a
void now, hollow but for the overture to danger. They
felt sick. Their horses were like lions beneath them;
they were enveloped in the terrible music and the
waiting was like waiting for death. There was no
other importance in the fields and fences and rivers
and unfamiliar distance beyond the necessity of crossing
them. Their other reasons stood beyond this. The
mountains and the receding day were as distant to the
moment as houses seen on the way to a scaffold might
be.

Then at the unbearable moment of tension the scene
changed. A voice, shrill and surprising as a bird at
night-time, telling the fox was gone. Wild and in-
human as a hound's voice this call for pursuing. Then
the horn blown and blown again and the hounds' bodies
no more bright and heavy and separate but creatures
light as fire, hurrying and flying together like fire in a
wind, settling heavy as bees an instant then gone as if
winged.

Pursue them. Pursuit through the hour that has no
time, over fields that have no name, a spirit purged of
itself pursues. A horse that was a lion is now a dove,
a part of self, but braver and more skilled for the brave
time. Green fields, dark fences, bogs that had no
meaning yesterday, dark alders and bright willows
let us through. Oh, kind little gap, be thanked. White
seagulls following the plough slip behind, forgotten.

Little arrowy snipe darting from white grasses, what is it to a mind and body so hotly set forward?

Before the chase is over the lightness and the thrill are suspended by moments of fear and heaviness and lack of quick choice. Fatigue of spirit and body weigh despairingly. A tired horse, staggering and recovering, is more sickening than a fall. The wild feeling of conquest is gone and yet the pursuit goes on. The hounds pursue, fearless, merciless, tied to their intention, untiring. They must be followed, with tremendous effort, vast expense of energy, absurd endeavouring. It is the phase of unhappiness which endures till fear is strangely purged again from the spirit. Thrill is passed and fear is passed and a period of reality survives in which the most jealous rider will take grateful advantage of another's decision, throw shillings to the gate-opener and desire only to endure until the end, not to surpass in brilliance and bravery his fellows.

An insufferable thing, a treatise, a homily on hunting. No, but a day in the spirit of those two who feared and loved it, Susan and Simon, would be like that. Composed of sickness and wild love and the real pain of fear and, that cast out, content to suffer or enjoy and then at the conclusion again exhilaration. Relief? Physical pride in endurance? The held memory of lovely things seen from a strange angle? The heron that rose from its fishing, the little fox escaping. The swiftly flying beauty of a pack of hounds, that dream-like quality they possess, the quality that goes behind the horns' notes too like ghosts of crowns on the air. A magnificence, a royal tradition, lies behind all hunting.

But such a day was done. They sat together in the changed dining-room of Garonlea. Diana for one could never overmaster the feeling of drama which possessed her at Cynthia's conquest of this room. Now she sat where Lady Charlotte had kept her place for so long, but the change, it was too far-fetched, again too dramatic to endure.

Lady Charlotte in jet and mauve satin, diamonds sprawling on her great square bosom, possibly an aigrette pinned in her dignified puffed and greasy hair, directing such conversation as there was. The first primroses. The Kaiser must be hanged. Surely I know a Cox's Orange pippin. Well, Diana, have you nothing to tell us? Nothing? And I think you spent the day at that enchanting house, Rathglass?

"And Ambrose: What's this Coulthwaite? What's this—Sago? *Really*, I cannot eat sago. . . . Where is the meet on Thursday? You didn't find out? No, it's not a Cox's Orange, I'm convinced it's not—Diana, try a piece—— One of those terrible nights of tears and unknowing frustration came back to Diana. How cruelly, sharply, the past can live even in contentment. Her eyes went down the table to find Cynthia and all reassurance in the warmth and beauty she had made. In the delicacy and plenty of the food, in the luminously pale room, and in the glow of wine. Cynthia in Lady Charlotte's chair, her black velvet dress with its great sleeves was winged and shadowy. It melted into the gloom the coarser fullness and thickness of her body. She seemed melodiously calm. Satisfied. Still beautiful. More beautiful because her lover found her divine.

Divine was what Gerald actually found her. Beyond him. Goddess of the chase he loved so much. He could not give her enough or love her enough, and he was generous and wholly loving. He had lived with Cynthia for three months now and his whole being was centred in his desire for marriage with her. To

218

make her his own so that she could not escape him. He was very simple in his possessiveness. He wanted her— his love—to keep. He was not unsure as David was unsure. Artistically selfish, living only for the moment, holding his life and his love at a distance, powerful in his refusal to sacrifice himself. Uncertain, unpractical, romantic, full of understanding yet hard as steel.

Could nothing tell Cynthia that this was her hour, her last hour, her day was ending, her power would soon be past, and here, giving into her hand, was the super conventional happy ending that her life demanded?

She could not perceive the truth of this. Always beyond her physical contentment with Gerald were the wraith-like, uncertain memories of her affair with David. The uncertainty and delight of those days of last spring, the joy and laughter and delight of her love with him. Other things, Simon and Susan's pleasure in his company remained with her. She compared their attitude to Gerald, half-scoffing, half-pitying, to this and it almost caught her to the same attitude. The laughter of the young is such a sincere belittling. Again, David's almost cruel independence contrasted with Gerald's heavy reliance on her will and choice in all matters. He was too trusting. Too generous. Too loving. Too loving was the worst offence. The pity of the mistakes people must make when they love sincerely. But it was not through too much loving that Cynthia made her mistakes.

Simon said:

"How lovely these peaches are, yours, Gerald? Sue, ah, Honey, try one. Just to please me. They really are very delicious."

"They're marvellous, Gerald," Cynthia said. She felt almost as earnest and unimaginative as he seemed

219

in the light of the children's quick mockery. His peaches and his orchids and his carnations came just too often. Not that it mattered. Other things about him really mattered far more. She poured out another glass of cointreau. On top of champagne, too. But nothing ever made her feel ill to-morrow.

Sue was sitting, affectedly enough, beneath the portrait of that ancestress whose dress Simon had copied. She was giggling a good deal and wiping her enormous eyes, sometimes on her napkin sometimes, on her handkerchief. Even a sip of wine made her like this. Anyhow, she always giggled for hours before she made a joke—"My jokes are so poor, people must have time to prepare their laughs."

Now she leant back her head and pointed to the portrait of her dress with an upflung hand, not looking.

"Pearls, darling. Not that I want to fuss or complain. But isn't it strange there are no Garonlea pearls? There ought to be, shouldn't there? Where have they all gone?"

"I can't tell you that, Sweetie, but I know where they ought to be and know where they will be, next time you and I make Bond Street together——"

Sue yawned, her laughter all gone before the joke was over.

"We're so unkind," she said to Simon, and then to Gerald:

"Dear Gerald, let me ride your lovely mare, ' Kiss Me To-night,' on Tuesday. And why did you call her that?"

"You certainly shall, Sue. She's a perfect performer. Beautiful mouth. Beautiful manners. Yes, I called her ' Kiss me To-night' because she was by Presto out of a mare called Gipsy Love by Poets' Corner out of Spanish Lass. See? It all fits in as neat as pie."

"Oh, it certainly does, Gerald."

"——How does she endure him at all, really, Simon. Tell me——" Sue said to Simon. She was sitting up in bed, the collar of her pink, quilted dressing-gown turned up like a shell round the back of her round head and little neck. Her lovely frog's or cat's face was polished and clean. Simon liked to see her turned back purely into young, virgin woman. There is something exciting about this sudden stripping of artificiality. It is each night a stranger's pale oval of real face. Pale lips that were red. Pale lashes that were dark. Wax flesh. Smooth hair that curled. "Will she marry him, Simon?"

"Oh what a disgusting thought. Is he to be always with us? Saying, Honey. Saying, I want you to know. Eating in that French-American way with his mouth half open. Did you hear him say his horse had very good 'gaits'? Action, he meant. It's a word."

"We shouldn't listen to what he says about his horses. It would make us give up our good food."

"He's clever about food."

"Will Mamma marry him?"

Marriage. Simon always remembered that love meant marriage to Sue still. Their mother's friends all wanted to marry her, of course.

"Perhaps she will, Sweetie. Do we want her to?"

"And us to live alone here with Diana? Oh, yes, Simon."

"It would be nice. But how soon would you marry and leave me?"

"Never, Simon, never. I have no need of Love and chaps when I have you."

"I think Diana would stay with Mamma."

"Oh, no, we must have her."

"I don't think we shall get her."

"How tired I am. Please put my dressing-gown away and open the window as little as possible. My horse did frighten me to-day, and oh, it was exquisite. Good-night, dearest Simon. Give me my little dog to keep me hot. Good-night."

Simon knew she was asleep before he was half-way down the wide staircase. She was the most complete creature.

That staircase was the wrong background entirely for Simon. He was too olive and oblique for its space and clarity. He required a background of red wall-paper and too many pictures. He was wrongly cast for his part in Cynthia's new Garonlea.

In the library he read the paper. The atmosphere of the room which he had broken left him feeling acid and miserable. His mother's idleness and silence, her palpable mood of waiting were obvious to him. So was the fact that she had, quite gracefully, had too much to drink. Gerald he hardly regarded. For him he only felt an absurd, enveloping contempt.

Active anger stirred in him towards Cynthia. He thought of Susan and he thought of Garonlea and he was furious that she should so slight the importances of either by this greedy intrigue. Simon was a person who would soon be very fully conscious. Even now, although he went to bed soon, muttering pleasurably to himself, "call you his mouse—or paddling in your neck with his damned fingers——"—he was completely aware that for his mother's sake he did not care two-pence who called her what or who paddled where.

XXXI

A WINTER's morning. A sky like a dirty old slate.
Trees untidy, lead-coloured brushes against it. The
air full of snow and rain and the hour full of the
absurd necessity of going out hunting. At such a
moment there is one thing impossible and that is to
see a summer's day. White flowers in the evening
seem the only terms in which it can become remotely
visible. Say, stephanotis in darkness.

Ten-fifteen at Garonlea on such a morning. Every
one felt cross and unnatural and a little late with
everything they had to do before they faced the
outdoor day.

Sue could not master her hair or her stomach. Her
hands were cold. She nearly cried.

Simon complained about his new boots and the bitter,
bitter, unfair weather.

Diana had to hunt to-day because it suited Cynthia
to provide her with a horse on Tuesdays. She had no
complaints to make except that her hat gave her great
pain.

Gerald, standing by the fire in a coat with a fur
lining and a fur collar and boots beyond all dreams of
brightness, was the only really comfortable person.
He was drinking a large portion of port and brandy.

In a moment Cynthia would be drinking too and
feeling better. Up till now she had been flying noisily
about in her boots and writing out stopping cards to
catch that day's post. In the snowlight her face, under
the peak of her velvet cap, looked old and faintly
sagged and terribly preoccupied.

Susan did not drink, of course. Simon would not have allowed it. And Simon did not drink, nor Diana. They stood round the fire, looking unspeakably sullen, their minds tortured with their own immediate problems. They could not have felt less genial or looked more dreary.

Cynthia came across the room, a bunch of cards in her hand. She took her drink from Gerald and swallowed half of it. Better already. Definitely better.

"Almost time to start," she said. "Don't I hear the car coming round?"

"I left the car at the door twenty minutes ago," Simon said.

"Oh, then who is it, I wonder."

It was David. He walked in among them—through them—past them. Coming to Cynthia. There was a faint colour of dismay and delight in her cheeks. She made a tremendous effort to appear casual.

"Why didn't you let us know you were coming? We'd have found something for you to ride."

"It's all right. I came over to try a horse of John Hearn's."

"Oh, I see. Would you like a drink?"

"No, thank you, Cynthia. A little early for me."

Cynthia and Gerald had another glass of port each, just to uphold each other.

"Well, hadn't we better start?"

It was all so broken off, so without meaning in that horrid hour. It was the sort of awkward moment David would choose to return, inconvenient, unromantical. Putting himself into his old position of ascendancy with this not telling, hardly caring. Riding a dreadful horse of John Hearn's when Cynthia's perfect hunters were at his wish. Contradicting all his uncaring by coming twenty miles out of his

224

road to see her, not for a drink nor any convenience of time or place. But to see her, though he claimed nothing, wanted nothing——

She did not ask him to drive on to the meet with her. He took his own car and Susan and Simon and Diana in it. Cynthia and Gerald drove together. At the back of the ease after her drink she felt bored and rather squalid with him. All her content was gone. But she was nearer to tears than to happiness in seeing David again.

She said to Gerald:

"Do you know David well, Gerald?"

He said:

"Quite, Honey. He's not the type of man I care about. I'm sorry he's a friend of yours."

Cynthia felt as unaccountably angry as she had done on the day when, she having complained at great length of Simon's tiresome and affected ways, Gerald remarked weightily, "I'm afraid he's just a great big Cissy."

Both Simon and David were really so much beyond Gerald's praising or blaming. It was a gross impertinence from him even to agree with her own complaints or condemnations of either. Now she was furious. She kept an elaborate silence and lit a cigarette for herself.

"Please?" Gerald said.

"Yes?"

"Cigarette please, sweetie."

"Oh."

She lit one for him.

"You're not mad with me for what I said about your friend? I know he's not your type either."

Oh, God. Really, what an uncivilised man. What a way to speak. What a pity simple people ever try to put their emotions into words. They so destroy any effect they might have just as emotions. Looking at him, Cynthia saw how deeply troubled he was. If he had not spoken she might not have had her pity made impotent by irritation.

"I really didn't pay much attention, I'm afraid. I'm wondering if it would be better to draw Lara Wood first or go straight to that little wild gorse where we found the last day."

Gerald, who had a simple and uncomplex temper, did not answer her. He sulked like a bear all day and jumped the most preposterous fences on the faintest provocation. That evening he got very drunk and was insultingly possessive with Cynthia.

She was clever about this, affecting a lightness and tolerance to hide her furious displeasure.

People came to dinner and stayed for the night. It was one of those parties where every one talks a very great deal, conscious of difficult currents in the atmosphere. The food, the wine, the warmth, the flowers and colours of Cynthia's Garonlea endured. But behind the outward ambience it was as though feet stirred and rustled in an old darkness. In trouble there was something behind—something more powerful than Cynthia's new Garonlea. Contentment and happiness were necessary to her house. To-night there was a faintly mask-like importance on her decoration of Garonlea—a little ghastly, as masks are.

Part of the mask was in her friends who really loved her and lent themselves now unconsciously to tide across this dangerous moment. Although, like Cynthia, they all preferred David, they were sweet to Gerald, indulgent to his drunken reminiscing over his horses. Perceiving, as who would not, the trouble in the air, they talked to Gerald so that he should not quarrel with David, and never let David alone for a moment so that he should not be disastrously and pertinently rude to Gerald, as he seemed on the edge of being all that long evening.

Beyond anybody, Diana felt the alarm and the quiet possessiveness of the unhappy past creeping back through all the kindness of people and the rich and fluid ease of present life at Garonlea. It had happened and she was helpless. Unhappiness was here again. She could not say to David, Go away now at once. Say no more. Never come back again—— He had come back and it was too late. Dear David whom they all loved. She herself and Simon and Susan were nearly as much in his power as Cynthia was. He was back again—tuning them each to a curious pitch, un-enduring for its very level. He did not wish his influence to endure. He wished for nothing beyond the moment, and for him the moment could not be fragile enough or short enough. Too selfish, too romantical, he would not see his moments endure and grow worn and sordid with use and time. To him there were no such words as dear or accustomed—the gentlest Pirates do not know such words.

XXXII

SIMON saw what would happen too, and Diana knew
that he was excited and pleased far within at the idea
of Cynthia's mistake and disaster. She knew it as she
had known how it satisfied Cynthia long ago to rule
and frighten Susan and Simon when they were pale
and cowardly children.

Susan did not know what was happening or mind.
It was lovely to have David back, that was all. And
cruel fun to see old Gerald sweat about it. Susan did
not play cards to-night. She hardly ever did, or smoke
or drink. She had a work-bag as big as a sack, full of
pieces of tapestry which she had worked at now and
then through the evening. She was very partial to
green tulips and then paler green tulips in her designs.
She was never cross and liverish from doing nothing.
If she thought about anything, she thought about
birds or Simon or how terrifying and lovely hunting
was, or she might wonder if her stomach would ever
allow her to ride in a race. She thought not.

"David, you're not going away to-morrow, surely?"

No one had said anything about his going or staying.
It was a cry of sudden pain from her distance and
wilderness of green tulips.

Simon felt so pleased. What drama the child called
forth, sitting a sort of Sabrina among her cool tulips
away from the men and women and wine and falling
cards.

"No, Susan, of course I'm not going. I want to stay
for weeks."

"You will stay here with us, dearest David. It will be lovely. Like in the Spring."

"Yes, I would like to. Last Spring was lovely, wasn't it? We were all so good-tempered and we thought of such good jokes."

"It was lovely."

"And do you remember——"

Across Sue's thin netting of talk, clear behind its meshes, Gerald, spoilt and simple and exhausted with the heavy pain of the evening, was staring at Cynthia, a little drunk, saying silently, clearly, with all his heart and strength, "You must not ask him. You can't have him here. Hell! am I to share you? I love you so. I love you so much. Honey, believe me, darling, I'd give you the earth."

And Cynthia, looking at her cards, knowing each word in Gerald's mind, looking at no one, hard, considering, eager, greedy, helplessly enthralled to David, made her choice and her mistake, thinking, "I can have David now. I can have Gerald back again."

She lifted her head and gave Gerald a long, strange look, surprise in it and dismay for his bad taste. A slow withdrawal of her eyes and herself. It was a piercingly cruel and stupid thing to do. Stupidity and impatience and greed were in that dismissal.

They ran riot in the ugly scene that was bound to follow it.

"But I don't understand, Cynthia. I may be dumb. Say I'm a big boob if you like. But you loved me, you said so. Isn't it so? Did you say so?"

Still the library, but changed. Much later. The hint of very early morning behind the close curtains and the low fire and the scent of flowers reasserted in a room empty for an hour.

There was no wish in Cynthia to keep Gerald, even

229

his future return was not weighing with her now. She accepted it as entirely probable. Now there was no kindness in her towards him.

"I may have said so. But does anything I said account for your curious behaviour this evening?"

"Behaviour? Hell, I just want you. That's all."

"It's embarrassingly obvious."

"What is?"

"What you want. The only thing you do want."

At least Gerald did not want mink coats and rich jewels and horses as well. She thought it safer to hurry on.

"You'll admit it's embarrassing for me."

"Cynthia, listen——"

"Simon and Susan. All these people——"

"Ah, listen now."

"You pull yourself together and listen to me. I simply hate saying so, but this sort of thing is quite beyond me. It just ruins everything. One can't compete with it."

"You make it out pretty rough, don't you? And what have you got on me? Tell me, because I don't understand. I'm no different to-night from what I was last night."

"You rather sickened me to-night."

"I sickened you?"

"Flinging everything about like you did."

"I don't see your viewpoint. The folk here to-night, they knew last week I loved you. I'm proud to love you. Your friends know it. To-night didn't show them any different."

"If you'd had rather less to drink, Gerald, you might be able to see some difference."

"I was drunk too?"

"I said, ' If you'd had rather less to drink.'"

"Well, let that be. I've been drunk before and I will be again. You can't pick on me for that, can you? Considering how you drink yourself, don't you?"

Cynthia was rigid with fury, stammering for words which she could not find.

"That's all right, Cynthia. I don't quarrel with you for that. Forget I said it, but it's true, remember."

"Thank you."

"And listen."

"I'm getting rather bored with all this listening. I'd be rather glad if you'd say what you have to say in one minute and if possible leave the house before I come down in the morning."

"But you keep interrupting me, Cynthia. You make me feel kinda mad with you. Child, you've got to know how much I love you. Of course you know. You don't forget Tuesday, nor Sunday, nor that lovely week in London. I know you don't. You can't. Then what is it? What's got you to-day and to-night? What have I done to you? What's happened to us? You loved me."

"I've told you. If you don't mind, I'm going to bed now. Good-bye."

"It's not Good-bye and you've not told me. Listen——"

"Oh, God, if you say ' Listen ' again—Let me go!"

"It's the way I act and the way I talk now. No, I'm not letting you go."

He had pushed her down in the sofa by her shoulders and leaned over her, amorous, animal, nervously, desperately in pain. His eyes and his mouth out of his control—he was asking to be hurt. He could not see the pain that was coming to him. It would be pain beyond consoling.

"You think I'm dumb. Certainly you must think I'm slow if I can't see what's happened to-day. You changed to me. Yes, I felt it the minute David walked in at the door. Didn't you now? Are you in love with him? No. You can't be. Not when I think of yesterday. Excited maybe? What's the good? He's no use to you. I know David better than I said to you this morning."

"You do?"

"I do. You're not his type. Believe me, it's so. His wife holds him. She's a clever woman. Do you know her? Like Sue. Sue reminds me of her, I can't tell you why. That coldness they have. That's what he likes."

"Really? Perhaps now I might go to bed."

"Cynthia, don't move. Don't make me bully you. I would. I've got to talk this out with you."

Cynthia leant back, pinned in her corner of the sofa behind all this heavy talk and heavy shadow of love. Where was all her wish for him that yesterday had warmed and soothed her? Now her body felt like a wire cage against him. Her arms fleshless as wire. Her eyes looking on him, cold. Back and forwards through her mind through all the truth of his talking, walked the poisonous word against her vanity—"You drink too much. I don't quarrel with you for it." Cynthia could not think how to hurt him enough for that moderate and deadly thrust. It was so true, so much more true than it had been years ago when David said it, as true now as his assumption, his knowledge of her desirousness. . . . "Like Sue—She reminds me of Sue—that coldness——" It was insufferable. She could never hurt him enough.

"—This is what I want to say. I want you to marry me. My darling, realise things. You do love me. Perhaps not to-night? Yesterday and to-morrow though! Ah, Cynthia, we could do so much together. Everything you wanted I'd get for you, Honey—— Everything I could."

That was all he said about his money. This was Gerald, simple and very kind. Speaking of his love, asking his mistress to marry him as ardently, as gravely as though theirs had been the most austere courtship imaginable. And this was Gerald really, beyond his talk of horses and his smart, dirty stories. This was the man whom Cynthia beyond any one should have known after these months of love.

She was too blind. Blinded by her vanity and by her greed and her pathetic lack of imagination. She was going to hurt him so that he must be forever lost to her. She was not very quick of tongue or very clever, so the words she chose were blunt, heavy, bruising words, strung together in short, plain sentences, easy to follow.

"——Do you realise quite what a crashing bore you are? How you think any woman would have the nerve to live with you for the rest of her life—— Did you really think I would? Poor Gerald, you did? Well, I'm awfully sorry—— Rather touching, I thought, trying to reclaim me from drink. I'm not sure when I've seen you entirely sober. Or have I ever? Not that I mind. But it dulls the brain a bit, doesn't it? You find it difficult to see when people simply have had enough of you, don't you? When they've heard *all* your dirty stories and all about your horses and all the fences you've jumped five times each week."

Gerald got up very slowly from the sofa and stood looking down at her from the great distance of his height.

"One thing I never thought I'd do, Cynthia," all the possible weight and simplicity in him behind the words—He stopped and then spoke, "I never thought in this world to despise you as much as I do now. For what you've said." He went away.

It was an answer she did not expect. She was left

233

alone, still flattened back in her corner of the sofa. All the words she had said, ungoverned, cruel words, were clear and loud in her mind for a moment, their meaning now trebled from what it had been in her angry mind. She felt frightened and ashamed and minded curiously that Gerald should have spoken so finally. She wanted reassurance terribly. Through and through she felt shaken. Reassurance? She crossed the room to a glass and looked in it, moving away a little wildly. Again she had been shaken—tired, that's all, only tired. It was never age looking back from the shadows. What had she said? What had she done? I can't be left alone. I must discuss this with some one. Diana, she understands. Diana loves me. Should I have a drink just to steady myself? Nor fair to Diana to wake her up and go on like a nervous wreck. I suppose I ought to have one. Yes, I will, it may do me good—— Ah, better—already much better. And a tiny one for while I talk. I've had so much talk to-night. It's so exhausting.

XXXIII

CYNTHIA went upstairs a little awkwardly with a glass in one hand and the front of her dress caught up so that she could not stumble. The picture of Lætitia, wife of Desmond McGarth, seemed withheld in a wintry isolation from her successor. All that Cynthia had done for her importance made the division clearer. Cold, honeyed mouth; hands preposterously unlifelike, inamorous as sea-shells. Oyster-pale dress. Powdered hair, and a crisp posy of white, moss-rose buds and tiny, dripping fuschias—such sanitary flowers to pin between those reserved and symmetrical breasts. Cynthia, groping a little stupidly for the light switch at the top of the stairs, was remotely aware of a chill breath between that piece of lovely decoration and herself.

Diana was curled in bed like a little dog, absolutely fast asleep. How strange that always she should have slept like this by herself. I can't think why it doesn't make more difference between us, Cynthia thought. She felt a surge of trust and gratefulness go through her as she put her hand gently down on Diana's shoulder.

"Diana, darling, I'm so sorry to wake you. Listen, I've had such an awful time——" Cynthia began to laugh. She certainly was not very steady.

"What is it? Sit down, Cynthia." Diana had woken in the quick, untroubled way children do. "Mind, darling, give me that whisky. You're spilling it all over your dress."

"Don't give a damn about this dress. Never suited me. I'll give it to you, Diana."

"No, darling."

"Yes, darling. You deserve it. You're such a comfort to me. Give me my drink and I'll tell you about it all."

Diana looked at her mistrustfully, but she handed over the drink. One could not argue with Cynthia at this hour.

"What an evening! Do you agree with me that Gerald was behaving shocking—preposterously, or am I wrong? I don't want to be unfair. But it wasn't fair to me, the way he was going on, was it? Or to you or Simon or Susan or anybody. Was it?"

"Cynthia, darling, I think you ought to go to bed and discuss it in the morning."

"No. Now. I came to tell you now." Cynthia seemed to gather herself for a moment out of her absurd drunken importance. "I've been simply brutal to him, Diana. The things I've said." Then she relapsed again. "Can't have it, I said, think of Susan, I'm always thinking of what she may think. I said, you must go away, Gerald. Stop boring us all about your horses. He said, I love you. I want to marry you. Give you anything, he said. All men are the same, Diana, about me, aren't they? Look at David. Back again."

"For how long?"

"I don't know what you mean. David's come back to me. Then he said I was drinking. Pulled me up for drinking. He did. I don't think that was fair, do you? That made me very angry. Wouldn't you have been? I do so hate unfairness."

"I suppose you refused to marry him?"

"Yes, David's come back. Wasn't I right?"

"How can I tell what you should do. Oh, darling, I wish you'd go to bed."

236

"Don't you want to talk to me? I want to tell you about this. You see, I've been very unkind to poor Gerald. He's off—he'll come back though."

"Probably."

"But I took a tremendously strong line. I thought it was only fair—don't you know, Diana? Finished, I said. Two months with you enough for any girl."

"I'm worried, Cynthia. He's been so good to you. He's so absurdly generous. How unkind were you?"

"I was very. I couldn't help myself, Diana. To have him here going on so unfairly like to-night when David came back—what could I do? I ask myself, I do really."

"Is he going away?"

"To-morrow, I think. That's to-day."

"Finally?"

"I hope so. Do you know he dared to say to me he despised me—— Oh, what *am* I saying? But he did say so. Fancy Gerald saying so! It's almost funny. I could only tell you."

"You must have been very very unkind."

"Yes, I think I was. Was it very cruel of me?"

"What are you going to do about all his presents! You'll have to charter a whole train to take them away."

"I forgot them. I like people for themselves. Not because of what they give me. You know what I mean. You see how I quite forgot his presents. Isn't it funny?"

"What are you going to do about them?"

"I can't pack up that horse and send it back."

"I thought you hadn't quite accepted that."

"Yes I had. And I may be unkind but I couldn't hurt him as much as to send back the horse he gave me, could I?"

"And the mink coat?"

"It wouldn't be fair on him to give that back. What could he do with it?"

"And all the diamonds?"

"Oh, the vulgarity of those diamonds! Poor Gerald. I can't throw them in his face. There are limits, and honestly, would it be fair to him?"

Diana got out of bed and tied the cord of a blue dressing-gown firmly round her waist. She looked very tired suddenly. Her hair and her face seemed faded and sad.

"Cynthia. Bed for you, my dear. Come on, I'll help you to undress and tuck you up. To-morrow we'll talk about it all."

Half an hour later Diana got back into bed. She slid down between the chilled sheets without resentment. She was curiously detached from the comforts of the flesh, enjoying them as presents, not as rights. Now she lay very straight, curled no longer like a warm little dog, while thoughts and speculation chased each other in her brain and she turned things backwards and forwards, puzzling and aching over the problems they presented. Cynthia, her dearest, whom she so loved, how was it in her power to help her? Could any one help and steady her now? This year had done so strangely by Cynthia. In it she had broken the abracadabra of her faithfulness to Desmond. She had found delight and disaster and content. She had loosed her hold on content, to grasp at stars again; and what now? But it was not like that in Diana's mind. It was not a stated situation. It was all in half-spoken thoughts, in condemnation caught back for love's sake, in old gratitudes and memories of the glamour and tragedy and drama of Cynthia's life that had so strongly caught her own into its flow. There were new thoughts too—thoughts the depth and truth of which she had hardly learned to own—thoughts in which fear and pity for Cynthia were equal. To her it was as if Cynthia had suddenly lost hold of herself and of integrity towards others. She could not see that this year of yielding and greed was only the outcome of all the long years when she herself and all who touched her life had deified Cynthia beyond reality. Cynthia in their eyes and twice magnified in her own eyes could not suffer change by what she did. Contempt was the one thing not alive in the world towards her, and could it now be coming to her? Contempt from the lovers she took too easily. From the friends towards whom her charm broke and failed. From Susan and Simon whom she had tortured yet cherished. Children who had made her ashamed and who filled her now with a sort of pride and distant respect. There was a pathetic nascent awe in her relations with them.

To-night Cynthia's cowardly refusal to give back those too rich presents of Gerald's had filled Diana with real dismay. Static little codes like that, of which she had read in books, were very true to Diana. She detached it from all she had known of Cynthia's tremendous power of taking. It seemed to her a real loss of integrity. Part of the year's undoing of Cynthia. Part of the lessening of Cynthia's self as Diana knew her.

Another question which tormented her about Cynthia, a question which she avoided in her own mind because she found it so unanswerable, was the problem of her drinking. Since Diana had known Cynthia first she drank far more than other women, but she did it in a virtuous, necessitous sort of way—a duty towards herself and others, almost as if she was going to communion oftener than others (and in this matter too she had for years been the parson's pride and pleasure). Diana had grown so used to seeing Cynthia always drinking a good deal with a great air of moderation that she could not have told when it was that it had first come home to her that there was nothing moderate about the way Cynthia drank. That it was a complete necessity to her. There had been a time, in that first wonderful freedom from Garonlea, when her life changed in the anguish and joyfulness of Rathglass, when Diana had attacked whisky (that was her drink) with defiant vigour. But this was only as a testament to freedom. The phase did not endure. She did not really like whisky.

Perhaps it was through Simon that she had grown to realise the extent to which Cynthia's drinking had developed. He seemed so sharply aware of it.

Coming home from hunting, "Oh, let's go the other road—Mamma will sit for hours if we pass Carney's Pub."

Cynthia's changed face of disappointment and her hurried search for an answer:

"No, we must go by Carney's. I want to give the men a drink. They're soaking wet."

It was Simon's quick look of contemptuous dislike, his restraint from argument, that Diana remembered. And the brazen furtiveness of Cynthia's excuse.

Of course she could have brought drink out in the car on hunting days, but she preferred to sit for hours in the gloom of one of her favourite bars, warming her feet at the low fire, her fur coat open, her face lost in the darkness of walls and early evening outside. Drinking bad whisky, slowly, contentedly. Buying drinks for any one who came in. Often, and so accurately, putting a name to the faces that were only a pale glimmer in the half darkness and half silence of those little public houses. At such times she was truly great, a great personage whose people were her friends. She would sit there, not making conversation, not condescending, but with them. They would drink her health gravely and she would nod back, raising her glass to the eyes outside the circle of firelight. She would blow up the fire and stretch her ringed hands to its heat while she listened to long stories of foxes and their whereabouts. She never showed her friends the discourtesy of hastiness. Often when she rose to go she would sit down again, held as it seemed by her wish for their company. And at last she would step forth reluctantly into the winter evening and drive away in her powerful, gloomy-looking car. . . . How many of those hours had Diana seen before Cynthia's need of them and Simon's hate of them were known to her? She had looked on them as part of Cynthia's being a Master of Hounds, not as part of Cynthia herself. She had often endured, uncomplaining, hours with cold feet and wet clothes while these royal drinking parties were in progress, because she understood they were so important.

When Simon and Susan were younger they used to eat biscuits and drink ginger beer contentedly in the back of the car. Now they restrained their more educated appetites till tea time and sat wrapped in coats and rugs, a complete accord of disagreeable condemnation between them. Refusing to be young and hearty, or even slumberously royal like Cynthia. Just lately they had shocked Cynthia very deeply and quite disgusted Gerald by bringing books with them which they read when they did not wish to talk to each other.

"That must be an enthralling book, Simon. You

couldn't wait till you got home to finish it, I suppose."

"No. Do you mind me reading while you drink?"

"It keeps our minds off our hot baths," Susan said. She read a little more with easy insolence before she shut her book.

Cynthia was not able for them. Whenever possible she sent two cars out hunting or arranged, which indeed they did not mind, that they should ride home.

All this Diana knew. She knew the tenuous strength of the discord that existed between Cynthia and her children. Nothing could bridge it—nothing solid and actual that Cynthia did for them, and she did a great deal, could overcome that familiar loathing for the poses they knew in her so long. The poses they had seen through so long ago. The poses they knew she could get away with over and over again. They could not possibly look on her detachedly. They could never forget, and they were too young to understand, her cruelty to them as children. They did not thank her because this very cruelty in its strongest manifestations had given them so much that they now had, their great and passionate pleasure in the chase. They thanked her for nothing. Not for all that she had at Garonlea. Not for their physical well-being to which she had attended so scrupulously. In everything she had done for them, in everything she still did, they could only see reflected back her own will towards power, or a further uprolling of that sickening full-sized cloud of glamour behind which they knew her for what she was. Or for what she was to them. How did they know what she was? And by what right of sorrow could they understand how terribly she had suffered and changed?

Diana could not understand it. She was far more

childlike in mind than they were. She only saw the danger of their bitterness towards Cynthia. She felt as though some wheel in life was coming slowly round. As in a dream she could not see the wheel or the circle it made, but she knew. She had been told before. She had the dream fever of panic and impotence in all her thoughts of Cynthia and the children.

How quickly they had become such entire and dangerous creatures, these two children. How flattered she was now by their love. And was this new, or had it always been so? Even in that other world, that strange dark young world of make-believe and escape, world of vast extremes in which she had been their friend? Now she remembered a shop they had once in a dark loft—their own loft. Their very gentle small hands, animal gentle, as they caught her hand to bring her and show her. Simon's simple embrace turned at the last moment into an effort to see if he could lift her. The elaborate mounting of a ladder. Then the telephone. Shut down the trap-door so that you can *only* hear by telephone. Speak softly. You will really hear. The exciting unreality of it—thinking of different things to say to each. Their pleased, eager replies. Her wonder. Then—the shop. Melted boiled sweets in little pots. A mouth-organ. Book markers— paper cat's heads. We can make you more *easily*. Tiny lavender bags squeezed into tight little stomachs. A tooth paste carton. Post cards. Look, isn't that funny? "LOVE FROM THE WHOLE D—— FAMILY." A frieze of postcards and cut out pictures running round the wall of the loft. But these are not for sale. The elaborate descent of the ladder. A pale donkey below. *How* do you jump up on her. It's quite easy now, but first I tried and tried and tried the whole summer through. The thing was that she still felt with

243

them as she had then. To her their present world was
as full of fantasy and sustained pretence as the time of
that shop had been. It was as real to them and she was
as much outside it as she had been then. Although
she could not understand she respected the reality of
their attitude. She did not ask to have it explained
any more than she would have asked them to explain
to her the use and sense of that telephone. She was
entirely detached from them and because of this there
was no bitter familiarity between them. She was
content to love and fear them a little.

Dear Cynthia—beloved creature. . . . Diana turned
in her bed to face the open window opposite, and the
full dark, most despairing hour of winter and morning.
In her thought of Cynthia tears came piercingly behind
her eyes, rolling their shaped and definite course
down her cheeks. . . . Ah, but these tears are pearl
that thy love sheds and they are rich and ransom all
ill deeds. . . . The pain of such tears if they can be
shed for another, the helpless fear for which they
spring and fall, the darkness they accept, the love
they spend as on the idlest air, from this, their very
quality of selflessness, they must go uncomforted,
their pain unrequited.

XXXIV

THERE was to be a party at Garonlea for Simon's
coming of age. Cynthia was far more excited about it
than either Simon or Susan. Simon froze quite still and
cold and then turned hot and fluttering inside at the
thought of making speeches, this in spite of all his
grand airs of detachment and toleration. But this
could hardly be called excitement. When Susan thought
of her cousins Nancy and Cecily who were coming to
stay (they were Enid's daughters) and of the smart
young men that Cynthia would ask, she felt quite
despairing.

It was curious how easily the young were overcome.
Now the thought of Muriel and Violet and Enid and
their husbands all being at Garonlea for nearly a week
did not trouble Cynthia at all.

Then Simon gave the party a sudden kick that
landed it in Drama. He opened a telegram that was
brought to him in the library after tea, and read its
clear delightful message to himself through and
across all the fragments of talk that were going on
round him.

"——Yes, I'll give you his address, Muriel. He
makes up one's own tweed for five guineas——"

"——I do think Simon has such a look of Mother
sometimes——"

"——A delightful little place. Quite unspoilt——"

"——Of course the exchange was in your favour——"

"——The waiter was so thoughtful. 'This cream
bun *can't* hurt, Madam,' he said——"

245

" ——I followed him through some long grass. It was ticklish work. Now and then you'd see a spot of dried blood on a leaf, or a pile of fresh dung——"

" ——Give me a mashie, I said. Not that club, you fool——"

" ——It's worth knowing, if you ever strain a riding muscle, you cross the handkerchief like this, knot it and go once round your leg and once round your waist——"

" ——These *lavages* are tremendously important, he says, so I go to this woman twice a week. Unpleasant yes—but not really painful——"

"Look, Sue," Simon said, showing her the wire.

"Oh, Simon, what a lovely saviour."

Simon said to Cynthia, "My friend Sylvester Browne would like to come and stay."

It was exactly as if he had said, my friend John Gielgud or my friend Noel Coward would like to come and stay. It was all very well for Sylvester's intimates in London to call him old Sylvester and say between themselves that the poor old thing could be a bit embarrassing at times. Or, "We *can't* be very pleased with Sylvester for that." But they did not deny his moments in which the acidity and spark of life, the depth and the laughter and the truth and sentimentality that were in his plays and in himself went beyond even the most ungenerous praise. So when Simon said in that after tea hour, "Sylvester Browne is coming to stay," reactions of all sorts went rippling round the room.

Cynthia thought, "How old is he? Quite old enough —thirty-seven? Thirty-eight? I'll have that lovely brandy up. Wouldn't have wasted it on Arthur. My white dress with the hood—pearls."

Simon was quite stilled with pleasure. He had

always wanted to have Sylvester at Garonlea, and now in this early autumn seemed the very best time. He had not asked him either. He had never thought he knew him enough.

Sue thought, "Perfect! How lovely for Simon and me."

The others talked.

Uncle Arthur said, "I went to a play of his once. Cecily took me. Thought it would be a suitable play for Father. Never was so bored in my life. I like a good musical show. Nothing highbrow. What was this about, Cecil?"

Cecily was furious. Sylvester was one of her shrines and she had very few. She said loudly, "Incest and adultery. Don't you remember?"

Quickly quickly Violet, the trained and gracious lady, said, "It always seems to me such a pity all these unpleasant plays. And why do they write them, I wonder, when a good clean play runs for ever."

"But we went to a charming play of his together, Violet. All about the Tyrol." Muriel spoke up for her shrine.

Simon said, "When a faded lady found love among the Eidelweiss. I always thought that was rather naughty of Sylvester."

And Cecily's play had been about a horsey outdoor girl and her romance with a French Vicomte. Cecily and Nancy kept a riding school near Cheltenham where they had been to school. So they were all busy making little arrangements in their own minds about how they would impress Sylvester and how they would see themselves one day in his plays. Every one except Sue and Diana. Well, Sylvester was used to it and anyhow Simon had not asked him to come to this dreadful party.

247

He arrived by train because he hated driving his car. Simon and Sue did not go out cub-hunting because they both wanted to meet him at the station. They were not a bit shy or compressed in their greetings and roared with laughter when it appeared that he had no idea there was a party of this sort going on at Garonlea.

"We thought you came to help with the speeches to the faithful tenantry," Sue said.

"No," Sylvester said, "I really came more to escape from my cousin, Piggy Brown, than for anything else. I thought you would make a nice change," he said to Simon.

"Yes. But there are lots of cousins here."

"We won't have to contend with them. I want to see your house."

"The house itself isn't good, but all round is nice," Simon said vaguely. "Isn't good" was the worst abuse he could bear to level at Garonlea's spread of battlements and turretings.

"The house is heaven by me," Sylvester said when he saw it. "It's my architecture of the moment. I hope the inside décor is in keeping."

"Well, no, Mamma has made that rather comfortable and modern."

"Oh no. Cream distemper and Flower Pieces and hunting prints and vellum lamp-shades by Lionel Edwards and trays for drinks?"

"Very nearly."

"Tweed sofa covers, or gay chintzes?"

"Still gay chintz."

"I expect you have black floating bowls with two roses and frog on the dining-room table."

"No, that's wrong. Diana makes nice plans about the flowers."

"Would she be a cousin too?"

248

"No. An aunt who lives with us."

"The only thing against her is that she thinks Mamma is too superb."

"Does she really, Sue? Or does she just try to defend her from you?"

"No, it's always been like that, hasn't it, Simon?"

"Yes, but we don't mind. Mamma is her romance in life."

"This is lovely," Sylvester said, "I could sit here for ever in the sun surveying this gorgeous piece of well-cared-for Gothic."

They had stopped opposite the hall door but none of them showed the smallest intention of getting out of the car.

XXXV

It was the same sort of afternoon as that on which Cynthia had come to see Lady Charlotte dying or dead and to make her first act of possession at Garonlea. Now the house seemed fatter and sleeker. Its mullions and turrets more tigerishly striped and spread with red creepers. The falling terraces below richer far with flowers. Behind the windows and within the open door there was a great feeling of opulent and spacious occupation. On that long past day the trees and laurels and rhododendrons had pressed dark and close round the house. Now they had been cleared back and back—cut down and entirely subdued from forest. Near the house sunlight poured on flat grass and on groups of blue hydrangeas and thickets of red-hot pokers. It lay the length of the opened bank of valley as hotly as in July. Black cattle standing close together in a ring of chestnut trees looked as if they were all carved from the same block and not yet unjoined from it. There was a shaken air of blue where the half turned bracken and the woods sloped down and up.

"Sumptuous—that's what it is," Sylvester said. "Your Mamma must be a remarkable woman. I came here once, you know, in old Lady Charlotte's time."

"You never told us that."

"But I was too young to enjoy it properly. I wish now I could remember."

"Perhaps it will come back to you."

"I don't think somehow there will be very much to suggest the house as it was then."

"We don't remember. We were too small."

"We lived at Rathglass across the river."

"Nice?"

"Very nice, I think, but we were such dreadfully unhappy children."

"Always sick."

"Always diarrhœa."

"Hated our ponies."

"Hopeless with our dogs."

"Starved our rabbits."

"Obeyed our governesses."

"Implicitly."

"I was a queasy sort of child myself. I don't think it counts. Look what lovely people we grow to be."

"I suppose we must face going in some time."

"Sometime always comes."

Sue sang:

"I shall never forget when the big ship was ready,
    The time drew near for my love to depart.
I cried like a coney and said Good-bye, Teddy,
    A tear in my eye and a stone in my heart."

"Yes, sometimes always comes," Sylvester said when the song was over. "Well, shall we?"

They went into the library full of a quantity of things. Pink lilies arranged with a shrub that grew pink marabout brushes. The smell of beastly Turkish cigarettes. The smell of expensive hair stuff. The smell of new *Tatlers* and *Bystanders* and *Sketches*, (for it was a Wednesday), and, to forget smells for a moment, the presences of Violet and Enid and Arthur and Lord Jason Helvick.

"Yes, really rather wonderful, five guineas to make up your own tweed——"

"——Twice a week. Unpleasant of course, but *not* really painful."

Violet and Enid were going on again about their little men who made up tweed and their little women who did unspeakable things to you twice a week, unpleasant but not really painful.

Arthur was telling Jason about a mahseer he caught in a river in India. Jason was far away in mind thinking about the habits of the lesser Plaitabils and of a photograph he had once almost succeeded in taking of the parent bird at feeding time. Everybody else was having a late meal after cub-hunting.

Violet began very capably to Sylvester about people they knew. Arthur went on at the top of his voice with his story about the mahseer, hoping the actor johnny would somehow be a little impressed.

Sue said, "Uncle Jason, you aren't listening to one word Uncle Arthur's saying to you." Lady Charlotte could not have spoilt poor Arthur's anecdote more completely.

Presently Cecil and Nancy and two indefensibly smart young men called Acres, John and Tony Acres, second cousins of Cynthia's, came in—still very much in their breeches and boots. They were introduced to Sylvester and at once began to read the *Tatler* and *Sketch* and the *Bystander* to show they weren't gaping at Fame or being unduly impressed. Cecil was particularly casual. She was a romantic secret young creature behind her extreme love of dogs and horses. Quite as passionate as Enid had been in youth, but with more possibilities of sublimation. She had Enid's pretty forehead and her eyes but mercifully none of those unhappy spots.

252

Into the extreme heart of this room came Cynthia. Her life and distance from these older peaceful women, fulfilled with their acceptance of triviality, satisfied by their small concerns, and from those two younger ones, choice examples as they were of England's outdoor girlhood, and full of rather pathetic striving toward a complacency they had not achieved, was at once real and remarkable to Sylvester. Compared to them it was as if a ring of fire burned round Cynthia, so complete was the division. Then she was more swift and easy and vague in speech. Unlike those girls, she had found time to change after hunting. Her hair superbly waved, the expensive flattering clothes—so far from the natty things the little men ran up for five guineas—emphasised all this difference. Besides she had had a couple of drinks and been kissed in a wood this morning by a daring and common man who had come from Yorkshire to buy horses. "I'm absolutely on the bit this morning, that's how I feel," he said. Cynthia gave him a sort of hard detached encouragement, agreeing with herself almost impersonally that he was most attractive, that was how she had grown about men now. She had other lovers since David had really left her, but at least she did not often romanticize her relationships with them.

——Simon might say sourly, "Mamma's admirers scarcely improve, do they? I mean I really am rather a snob at heart and I do mind."

But Sylvester said, "In a way, you know, I think it's rather royal of her. She could so easily stick to her own class."

"But I do mind," Simon said again. "Has she made passes at you yet?"

"You embarrass me, Simon. How could I compete with a gorgeous horsey creature like that?"

It was the evening of the second day of Sylvester's visit. He and Simon were walking over to Rathglass to meeet Diana who had been gardening. Over the river they were in that different air of Rathglass. Here the autumn seemed less opulent, of a thinner quality than at Garonlea. It was a slight shock to see trees taking shape again. A group of ash trees posing their new nakedness in an extremity of grace and affectation against a still dark wood. Limbs turned divinely. So much, so far too much has been written about the autumn. Fur and flesh. Ducks and shrivelled leaves. Level winds as flat as ribbons. Seagulls sitting on little fields—domestic inland birds heavy as pigeons. New grass, greener than spring time grass. Food and death for fat little birds. All these curious autumnal contrasts were in the evening as Sylvester and Simon walked on together towards Rathglass.

They did not talk any more about Cynthia. Sylvester guessed that Simon was thinking if he said any more he would become over confidential and perhaps a little embarrassing. Sylvester was feeling his way about Cynthia. He thought she would perhaps be good in one of his plays but he was terrified of encouraging any of her advances. Already she had made what Simon called a pass. A pass for sympathy it had been. A sort of Queen Lear pose about Simon and Garonlea. He could see that Simon was going to have hell if he wanted Cynthia out of Garonlea. If he wanted her to take herself and her horrible boy friends and that perfect dear Diana off to Rathglass and leave him and Sue to live together at Garonlea, marriage was really his only chance of avoiding unbearable scenes and woundings. But Sylvester did not think Simon or Susan would marry for a long time. They seemed to him as sexless as two jade doves swung in two silver wire rings. He wondered how much Susan understood about Cynthia's really rather skilfully decorous rioting with her lovers. Almost as much, he supposed, as Simon could understand the agony of potency in an old woman. Old? Not old. But however smooth and full the skin and restrained the figure, forty-five to fifty is old, horribly old to go looking for love—to have still unfulfilled that fever in the blood and through it to encourage the advances of tough men like this present admirer whose name Sylvester could not quite recall. How many like this had Simon seen? Just how arid and revolting did it all appear to him? Knowing a little of his feeling for Garonlea and for Sue, Sylvester realised that Cynthia would be quite beyond his pity.

They had a nice walk back to Garonlea with Diana who was particularly satisfied after such a time at Rathglass, warm and companionable. Talking to her and looking at her thin happy face Sylvester felt very glad that there were women in the world who found such contentment in things and places. Among the happy old maids he found some of the people he liked best in life. There was an unspent sweetness, an ungiven power of loving in Diana that was not sad and ingrown but within and about her, a thing both romantic and attractive. She had none of that sourness so often evident in those who lead unselfish lives. She had dignity and balance and no affectation. He was not surprised that Susan and Simon should be so fond of her. And he thought her devotion to Cynthia both natural and uncomplicated by any curious inhibitions.

In the course of his career as a playwright and player, Sylvester had seen many women fall for him. In the way old women worship doctors, he would say; a *little* difficult, he would say. But he was not often as unkind to them as he sounded. Impelled possibly by his morbidly simple sense of humour, partly by curiosity and by his perpetual discovery or fabrification of drama, he was given to the collection of rare and rich and if possible, eccentric old women. He would dine with them and listen unwearied to their stories of dead loves, dead injuries, unforgotten spites, old wills. He revelled in their rich clothes and food and wine.

And to his old ladies naturally he was the luxury of romance personified. They idolised him, went to all his first nights, struggled to understand what his plays were about, and never told him how much they preferred a nice costume piece. But all this he knew and would put the right words about his work into their hesitating praise, and if he took them to the theatre, would select one of their favourite dramas in which the gentlemen in the cast could be relied on to wear tights and the ladies to have bosoms. He was very valuable to them and it was only fair in return that he should be able to entertain his friends by the faithful repetition of their shattering aphorisms and the curious fetishes they would discover to him when a little in wine.

Such ladies he could contend with and enjoy. But the Cynthias were in another category. Try as he might he could not find them very amusing. Even as copy for some reason seldom useful. It was the more distressing to him that Cynthia with all her obvious and shockingly frequent moods for love should have about her that spark of reality, that undeniable fire of life which compelled his attention. He could not avoid his own interest in her. She captured his imagination. And he felt the pity for her that Simon obviously would never feel. He profoundly admired her attitude towards Simon's loyal tenantry and all the men at Garonlea of whom, what with servants' balls and presentations, they saw a good deal those days. Her quite royal absence of graciousness or any Lady of the Manor tricks, seemed to him most estimable. He thought, seeing the strength and dark good looks of some of the young retainers, that it was perhaps a pity that the *droit du seigneur* could not be recalled and reversed in this instance. It was a curiously wild thought but he imagined with Cynthia such a situation might have tremendous simplicity and success. . . . Possibly his inspiration came from the lovely brandy, or from the showery falling fireworks in the still first cold of the hollow autumn skies, from the bonfires and barrels of stout, from all the feeling of inheritance and forgotten powers that Cynthia seemed to personify so much more really than Simon.

XXXVI

THE festivities exhausted Simon and made him feel taut and nervous, so that in the intervals he took no ease and found his uncles and aunts and cousins quite hateful, and his mother's friend, Reuben Hill (he was staying at Garonlea now so they had been forced to know his name at last) beyond his endurance. It seemed to him past anything. That she had not been able to resist having this man here now was the crown and summit of the reality and unreality of her queen-ship at Garonlea. The superb insolence to Garonlea and its two assembled generations. The superb success of the insolence, for the Uncles and Aunts thought him charming and those two girls of Enid's and the smart chaps talked to him ceaselessly about their horses and had lots of jokes with him all the time. He was easier than Sylvester. None of them had begun to get near enough to Sylvester to bore or aggravate him.

But it all got tremendously on Simon's nerves.

So many things at Garonlea happened in the Library. Now the room was full of easiness to such an extent that the air was squalid with it. Everybody said what they liked there after they had had a drink or two. Cynthia arranged so that even Muriel should drink up a little before dinner. Trembling inside her silvery velvet, Muriel grew daring enough to say to Sylvester, "Of all your plays I think perhaps I like best 'Grey Morning.'"

"Ah, did you?" he said. Then finding geniality, for

258

it did seem a shame to distress her, "I'm so glad. Very few people really enjoyed it. I liked it myself quite a lot——" ——What was he saying? That awful unmeaning bit of success.

"Indeed I did. It seemed so kind and so true."

To the women who have little flats in big buildings? To the women who have had no love at all? Yes, they had been a good and paying public. And this gentle creature who had been turned away from everything all her life thanked him for them. Sylvester felt rather ashamed. Perhaps a little crucifixion would do his soul good.

"Tell me which of my other plays you liked," he said gently.

Cecil, the one in blue taffeta—a girl is always safe in blue if she has blue eyes—was trying to talk to Simon. She could only talk about horses so far as she had gone in life, and now she was saying, "I'm not really a good judge of a horse in the rough."

"Aren't you?" Simon said. Behind all such nonsense the room was full of the scent of lilies, heavy yet coming sharply. He tried to keep himself sensible of it.

"That was a pretty gay tie you wore to-day," she said. He didn't realise she was trying desperately for something to say.

"Yes, isn't it amusing? It's a ballet colour." That will put her off. Her and her nasty rough horses, he thought viciously. He was right, it left her with nothing to say, mute in her blue taffeta.

Enid, who was telling somebody about the only efficient electric toaster, saw her child's embarrassment with Simon. As across the world the ghost of a sensation came back to her. Raging impotence to make oneself understood in this room. "Yes, I'll give you

259

the address, it really is worth knowing. One just pops the slice of bread in and it's really such *fun*——" but the draught from elsewhere had chilled her warm and ordered preservation of content. She called across several people to Cecil:

"Darling, did you like that horse you rode to-day?"

Cecil shot a furious look across the room. They never allowed their mother to talk about horses. There was hardly a subject in life in which they did not find her ignorant, confused and tiresome, but about horses most of all.

"So next season you are to be presented, Susan? Won't that be thrilling?" Violet's placid voice would have made an expedition to Tibet sound tame. The pomp of courts was as nothing. Of them all she who had least suffered in that room seemed the one most steeped in its past air. "I was so excited when Granny presented me. Queen Alexandra looked so beautiful and she gave me quite a little smile, I remember. And I smiled back, which Granny said was rather a little breach of etiquette."

"And did you go to a lovely party afterwards?"

"Oh, no, dear. We went home to bed. I was quite tired out by the excitement."

"Simon and Sylvester are going to take me out," Susan said grandly.

Simon heard this and smiled to her. "And Diana too," he said, "she's to have a new dress, the smartest model in London."

Diana went over to him, and Sue, thinking that Aunt Violet had deadened her enough, came and sat on the other arm of his chair and sipped his sherry. The three of them made rather an insolent and secret looking group.

Cynthia came in late, partly to make a little stir,

partly because she was the person behind all the smooth running of this party and these celebrations. Her evident beauty and evident content seemed very strong in comparison to Muriel's pathetic tinkling vicarious ways of talking or living.

"I'm terribly late. I must be forgiven. It's not really my fault. You see, people come to me to arrange everything," she said to Sylvester. "Something had gone wrong about the fireworks."

"Are we having more fireworks? How delicious!"

"Oh, of course that was last night. No, it wasn't the fireworks, but they make plans and then when they are failures I have to do it all."

Extraordinary the illusion of youth and helplessness that she maintained with all that efficiency. And boasted efficiency. She was wearing a white satin dress with a sort of hood at the back and lots of pearls. Her face was painted so smoothly, it looked as clean as ice. Her vitality made all the people in the room seem less than they were. Because she only wanted to speak to Sylvester he felt quite alone with her. That frightened him of her strength. And she was so interesting.

"May I ask you something, Sylvester?" she said.

"Please."

"I have wanted to ask you so much how this house seems to you?"

"It seems just like you."

"You knew it before?"

"It was just like Lady Charlotte, I think. But I was so very ignorant and young."

"I'm terribly interested to know how it feels to you now."

Praise, that was what she was looking for. Only praise. Should he say, "It feels like an expensive country club?" He said instead, "It's the Ritz for comfort."

That was enough. She didn't see much beyond that.

"Shall I tell you what it was like when Diana and I came here? No lighting or heating. Tepid bath water at the best. All the wall-paper dark green or dark red. Festoons of red velvet curtains, tassels, fringes. In this room seventeen 'occasional' tables besides big ones and a vase of flowers on each one."

"Fancy!" Sylvester felt that his exclamation was conjured from that same past. An emanation from the ghost of an asparagus fern.

"That was what it was like. Really. And the atmosphere, I can't tell you."

"Mutton chop whiskers, family prayers, dead wishes." He gave her the three rather obviously.

"It stank of all that. And unhappiness." For a minute she became less boastful and sank her voice down so that they seemed more alone. "They were all so unhappy. They could not escape from it—Muriel and Enid and Diana."

"They seem easy enough now."

"Yes, you see that? There was triumph. She had partly made and partly been given her point. "And because the house is changed. I said I could do it and even Diana said it was impossible."

"Diana hated it most, I expect."

"I took her away from here. It nearly broke her heart making her come back."

"And now, even Diana?"

"Even Diana—yes. Even Diana is happy."

"To change a place—a whole world of tradition! Weren't you afraid of so much power?"

"You see I did it when I was too unhappy to be afraid of anything," Cynthia said.

262

And he had thought she only minded about the comfort and efficiency of it all. The warmth and endless bath water and soft beds, flowers, wine and cream distemper. But there was more than that. It had been a spiritual contest. She had not been unconscious of the animosity walled and closed in Garonlea.

"Why did you want to change it so much?"

"For Simon and Susan, you know. I couldn't endure for them to grow up in the house as it was."

"Really for them you wanted to change it so completely?"

"Well, I hadn't much other reason, had I? Not much left then——"

She moved away from him on that unemphasized note and took up the thread of a joke with Reuben Hill—a good joke probably. What a victory hers had been over Garonlea. Really rather superb.

XXXVII

AT dinner time he sat beside Diana.

"Cynthia has been telling me how you first came back here,' he said. "I'm enthralled. I want to get hold of it all. I've never seen any one so completely at the top of success as she is."

Diana said, "You know I adore Cynthia, but I'm going to say something too extraordinary. I'm more frightened for her than I've ever been."

"Frightened?" He looked up and down the long table in the warm emptied room. He saw everything of her making; space, warmth, delicate and luxurious food; the party she had called into being sitting round contented and well entertained; her children, interesting and decorative. Last he saw herself. She seemed a long way off at the head of the table. She had gathered that white hood round the back of her head. It was a sort of nimbus behind her strong, her pathetically endless beauty. That was it. If only she could end her beauty, and the life behind it, she might come to some sort of peace. As it was she could never put the power of living behind her.

"Why are you so frightened for her? You must tell me."

"Because I think she's done everything. I think she's come to an end."

"You must tell me more."

"I can't now. You're the only person I've ever wanted to discuss it with."

"Would I be any use?"

"Oh, I think so."

Diana did not say any more then. Once or twice during dinner he thought she looked across the table at Reuben Hill with an expression of fear and dislike. Probably that was how she had doubted and disliked all Cynthia's lovers. There of course he was wrong.

Later in the evening he listened to a curious three-sided argument which took place between Cynthia and Reuben and Enid's Cecil.

Cecil and Reuben were having one of those dreary arguments about whether women should or should not ride in point-to-points, other than in Ladies' races. Reuben was hotly against it and Cecil spoke up for it as though it was her only creed or idea of value in living. He had not observed that spark and determination about her before, only that she was a pale lengthy creature with dark hair and rather surprising blue eyes, whom he avoided because he knew how endlessly she talked about horses. Horses in sickness and in health, their bits, bridles and accoutrements, their habits and peculiarities, engaging and otherwise. He did not wish to speak to her himself but he did not mind listening to any passionate discourse addressed to some one else. Another thing he observed, that Reuben's arguments were produced and upheld only to spur this furious child towards further self-revealings. It was when things were going really well that Cynthia joined in, sounding her lazy queen-like note of authority on Reuben's side of the argument.

"But, darling, aren't you a tiny bit talking nonsense? I mean really it's a question of sheer physical strength."

"No, not entirely, Aunt Cynthia. I mean, were all the good jockeys you've known strong men?"

"I maintain you girls mustn't ride against us boys

racing. Because you might defeat us and that would be bad." Reuben lit a cigarette and Cynthia went on:

"Really darling, I think, jokes aside, we must leave the chaps their racing. We can't compete."

"I don't see a bit why not."

"Perhaps when you've had more experience——"

"How much experience of riding point-to-points did you have, Aunt Cynthia, before you knew you were no good?"

It was an extraordingary rude little speech but delivered with so much earnestness, a direct question to Authority and to the Past, that it was stripped of insolence. It was simply cruel. Sylvester saw Cynthia grow very slowly red—an old and unbecoming red— before she said:

"In my day we found hunting five days a week just about all we wanted."

Then he realised that Cynthia had never competed in a race of any sort and, most strange most pitiful fact, was ashamed of this. And was it necessary for Simon to slip himself lazily into the conversation to ask:

"How many delicious Ladies' Races have you won, Cecil? Dozens, haven't you?"

"No, only seven."

"Well, that's seven more than most of us. But apparently not enough to prove to you you're silly to try."

Reuben looked at Cecil quickly. "You sly little devil," he said, "I didn't know you were an expert."

It was a slight matter enough to bring that look of complete animal fear across Cynthia's face. Sylvester thought that never before had he seen fear of what must be written so desolatingly on a face. That and a blind refusal and lack of understanding or acceptance.

Was this the sort of catastrophe that Diana feared? Had she heard? Had she seen? Did she know those other lines, "Lady, the bright day is done and we are for the dark——"

Diana had seen. She had seen what she had been watching for all that day, ever since a blush and a giggling whisper of Cecil's had set her wondering. She had seen what for years she had known must one day happen. Cynthia would lose a man to youth. She would be defeated by that lost thing strong and present in another.

"Why hasn't it happened before?" she said to Sylvester.

"Are you sure we aren't wildly exaggerating?"

"Of course it is a trival thing, looked at coldly."

"But who's looking at it coldly? If you look at it at all it's a shattering thing. Don't say excusing things like 'looked at coldly,' they mean nothing at all, do they?"

"I was only trying to be moderate."

"Need you be moderate with me?"

They were in the old schoolroom, the room where so many thousands of years ago Enid's drama with poor Arthur had filled the air surging full. The little piano (not even its ghost left here) had shaken to her vibrations. The air had stilled and frozen about her while they fitted that trousseau on her unresistant body, locking all the desperate doors in her mind. Everything was changed. There was no single object to recall the room as it had been. Not a book of girlish adventure—not a photograph of Violet in her Court feathers. There were now low, long chairs and a sofa on which you almost lay upon the floor, pale grey walls and one large and lovely flower piece in a white frame. There was an urn-like vase of different purple flowers of contrasting textures, the feathers of Michaelmas daisies, the flesh of dahlias, a well-shaped decoration.

Diana said:

"It's very curious your being here now to help me. I've never had any help about Cynthia before."

"I've gathered so little from Simon. I think he feels rather violently about her."

"You see she was so cruel to them when they were little. I must speak the truth about her."

"Yes. If I'm to be the smallest good."

Diana leaned forward, looking into Sylvester's face, who sat in the low chair opposite to her. He seemed longer even than its length allowed for, leaning back and stooping his head forward towards his eternally long fingers, fish bones picked by sea-birds they suggested. His face was very gentle and interested. She knew that it would be completely safe for her to abuse Cynthia to him without apology. He knew and understood enough of her love.

"It seems simple enough to me now," Diana said, "though I couldn't see it as it happened. When Desmond was killed she had nothing left. She simply was in a desert—salt water to drink. She had the hounds. She always has been marvellous about that, but what I am saying——"

"You said, 'she had the hounds.' I thought it sounded rather a far-fetched substitute."

"Yes, but it was occupation, exhausting occupation."

"Ah, yes."

"Of course, she should have married again. But really, she couldn't then. No, it wasn't all vanity and thinking herself a queen. Not then."

"I do easily see that."

"And having the children so unlike her or Desmond

was somehow terrible to her. I think almost as if they were a little monstrous. Can you understand that, Sylvester?"

"She had to torture them into some sort of likeness of the children she wanted for Desmond."

"The poor little things. They were so frightened. I used to feel ill myself I was so sorry for them. And the best I could do was to do nothing."

"And outside that—all the time her queendom kept swelling and growing?"

"It was—it is—extraordinary how people do really worship her."

"And when she got here at last—was it interest or only revenge, all she has done here?"

"First one, then the other. She could not work herself hard enough for both. She just went on in a way that would have killed any one with less strength and vitality."

"And she won?"

"In an outward sort of way, yes, I think she won. She imposed her will on the place, and it's the strongest will that has ever ruled here, except perhaps my mother's."

"Do you ever feel, Diana, that there's still a sort of contest going on between them? Between Cynthia and your mother?—Child, don't look so terrified."

"I'm not frightened really. I'm so used to the idea. It's always behind everything. And the tremendous will of the place against her too. I do know about that because all my life I've known it and hated it."

"But what a victory! One can't get away from it. And she has been happy here with other men, I gather. And avoided fatalities?"

"I don't know. I think the first—David—she really minded about. But he was so hopeless and uncertain."

"Did you like him?"

"I couldn't have liked him more. The children were devoted to him too. He could have made Cynthia's life all right again. But he just didn't want to enough. He simply left her. Once she recovered from that. I thought the—what was the word you said?—the fatality of Garonlea would get her down over that. But it didn't. Quite soon she was happy with some one else, or seemed content. She'd have stayed content enough if David hadn't come back. Then her vanity got the better of her; she thought she could have everything her own way. But David didn't last and her American never came back. She'd hurt him too much, I expect."

"Since then?"

"Since then she's been just hard and silly and terribly vulgar. *How* she gets away with it!"

"Doing things like having that present fancy here with Violet and Muriel."

"Hundreds of things like that. And always thinks she can't be detected or defeated."

"It's a distressing situation. Ageing beauty. Sad enough when cold. Terrific when incontinent. Can no one tell her how unattractive and hopeless it all is? But of course not. She's very beautiful really and as strong as a horse."

"I think before long Simon will tell her," Diana's voice was strained and small when she said this.

"You'd like to spare her that if possible."

"You may think me altogether too romantic, but, you see, Cynthia has given me all the happiness I ever had in life. Really she has. I'd like to spare her that if there was any way I could."

The fire in the little room had died down long ago.

It was cold with the weak, enveloping chill of very early day.

"What would you say she feared most, Diana? Could anything make her see herself truly? Even for a moment?"

"She's afraid of age and she's afraid of defeat."

"If the possibility—the certainty of both, were brought home to her once, really cruelly and consciously, what would she do?"

"Immediately, I don't know. Presently I think she and I would go and live at Rathglass again."

"Rathglass again." Obviously Diana had no truer idea of happiness. He understood it and envied her. It was strangely moving to find a person capable of almost hysterical love for a place. Blessed in its air and earth. At peace in its service.

"Oh, I do believe, Sylvester, if I could get Cynthia away from Garonlea, that even Simon might like her better."

"Yes, it's possible. And Susan?"

"Susan is simply in Simon's hands."

"I think those two have a queer story in front of them, you know. And Sue is ravishing—or going to be. No, is."

"Simon reminds me of Mother. Sometimes—about Sue——"

"How dangerous."

"Dangerous! If you knew Garonlea half as well as I do, you'd know there was danger round every corner and behind every shut door in the house. And not asleep either. Look! This room—goodness knows it is changed from what it was when we were young, nothing the same. But Enid came in here to-day and she couldn't stay in here—even Enid—it's too full of her own unhappiness and despair."

Enid and her electric toasters and her wonderful little men, had she ever despaired? It must be so or Diana would not have spoken as savagely of such a time. Sylvester thought if he had ever seen truth unstrained and unexaggerated it was in Diana. She seemed to him to be altogether truthful and sane.

XXXVIII

BUT it is the wet afternoons in life that are responsible
for so much with their interminable dangerous leisure.
It rained at Garonlea on the day after Sylvester and
Diana had talked so long. It rained and stormed and
blew quite preposterously. Swollen leaves as big as
birds and birds as big as boats were flung along the
wind across a sky full of rain. There could be no
question of going out of doors. Every one sat about,
growing more cross and liverish with each paper they
read, and each hour that passed. Those who could have
a nice drink now and then were the only people who
had anything to look forward to. Those who forbade
themselves this relaxation from principle or because
they didn't like drinking or its effect, grew waspish as
well as liverish. In the hour succeeding luncheon, the
library was full of edgy people, stuffed with food and
inclined for either sleep or argument, but not for
pleasant intercourse with their friends.

Only Violet and poor Arthur seemed to be enjoying
themselves. They were playing a game they both
understood—its point lay in looking up people in
Debrett or *Who's Who*, that they did not think could
possibly be there. Violet stitched away placidly at her
tapestry, for it was a long game and required some
such occupation. —"Look up so and so. Can't you
find him? Isn't that rather odd?"

"How old is her eldest daughter? Look her up.
She's sure to be in it. Really not?—How very funny."

"I must look up old Johnny Hood—dearest chap—

don't you know him?—Now that really is odd, isn't it?"

Simon said presently, "For two such awful snobs it is surprising how many common people you know. Are none of your friends socially O.K.?"

Enid's Cecil was writing a letter. She said to Reuben Hill, "But how shall I ask him?"

"Write and ask him nicely."

"How do you mean—'nicely'?"

"Well, you might say, 'I'm coming to stay with you.'"

"Is that nice?"

"I'd think so."

"Who minds what you think?"

Sue was telling one of her stories about a party: "He took me out to dine at Boddinino's and as it was Sunday they shut down too soon. So he collected a party to go on to a night-club. There was the Pro dancer and the dancer in the cabaret whom he was in love with, and her partner—a Russian peasant—and the head waiter and a man who was in love with the dancer in the cabaret too. They were the most boring possible collection of people. I had no latch-key so I had to wait for him to want to come home."

Cecily's sister Nancy was telling Simon a long, long story about a horse and the queer things she did to its hocks.

"Where are its hocks?" he said at last, as if he said, "Whereabouts is Bokhara?"

"Oh, those things with knobs. Yes, yes, I know. We ought to have dummies really that we could talk to about our horses and bore as much as we liked."

"Simon," Sylvester said, "you do keep this party in a roar, don't you?"

"Very well, I will," Simon said, "then every one will

274

hate me more than they do now. Listen! Who votes for a nice game of Hide and Seek?"

Sylvester said, "I do. I feel wonderfully light on my feet." Some words of Diana's, spoken the night before, came flying into his mind—Danger. Behind every shut door in the house . . . and not asleep either. Where so well as in the course of such a game could one see and feel the atmosphere in unused rooms and stuffed, airless attics? The very fact of being fugitive, or seeking, hunting or pursued, lent an antic isolation to such a spying out, such a ghost-smelling as this might be.

Other people thought it might be fun too. Cecil's eyes sparkled with old memories of turret-rooms and little dark staircases known to her as a child. Reuben Hill was as keen as a kite for any hearty games. Arthur was delighted that the young should be removed as there were at least three picture-papers he hadn't been able to see. The two smart young men said it would be exercise. Sue said it was the most boring, awful game possible, but she knew an attic with a good key that was full of old *Punches*, "and there's a very nice book there, too, I often read," she said, "called *Till the Doctor comes.*"

Enid did not raise her eyes from the rather daring novel she was reading, neither for nearly ten minutes did she turn a page.

There was one person who wished with raging impotence that the game had not been thought of— Cynthia. She was not going to go bouncing and screaming about the attics. That was not her type of youthfulness. At the same time, she was not going to be out of things. "Well, I thought a nice game of bridge," she said. She put out her cigarette and considered the party without haste. "You'll play,

Reuben, of course. And Arthur? Violet, would you care for a game?"

It was so little. It was so much. A tremendous misjudgment. A gross imposition of will. Or simply the slightest demand of a hostess to a guest whom she knew to be her best and readiest bridge-player. But Sylvester thought she looked rather tight about the lips as she went across the room. She opened the lid of a big lacquer box and stood staring into it for longer than was necessary to take out cards and bridge markers. He was quite right. She was shaken and hardly knew what she was doing.

The rest of the party trailed rather drearily out of the room. Soon the game was in full swing, which is to say that those who had not locked themselves, nicely provided with books and cigarettes, into bathrooms and other places, were sitting in Sue's attic, taking it turn about to read aloud, *Till the Doctor comes*.

All but Sylvester who, filled now with delight now, with horror, now by a collector's frenzy for possession, took his fill of prowling in and out of those rooms, strange, full and silent, that harboured the abandoned furnishings of Garonlea. They were indeed like harbours full of old ships and ghost-ships—those rooms. Dark harbours where the sun did not shine nor the winds blow, and old ships rotted and mouldered in excruciating quietness. Once Sylvester had seen the saddest possible sight—a boat lying on its side in a shallow green estuary, its sides rotted away. He could see the water flowing through its staves like open windows. But here there was no water—only time, to rot and make an end.

But how right Cynthia had been. How terribly just and right to strip and change the house as she had done. The air of melancholy in these abandoned places

276

seemed to him to crawl even in his hair. He sat on red plush sofas and smoked cigarettes. He counted a thousand little tables. He touched mouldered piles of velvet that gave him the recoil that the sudden touching of a still but living bat might give. He saw things that others before Cynthia must have cast away. Some gorgeous and unfaded wax fruits beneath a glass dome which he coveted. They seemed so seemly and bright here. An astrakhan cloak, moving with moths and their worms. An unbelievably elaborate and ornate bird-cage hung rusting on a wall, with all that it meant of prisoner and dead prisoner too.

And at last—it was Diana who found him—still with horror and inspiration, standing spellbound before two life-size black boys in Saxon porcelain. They wore curious, bustling sort of pale-blue kilts and white boots with tassels. They carried torches. They stood on tiptoes on heavy, elaborate pedestals.

"Diana," he said, "tell me all you know about these superb pieces. I stand simply abashed before such marvels of Rococo. At the same time I want to giggle a lot."

Diana said, "Aren't they hideous? My grandfather brought them back from Germany, I think."

"My dear, he must have had the strangest fancies."

"I don't quite follow. I think he just had very bad taste."

"Yes, of course. Exactly."

"They were almost the first things Cynthia whirled out of the library."

"Naturally. They would be. Diana, you haven't forgotten our talk last night?"

"Well, could I?"

"You may think me fantastic and mad. You may

277

think my idea cruel and dangerous and useless as well—perhaps you would rather I left you out of it?"

"No, I wouldn't. I'll play." Diana sitting down on a rigidly-curved, back-to-back sofa, its plush buttoned a thousand times over, looked strangely weakened, as if in this room she felt the strength of old currents almost beyond her.

"Play-acting, Play-producing, those black boys—I don't know what gave it to me—but this is my idea——"

"Those black boys? Oh, no, Sylvester. Don't say things like that in dark attics—people will begin to talk." It was Simon. He came in and sat down beside Diana. "But what is your idea? And what have you been doing for the last hour or so to make my party go with a swing? Nothing. Not one merry call of ' cuckoo.' It's a shame. You're not pulling your weight."

"Well, I've thought of something all the same. A very good idea—for a Good Time to be had by all."

"With those two boys? Sylvester, I'm not sure they're not pretty marvellous really. They give me the same feeling of complete vulgarity that one has from looking at Delysia. I'm glad they're in your idea. Look at their boots. I think they're swell."

"Well, listen——" Sylvester spoke, and as he told his plan the after-storm light coming so ekeingly through the narrow mullioned windows grew less, fumbling its way out of the unshuttered room where in the dusk the monstrous, cluttered shapes of furniture loomed and leaned, swelling infamously, dirty and bubble-like, stinking feebly of moth-balls and corruption. The two black boys in their petticoats leered unpleasingly, the last light catching an oily glimmer on their cheek-bones and on their fat, springing calves.

Diana, from her corner of the sofa into which she

seemed to have shrunk small and brittle as a bird's bones, said:

"I think it's a terrifying idea, Sylvester. Don't let's."

"Simon, what do you think? After all, it's your house, though we do keep forgetting."

Simon was shaking a little with excitement. "I think it's wonderful," he said. "It couldn't be more exciting. Of course we'll do it. And we'll ask everybody. On Friday night—would that give us time?"

XXXIX

"——But, Simon, what can we do about our hair?"

"What about our bosoms?"

"What about our behinds?"

"We have no corsets——"

"We have no bosoms——"

"Wear hats, girls. Pad yourselves into womanhood. Use your imaginations. Have a look at those bound copies of the *Lady*."

Not a doubt about it, Simon's idea (for it had come to be called Simon's), had caught on.

—— A Period Party. '95 to '07. Come as yourselves, your Uncles or your Aunts. Dancing 10.30——

The countryside ransacked their cupboards, delved in domed boxes, fought out bitter contests for the first services of the Little Women Round the Corners. They were all coming to the party at Garonlea. Simon's party.

Yes. It was entirely Simon's party. It was some time before Cynthia took it in. Before the, "No, you mustn't bother, Mamma"—"We've seen to that"—"no. I've had a marvellous plan with Mrs. Bryant about the food"—"I won't tell you, it's to be a surprise for you too"—"Sylvester won't hunt on Friday, and we want to keep Diana"—really got home to her. When she did realise that in this scheme not only was she unwanted but unnecessary she was left with a feeling of strange disorientation and more of wonder than of bitterness. It was so long since anything at Garonlea had been carried out except by her direction and inspiration that she felt now almost as in a dream, voiceless, purposeless, but terribly resistant. Yet through it all she kept up the magnificent pretence of everything being done according to her own wish.

"Violet, Darling, shall you mind terribly if we sit in the billiard-room and schoolroom on Friday? These awful children and their mysterious decorations, you know."

"I know you don't mind, Enid. You see, I've always encouraged them so tremendously to look on the house as a place for enjoyment—not as a sort of museum in trust. I mean, if they want to break the china, let them break the china——" She was, perhaps, the only person to whom the idea of Susan and Simon breaking china did not seem a little grotesque.

"Muriel, dear, I know you don't mind."

Nobody minded. They had all caught the fever of Simon's party and thought only of the impressions they were going to make of their youth and prime.

The importance of all that was happening at Garonlea was dwarfed in Cynthia's understanding by that other terrifying thing that was happening to her at this time—the thing she fought with every powerful yet futile weapon at her command. While she refused with desperate obstinacy to admit to herself any matter for contest, no other woman could have used with less conscience her power to keep Reuben Hill and Cecil apart.

It was not that her love was involved in the matter. She had been attracted, that was all. There had been an amusing prelude to a possibly satisfying love affair with an ardent and common young man. Cynthia's world was full of such. This defence and battle were for herself alone. She could not dare to admit defeat at the hands of a girlish and earnest young creature. She fought a whole generation. She might as well have quarrelled with the tides or argued her needs out with the winds. But from all her life there was no reason why she should be able to see this.

It was an ordinary enough little flirtation, Reuben's and Cecil's, and might have made its small progress and died its small death of inaction had it been let alone, or even encouraged. But, alas for Cynthia! she thwarted these two rather simple people at every turn and so blew up absurd fires between them. They had to make plans to meet and kiss. They knew a common enemy. Cecil giggled a little about Cynthia's age. Reuben, who was a great deal more sophisticated, did not think this so preposterous. If he had seen just a very little more of Cecil, this weapon might have turned dangerously in her silly young hand. There was in Cecil as much of that romantic incontinence as might be looked for in Enid's daughter.

After such times as that enforced game of bridge, there was terrific excitement in eyes meeting and falling across a drink at six o'clock, in small and precious plots and schemes for the evening, for the next day. Simon's party was a godsend, a milestone, a beacon in the week towards which they looked with disproportionate eagerness.

It was curious how excited everybody was over this party, as if a suddenly loosed power in the house had caught them all. Even the entirely phlegmatic Violet kept regretting that she had not brought her Madame Pompadour fancy dress with her, although everybody explained to her that she had got hold of the wrong idea. "Can't you remember what you wore in 1900, Aunt Violet? You can't have gone round as Madame Pompadour."

"Well, I don't know. I think we wore much the same sort of clothes as you do now."

"I thought you wore tea-gowns for tea."

"I don't think anybody wore tea-gowns unless they were *enceinte*." She used the word in the most delicately indecent way.

Enid said, "We always carried a fan for dances and wore gold shoes, of course, and gloves turning the elbow."

"Oh, Aunt Enid, how daring it sounds."

"Daring? I don't know. We were much the same, really, as you are now. Long skirts in the evening, you know."

"I would like to come to Simon's party in some very high corsets, to make me some bosoms, and a really tricky pair of white nainsook knickers run with black ribbon and frilled at the knees. Pads for my hair, and black silk stockings. I know I'd have the success of my life."

"I think it would be very unfunny, Sue," Simon said.

"Well, I won't, Simon. But I could have called myself a postcard from Paris. And do you remember the first bicyclists, Aunt Enid?"

"No, I don't," said Enid, sharply. Why should it appear grotesque to any one—that time of youth? These clothes that had looked so right in their own romantic period! All there seemed to say in their defence now was that they were not unlike the clothes of to-day. You could not say, "We looked lovely in those clothes. We had bosoms which attracted the gentlemen and beliefs to which we clung, we were not rude and unhappy and flat-breasted like our children." Not unhappy? The denial too was true to type and period.

Only Cynthia who had been beautiful and enormously happy then, felt a real reluctance towards this one evening stepping back into the past. She took it all seriously and rather childishly, saying, "It's an absurd idea—it's not long enough ago." She was frightened of the travesty this would be of what she was. She would have avoided it by any means, but the general will was too strong for her.

Enid and Violet and Muriel got together about
clothes and really enjoyed themselves at last, turning
out cupboards and trunks and putting every possible
garment into an empty bedroom. Soon the bed was
up to the ceiling with its load of gored skirts and boned
and frilly bodices. There were clutches of rolled silk
stockings on the tables, black and openwork. There
were boxes of gloves, and fans in boxes, and plumes
from old hats and lovely hat wreaths of frail pink silk
roses and forget-me-nots. There were petticoats with
tapes at the waist and finely-worked nainsook knickers
(the kind Sue wanted to wear) still gathered into a
frill at the knee by a faint and quite rotten piece of blue
ribbon. There was a terrifying engineering work in the
shape of a pair of black satin corsets. "Darling, darling
Mother's, don't you remember?" Muriel put them
sacredly back in their encoffering cardboard box.

For a day they were extraordinarily excited and happy
among those clothes so carefully put away by careful
maids, so old and clean and fustily sweet, so complete
and elaborate. With flushed cheeks and sparkling eyes
they shook out skirts and held them up, marvelling
politely at the size of each other's waists. It was not a
new thing with these three. They had always been
sympathetic and sweet to each other, but this was a sort
of autumnal reflowering. Roses in the brief heats and
chilled dews of October, seeing the June before in a
heavy, luscious haze.

"Violet, you remember the Viceregal lodge party

285

when you wore that hat? Those pink roses and the white feather boa——"

"Muriel, darling, these must have been yours. You had tinier feet than any of us. But these seem too ridiculous."

"Your blue, Enid, surely. You always wore blue at night. It would have been a crime to wear anything else with your eyes."

Cecil and Nancy and Sue came in. They were for a moment awed and intrigued by the breath from the past that swirled and filled the room like wreaths of roses and whorls of braid, as though invisible parasols, long-handled, had flirted open, and dexterous fans swayed coyly in the air, stirring the close scent of old sachets and of stuffs long ago grown brittle in their folds. They felt a little shy of their aunts' and mothers' gay memories.

"Thank God we girls don't have to unhook each other," Sue said, picking up a gold tissue bodice veiled and ruched and frilled by an apricot and diaphanous cloud. "I'll have this, I think. Or shall I? Oh, my nainsook drawers! I suppose Simon wouldn't be very pleased if I wore them. But I could, underneath. A nice bit of atmosphere. I'll have these anyhow."

Cecil said, "What a curiously cruel shade of blue. Truly electric."

"Not by lamplight," Enid faltered a little.

"With her eyes it was quite your mother's colour," Muriel's feathers, like an angry wren's, rose and buzzed.

They took off their skirts and walked about in their slick knickers, green and pink and red. They whirled skirts off the bed and struggled into them, unashamed of the vast contrast between theirs and their mothers' waists. Only on Sue would any of them meet. Her long frame was so very narrow. Even the folded belts,

286

pointed behind and buckled in front, she could just with a deep breath, clasp. They took their pick of all the clothes and ran down the passage to the sewing-room, their arms foaming and full of strange anachronisms, as any of the three left behind could have told them if they had been asked.

"*Deep* blue, I call it. Not *Electric*," Enid said in a wounded voice, gazing at that blue gown of '04 which had not then seemed such a cruel and persistent shade.

Violet laughed, not quite so calmly as usual. "The wreath of my garden party hat," she said, "has gone with a mauve ball dress. When do they think we wore them, do you suppose?"

Muriel said forlornly, "They haven't chosen any of my clothes."

Diana had been very firm. She would not exhume anything from her past wardrobe. Nothing, she said, had escaped from jumble sales.

"I'll have my dress made," she said, "I'll be myself by night."

Excitement ran high. Enid and Violet and Muriel recovered their setback by Youth and plotted and planned and praised each other busily. The girls, seeing their own shoulders suddenly from new angles, and curves where curves had never been before, grew enthralled by the change in their bodies and quite anxious to impress the gentlemen with their new selves. Sue even becoming something of a purist, abandoned her apricot ruchings and held grave consultations with Sylvester.

"But I'll look awful, Sylvester," she wailed, when he had given his verdict.

"No, you'll look most amusing."

"That always means wildly unbecoming. Don't I know."

Behind locked doors mysterious and noisy doings went on in the library and drawing-room. Sylvester and Simon, the carpenter and many helpers were there all day. Diana slipped in for half an hour when she could do so unobserved, and came out looking as if she had walked five miles in July. She lied to Cynthia about this too.

"I don't know what they're doing, Darling, but whatever it is, it won't be permanent."

"They seem to have ordered practically nothing to eat beyond fruit jellies," Cynthia said, "and I do so like my parties to be well fed."

"I think you're being very good about it."

"But it's just what I want," Cynthia said angrily. "You know I adore them to be independent. All the same, I shall see there's a good supply of bacon and eggs and sausages laid in."

"Have you decided about your dress yet?"

"No, I haven't. I haven't thought about it. It's a bit putting off the way those silly old things keep on about theirs. Rather pathetic when you've once let your face and your figure go completely."

For the last two days, as Diana guessed, Cynthia had thought of nothing else but her dress for the party. She had made plans and abandoned them for one reason or another as fast as they were made. She was feverishly anxious to look her best and to look different from those silly old things, her contemporaries.

In the passage outside Cynthia's room there stood a very long cupboard. Locked in it were all the clothes of half a lifetime, which she had thought when discarded too good to be given away. With all her sweepings and clearing of other people's hoards and rubbish, she had never found it possible to dispose finally of her own. She had always loved her clothes, buying them with vision and extravagance, wearing them with immoderate success, and cherishing them beyond any useful purpose. Because they had been part of herself, part of her beauty and glamour and powerfulness, there was a lonely and jealous force in her guarding of them. She would have been desperately angry had any attempt been made on that locked wardrobe of ghosts and memories. All the dresses were hers, her very life. Not a vague, general past. They had decorated her, been warm and light, part of her hours of love and dreadful loneliness and quickened content. There were dresses hanging there that she had worn before her wedding-dress which hung still like a Spanish queen's in isolated perfection. Not as much as a knot of orange blossom unpicked to fit it for lesser moments. Inviolate, unaltered it hung, a shapely thing in the darkness, its sleeves puffed still, as by the breath of romance, its sweeping white line a gorgeous full memorial to ripe virginity.

Often during the last two days Cynthia, suddenly and completely relieved of her responsibilities as hostess and entertainer by the occupation and interest found by all her guests in their preparations for Simon's party, had looked into this cupboard, comparing, wondering, strangely touched and plucked at by the past which she would deny and ignore and pretend to forget while keeping so many of its ghosts for torment and delight to feed on.

All her life had seemed so near her till now. Not ten and twenty and twenty-five years past. But yesterday, and to-morrow, and a thousand things done and to do. Troubles to surmount victoriously. Difficulties to smooth away. Sorrow to forget. Love to have again. Always horses and hunting to excite and anæsthetise. Always a drink if one felt low. Never an empty time in which to see fatigue or triviality or purposelessness in anything. Life for her never ceased to be a competition in which you won and won. You must never lose, never entirely let go your hold even for a moment. That was the way of desolation.

Now in her most triumphant hour was uncertainty to shake her? Was there to be a foreshadow of defeat? A week ago all tributes had been her right, everything had combined for her praising. There was truth in all the flattery.

As on Reuben's first day: "Cynthia, how can one believe that Simon is twenty-one. Seeing you together it just doesn't seem possible."

Those dramatic entrances and moments with the people on the place. She had felt a sort of wild power in her popularity. Simon was only a doll. She moved him about. The rejoicings were for her really. She knew it.

Even Enid saying, "I can never quite get used to being entirely comfortable at Garonlea. The contrast is too big—though we've known it for years."

And Violet, "How *do* you compel your hydrangeas to be so divinely blue, dear? We try everything in turn and it's no use. It seems so unfair because they never used to be blue here."

All nonsense, all ambrosial food, earned and merited by her years of unrelenting efforts and triumphs. It was true she was a greater and more lovely person than others. It was all true. She could not fail.

When did the draught of doubt first blow on her? Was it on that classical, that too nearly perfect autumn morning, when she had ridden on with her hounds for three miles between coverts expecting each sun-mellow minute to be overtaken? And Reuben had failed to overtake her.

Since then there had been other more obvious escapes, but she denied their reality with a fierce and terrified obstinacy, pinning all faith on a future moment when her beauty and her skill in loving should sweep him back to her for so long as she might wish. At this party of Simon's—that was another window into the unknown. The idea of it troubled her. It was strangely disturbing not to know precisely what was going on in Garonlea. To be in ignorance and too vain to pry and ask servants who must know of some of the doings afoot: to restrain herself from giving advice: to show the right balance of tolerance and amusement towards the unknown thing in store: to yield this shade of her authority for the first time: all seemed beyond her. But it would have been more cruel if she could have seen where the trend of it all led. Here her assurance and vanity for the moment were her saviours.

At six o'clock on the evening of Simon's party a maid was sent to summons Diana to Cynthia's room. There was something in the girl's quiet decorous face that sent Diana flying, her hands wet and iced by the arrangement of a quantity of floral decorations. A vast quantity.

"Darling, you wanted me? Cynthia, what is it?"

A fire burned in Cynthia's room. The air was warm and sweet. A big vase of malmaison carnations stood on the dressing-table beside Desmond's photograph. This morning Diana had put them there. The shaded lights lit wan ivory and gleamed in dark woods. But they lit the looking-glass as truly and baldly as mirror can be lit. The clever photographs of Susan and Simon and a hundred of Cynthia on her show horses or her best hunters stood low and high, on dwarfed tables and tallboys round the room. Light caught the superb

curves of the smooth rich bed. Over every chair except
the one where Cynthia sat, there hung a dress, many
dresses. They kicked their frothed skirts on the floor
too.

Cynthia's naked and most beautiful arms hung
straight and dropped over the chair arm where she sat.
She held a glass, her fingers pointing downwards
round its rim. The bottom of the glass just did not
touch the floor. Her head was bowed. She was pretty
drunk, Diana feared.

"Lock door," Cynthia said with pompous brevity.
Then she added, "Darling, please."

Diana came over to her, small and dark. Harsh with
anxiety and real loving.

"What is it, Cynthia?"

"I'm not feeling too good. I'm not feeling well.
Had to have a drink. Hate drinking between meals as
you know. So bad for the figure. Never do it."

"You've broken out for once, haven't you?"

"What d'you mean?"

"I mean, aren't you rather drunk, Cynthia?"

Cynthia dropped her glass (it was empty) and put her
hands up to her face. She was not crying. Diana had
never heard or seen her crying. She put down her
hands and said:

"Help me, darling—in despair. I've been trying on
my old frocks and—oh, God, Diana—Oh, God,
Diana——"

She did not say any more. Her appeal for help was
so royal, so defenceless. It was almost more than
Diana could bear.

Cynthia said harshly, "I look a joke in them. I
needn't be made to look a joke, need I? Do help me
about it."

"Darling, are you sure you can't wear any of them?

Couldn't you put some on and let me see. How can I judge?"

"There's nothing I can wear," Cynthia said fiercely. "Don't I know about clothes? Don't I know what they do to people?"

"Yes, you do know. I expect it's your short hair that makes you feel queer in them."

"Thanks, darling. Perhaps."

She seemed to have gathered herself up a bit. She was infinitely more sober and more unhappy. She looked slowly round at all the clothes tumbled about her room—all the lovely clothes that she had worn once. She, herself. Diana could not bear to think of her trying them on alone, flinging them off in despair, tearing at hooks and fastenings, perhaps crying. Alone, Diana thought she sometimes cried. Drinking herself stupid and sending to Diana for help. It was terrible to Diana to think she had entered into any plan that could bring Cynthia to this pass.

She was shivering now in this warm room. Chill and half-way to death, like a person coming-to after an anæsthetic.

"I give you too much bother, darling," she said. "I don't know how you can help me."

"I'm always here to help you. You know that, Cynthia."

Cynthia's eyes went hunting round the room again. Shifting from one thing to another.

"You know what they say about drowning? You know, seeing all your past life? I think these clothes make me feel like that. Perhaps I'd better have another drink, Diana. It might steady me up."

"No, darling, please. Isn't it silly?"

"Oh, *must* you be so tiresome?" the impatient demand, helpless, determined. "I'm not coming down to this party in any case. I'm feeling terrible. Let them see how they get on without me. I haven't a thing to wear."

"But you must come down, Cynthia, they'll be shattered." Unconvincing because perhaps this would be her last escape, her only escape. Then—Save her from this plan made in the first place for her salvation? This plan that seemed to outrun itself, already its effect was so overwhelming.

"Cynthia, you can't not come down. Or—have dinner up here and come down for the party. That might be best." How many drinks had she had before these two? How many would she have before dinner? Drinking alone in her bedroom in that horrid, reasonable way as though each drink was something she owed herself, these lolling, empty shadows from the past clustered thickly about her, her strong present hold on life had slipped. She was not trying to defend or strengthen herself in any way.

Diana said helplessly, "I'll be back soon." She must finish those filthy flowers for Simon.

Cynthia nodded. She seemed more contented and quite vague now as to why she had asked Diana to come to her.

"Why not lie down and take some aspirin?"

To her surprise the suggestion was accepted. Cynthia moved over to her bed, rather slack and ponderous, pulling on her soft pink dressing-gown.

"Put out the lights, Diana." Her voice was almost drowsy before Diana got to the door. She left her half-sleeping, half-stupid in the firelight.

XLI

Simon heard with some pleasure which he mildly tried to conceal that his mother would not be down to dinner.

"Sue, you won't be late," he made her swear it.

She came down punctually at a quarter to eight and found him waiting for her in the hall. He took out a watch, cabled across his stomach on a gold chain, glanced at it and at the clock—gave her arm a little pat. "Splendid, my dear."

Sue giggled weakly. "Must we keep it up alone? What do you think of me, darling? Don't I look awful—you do—or I don't know. Perhaps not."

Sue was wearing a smart afternoon dress of stiff brown silk faintly but steadily striped, it dragged back from her stomach and puffed hugely over her behind; just above the puffs a little coat of the same material flirted its tail like an amorous robin. Its shoulders were tremendously stuffed, its sleeves long and tight and rocky with braid. On a foreign and glorious shelf of bosom gold and turquoise chains and lockets lay almost as on a table. A gold and heavily-padlocked cable ornamented one wrist. She wore buttoned brown kid gloves, three lines of braid like heavy starfish on their bursting backs, and buttoned brown kid boots. She had washed all the make-up off her face, divided her hair in the centre and plastered a quantity on the nape of her neck inside a netting bag. Secured by an elastic to this anchorage and further safeguarded by long pins with butterflies trembling on spirals at their heads

295

was a small and curly brimmed hat in which feathers and flowers fought for supremacy. A hideous and wicked little hat, frightful to wear and frightful to behold.

Simon wore an almost parma violet suit with a striped green and purple waistcoat, a colossally high white collar which forced his chin into a position of unnatural arrogance and round which was tied a tight, small tie, very low down at the base of the collar, with a tiny knot and a big pearl pin. His trousers were tight to the leg and he wore patent leather boots with pale cloth tops. He carried dogskin gloves and a narrow, curly-brimmed bowler hat. He had contrived the slightest of side whiskers.

Their costumes were faithful copies of a photograph taken early in the '90's of Lady Charlotte and Ambrose. In the photograph she sat upright as a dart in a high-backed chair. He leaned negligently, even a thought rakishly, behind her, against a papier-maché balustrade. A palm in a well-clothed pot on a well-clothed table did its bit to lend an exotic atmosphere to the picture, which had been taken in Bath.

Simon opened the library door and they went in. It was very dark. Not pale and luminous as the fire-light could show it. There was no scent of lilies or cigarettes, expensive or cheap. No hearty voice telling of a good hunt or a good scandal. No laughter. No sound of glasses. The silence was not lazy and rich as any silence here had been for so long. It was of another quality, heavy and chill. It might have come from a place far-off in time. It was so remote. Yet it seemed dangerously near.

Simon put his hand up to the switch near the door and the room was lit.

"It seems so dreadfully real," Sue said, blinking in

the light and talking at once to break the feeling of silence.

"Diana was quite right about the lighting," Simon said. "We had to have it in the old places. I suppose it's a bit bogus but the effect is there."

The effect was indeed there. The library might have been waiting for a dinner-party at any period in its story up to the date of Cynthia's queendom at Garonlea.

Informed by the perfect memory of Diana's perfect hatred, not an ornament, not a sprig of asparagus fern, not a photograph was out of its place. The exact smell, cold, fragrant, a little musty—of chrysanthemums, absence of dogs or scent—that had been the library smell for so long had quietly taken the air again. It stirred in the length and fullness of the red velvet curtains that bellied opulently under the swags and fringes of their canopies. Even the Saxon black boys seemed less exuberant in its breath, more of the page and meek slave about them, less of the curiously pampered favourite. All sofas and chairs had found their own places and wore their clean antimacassars. There was a sort of glum smugness about the portraits. They sank back with a conscious sigh into their proper setting of dark red wall.

Sue sniffed at the air.

"I don't think this would have been quite my favourite period," she said, "what about you?"

Simon took a deeper breath.

"It has its points. Anyhow, it's a good purge for the room now."

"I do think you are brave not to let them drink in here before dinner."

"One couldn't drink in here, not cocktails."

"Perhaps you're right." She giggled at nothing and

shivered, moving closer to the neat coal fire. "Not even a glass of champagne?"

It did not seem easy to talk. It was not altogether the faint feeling of panic before your own party. Each word seemed silly and flat and yet was eagerly listened for as if the least word had importance in a huge dullness. Suddenly Sue felt herself in a panic for something to say to Simon—Simon to whom she said everything as naturally as she breathed or drank water, Simon with whom she examined the details of a pain in her stomach or a spiritual ecstasy. She was apart from him as if a great station full of trains was between them. She longed for Sylvester to come in and pick his way over to them, avoiding tables and sofas. For Diana. For some one to dispel this strangeness. Suddenly and strangely she thought of Cynthia and wished for her most of all.

It was Muriel who came in first. The door had been open. She turned round and shut it gently and came towards them easily, knowing the old tracks between the tables and chairs.

Sue and Simon were a little scared by her look. She wore an expression of extreme disquiet as if she sought for love within herself and only found fear.

"Simon, goodness me," she faltered, "you look so like dear Granny. And the dear old room. It's quite a shock." She looked round it distressfully. "Oh, Sue," she took her in at last with a little squeak of awe, but although the clothes were there she did not convey the same impression as Simon.

"Well, anyhow, I'm in time," she said, as if suddenly it mattered a great deal not to be late. "And how do you like my dress, Simon?" She stepped away from them and turned slowly, meekly, waiting for criticism or condemnation. It was more in the gesture than in the dress that the years of independence in the ever-thrilling flat of her own fell away from her and she was left poor little Muriel in pale turquoise satin, a tiny belt round her tiny waist, a white net guimpe and vest veiling her faint shoulders that sloped into nothing, softening the collar bones from which her neck rose silly as a cygnet's, still. She twirled meekly before Simon, and when she turned to face him again, one could have sworn it was not Simon she expected to see, all powerful on the hearthrug.

"Quite charming," Simon pronounced kindly. "Yes, quite charming. And so lady-like."

Sylvester came in and Sue flew to him. He looked almost normal except for his smoking-cap which he wore with style, and the blue silk, tasseled cord that supported his trousers, and of course, his collar was rather tricky too. He came out of *Trilby*.

"Simon has just told Aunt Muriel that her dress is so ladylike," Sue breathed. "Where have we got to now?"

Sylvester pulled her ear. "It's only a game for one night," he said, "but if you feel hysterical come to me."

"I will. Oh, look at Aunt Violet all in mauve."

"That's Violet for her name. She'll go straight to her photograph, you'll see."

She had swept with dignity and beauty back into a youth that made Sue and Simon the only grotesques. Her hair had gone back into its old crimps. Her figure obeyed its corsets. If the floor had been polished she would have floated across it like a dyed swan. The gentlemen should have held their breaths, emitting, "By Joves!" hands to smooth moustaches. She was no joke. She was gorgeous in mauve. Her eyes took their true iris from it. Sylvester was right. She went straight across to her photograph on the shawled grand piano.

"Gracious, what an old lady it makes me feel," she cried gaily, for she was fresh from her bedroom mirror. "Why, Simon," she crossed over to the fire, "you gave me quite a little start. You might *be* Mother, mightn't he, Muriel? It's so odd in trousers." Her voice was as fresh and flat as usual.

"It's tough on me," Sue whispered, still holding Sylvester's hand, "in all her clothes too. He's stolen my act away."

"Do you mind?"

"Oh, Sylvester, ask me do I mind again. Do you know, I could go wrong on one of those buttoned sofas to-night, just to prove to myself I'm not grand-mamma."

XLII

The smart chaps came in, a riot as themselves in '04—
white pelisses and lace hats and pretty disgusted not to
find a drink. Then Enid in Blue and too oppressed by
the library to speak at all. She glanced round it,
nervously warming her hands at the fire and whispering
to Muriel. Then thinking she caught a certain look in
Simon's face, suddenly raised her voice to a note of
strained uncertainty. It was as if she obeyed a voice
saying, "Speak up, my child. We should all be most
interested to hear anything you may have to say—
Nothing so vulgar as whispering in corners."

An early cricketer came in, Uncle Arthur, very much
in period. Jason had forgotten to be anything, but he
wore a paper hat to show he meant well. Nancy looked
really comically successful in a lady's cycling costume
of the wrong date. Cecil had lost every breath of
horsiness and developed a tremendously romantic
likeness to Enid. She vibrated within an apricot
brocade, a veiled, fringed and beaded model that
became her beyond any words. Reuben (a Bush
Ranger) grew every moment more dangerously
enamoured.

"Well, if we're all here," Simon said, "shall we go in
to dinner."

The door opened again and Diana came in—Herself
by night, she had said. Her costume was warm,
modest and yet in effect a sharply discordant note in
that restored room, among those newly resurrected
busts and naked shoulders and puffed heads of hair,
those behinds caged and basketed so formally. Herself

as she would never have dared to appear by night in those tortured years of her youth.

She wore a pale-blue flannel dressing-gown, very gored and full to the ground and pieced in the back, corded in blue silk, collared and cuffed in prim white muslin. Underneath it a fine nainsook nightdress, yoke a l'empire, buttoned to its round, close neck-collar, buttoned at the wrists, threaded heavily with blue ribbon. On her bare feet were quilted, heelless slippers of blue satin. Two dark plaits of hair were tied at the end by two blue bows.

She put a silver candlestick down on a table near the door and blew out her candle. The room applauded.

"Diana, you're marvellous."

"You look the cosiest, the sweetest——"

"I'd have chosen you——"

They clamoured round her, laughing, touching, examining as if she was a perfectly-dressed doll. Even forgetting that they wanted drinks.

Enid did not move from the fire. She had started obediently into life together with Muriel at Simon's summons to dinner. Now they stood side by side, different thoughts of the Blue room and the Pink room overcoming them. The chill, the candlelight, the white quilts, the curious, ornamental tatteries they had devised and found decorative breaking back on their minds like waves out of a mist. Enid could see Diana doing her Swedish drill in black silk stockings and knickers, undressing not always as modestly as she might and diving into such a nightdress. Cross, dark, un-communicative, unusually unattractive to gentlemen. That was Diana. Then she had none of the things that counted and now she seemed to have all that mattered.

Muriel whispered inevitably, "What *would* dear Mother have said to her?" Violet calmly aloof, expressed ably and without words that it was perhaps a pity. Simon, on the way in to dinner, paused with icy ceremony while one of the chaps relit Diana's candle. He wanted her to know that her joke dress did not amuse him a lot. They all shouted with laughter, thinking how brilliantly he played up to his clothes.

"Though I was always the one for the stage," Sue said. "The elecution mistress at school thought a lot of me—do you hear me, Sylvester? I remember she nearly cried once when I recited, *Play Up! Play Up! And Play the Game*."

"Do it now. It really would be riotous in that dress."

"My pretty gown. Look, Sylvester, is this party going to flop badly? Is there only to be one glass of champagne for the ladies? Game *and* a joint. Well, fancy! It's right for the date, I suppose. A white tablecloth looks nice, why don't we always have one?"

Except for a white cloth and an array of small flower vases arranged on a lacey and beribboned table-centre, the dining-room had not been changed. But as in the library, easiness had withdrawn. The party seemed to sit more straightly in their chairs—to converse rather than to talk to each other. If anybody thought of a good crack or a dirty joke it was made furtively and as likely as not suffered a change in the telling towards extreme unfunniness. Under the flowing tablecloth in real 1900 style, Reuben pressed Cecil's foot and sometimes touched her knee. She looked lovely and rather wild with excitement. Arthur felt a curious note of memory sound to him as he looked at her. Clean across all his present occupation with food and wine the sensation of acute unease with which Garonlea had inspired him in youth came fluttering indeterminedly back to consciousness.

Muriel pecked at her food and was quick in ready,

303

obvious replies again as though she knew some one listened. Some one waiting for a suitable moment to say, "Really, my child, you must try to make some impression. At least answer brightly when he speaks to you." The delightful stable thought of her own flat had retreated beyond consciousness. So with Enid, the little men who ran up your tweeds and all the domestic implements called gadgets had lost personal importance. She was a long way previous now to the Enid who had gossiped happily to Violet over the library fire, warm in the lily-scented room, pleased with her two attractive, horsey daughters who did not even treat her as a human being.

Diana, sitting between Sylvester and one of the chaps in pelisses, was the only person who kept up a real fight for gaiety. The chap in the pelisse was her child. They maintained a children's party-joke for about three of the endless courses. With Simon suddenly attentive it became in a moment quite pointlessly silly.

Throughout dinner, Simon talked with able dignity to Violet on his right and to Enid on his left. Each minute Sue thought he grew farther away from her, less tolerant of her giggling to Sylvester, more venomous in his maintenance of polite conversation with Violet and Enid. And neither was the easiest person to converse with after a week spent at close quarters. Simon was good though, he got Violet on to the McGrath pedigree with such success that by the end of dinner their blood (on the distaff side) trickled practically Royal in their veins.

Violet was all right. She was delighted with herself in mauve. She was undisturbed, calm, unaffected by memory, not haunted like Enid, who thought she had forgotten. Or like Muriel who thought she had loved. Nor was she in arms like Diana who knew how long and bitterly she had feared and hated all those things which had been brought back to shadow nightmare life at Garonlea to-night.

After dinner when at last the gentlemen joined the ladies, Sylvester said to her uneasily:

"I feel a little as if we had put on such a good puppet show that the Puppets had come to life."

Diana said, "No. Parts of it are too grotesque."

"The pelisses and the cyclist. But other parts of it? He indicated particularly Cecil, the virgin queen of outdoor sport, who sat now swaying at Reuben with tremendous effect. She made a world of separate romance behind her great, soft, feather fan.

"Yes. That's just how Enid used to go on, and then terrible scenes and tears with Mother."

"Poor, oversexed girl. Happily for her she married young?"

"Well—Yes."

Simon came over to them, a cup of coffee in his hand. "I enjoyed dinner," he said, "I think a little restraint must be good for one's digestion. I'm not sure we won't always have a table-cloth now."

"I thought dinner seemed endless," Sue said. She was beginning to feel a little cross and to resent her hideous brown gown. "Why did you want to have all that food?"

"You'll get used to it, my dear," Simon said. "We're often going to have dinners like that. They soothe me wonderfully. I feel myself after them."

"You don't seem like yourself." She looked at him, picking at the rolled gloves in her hand, raising her striped, uncertain eyes, dropping them, sulky and despairing.

Simon paused as though to consider whether she was worth answering. Then he turned to Sylvester. Diana said, "Come and have a look at the ballroom with me," because she thought Sue might conceivably cry for twopence.

The drawing-room had been stripped and garlanded plentifully with evergreens. Even the window curtains had veils of ivy running up them, producing the maximum of naturalistic with the minimum of decorative effect.

"I think it looks unspeakable," Sue said. "Why did we think it would be amusing to do this? Everywhere I go in the house I feel choked, and I don't know what's happened to Simon. What has happened to him, Diana?"

"I don't exactly know." So hopeless and negative. A lie to this child who had been imprisoned for a night in this parody of another age. And what could one say to her? It was all fantastic. Like shouting in dreams that can't make sound. For the moment as dreadful as that can be.

Diana remembered, Diana knew this sense of being blinded and choked by the sadness of Garonlea, the sadness that was evil, the thing that Cynthia had hunted out and kept at bay, but which they had loosed again, pillared and supported by unhappy memory. Herself, Enid, Muriel, the knowing and remembering past. It was not strange that the old vibrations should stir in the air for them.

But that Cecil should find herself no better equipped for love than Enid, in her youth, that Simon should seem possessed of a spirit that Diana did not name for fear, that darling Sue should be as near to tears as they poor, uncertain girls had ever been, all these things seemed an improper, disproportionate sacrifice and spending of emotion for the cause in hand. They had done too much, gone further than they ought on Cynthia's behalf.

XLIII

AND now the party was on them. It was in being.

"I asked lots of girls," Simon said placidly, "because I wanted the good old wallflower atmosphere. A stiff upper lip in the ladies' cloakroom."

He and Sue, posed as in the photograph that was their original, received their guests in the hall. They made a good group, what with their palm and their balustrade. This unkind remark of Simon's was the first he had made as Simon for a long time. And he would not let Sue leave him, though half an hour of sitting on a hard chair, being Grandmamma's photograph, had got her down and for nearly the first time in her life she badly needed a drink.

"There are far too many. I'll have to make all my chaps dance with them instead of me."

"Well, dear, your partners won't think the less of you for that——" He had gone back to it again. One could not get near him. One gleam of unkind humour and the mask was down—"Mrs. Hughes, I'm so glad you brought *all* your young people, Evelyn, Adela, Dorothy, Barbara, delightful to see you *all*——" On and on it went. Eyes and mind gaped wearily at an endless series of strange dresses and glaringly familiar faces. Plucked eyebrows and varnished nails were not affected by the leaders of society in 1900 but Sue grew grateful to them for their unreality—they were a faint hint that this party could not last forever—that it could sometimes be laughed at and forgotten.

But not yet. Not now, though she had escaped from

Simon at last and had her glass of champagne, the party seemed to her hopelessly stagnant. Clogged with failure and unhappiness. Ranged tragically round the rather chilly drawing-room, girls who could not hope to dance, sat with their mothers, or stood about in little groups talking brightly to each other of hunting, that glazed, hurt look about them which is so pitiful to see. Over the scene there was a horrid air of truth. Things had been like this and were again. Mothers looked anxious, watching. Over-polite to one another about daughters who danced or did not dance. The party was no longer a joke. It was real. The dresses had become familiar. Soon they did not even serve as subjects for conversation. The band played waltzes and polkas without any remarkable verve. Simon danced with the ugliest girls, other men danced only with the prettiest or the most amusing. Sue thrust her partners on her friends to the embarrassment of both. Supper seemed endlessly far off. The unwanted girls migrated to the library fire and tried to cheer themselves with weak champagne cup, which was all the drink Sue could procure. She found it a terribly stiff task entertaining them with her stories when they only wanted men. She felt almost wild with despair.

Enid had moments of some enjoyment selecting those old friends whose daughters were not dancing to talk to while Cecil and Nancy twirled continuously round the room.

Simon was in a curious state of satisfaction beyond him to account for. He demanded this unhappiness as a sort of blood offering in his purging of Garonlea. Beyond any idea of Sylvester's he had schemed and planned just such an evening as this, in complete contrast to every party he could remember at Garonlea. It entirely satisfied some instinct against pleasure in

308

his nature. He had never before to-night felt Garonlea to be his own. Now he held a strange power over it, or rather with it. The house itself was his mate and equal in power. Together they avenged the insults of freedom and coarse taking of pleasure that had filled the easy years.

After to-night there would be other parties. Parties for witty talk and restrained pleasures, he and Sue would give them, delighting their chosen friends, themselves cool and happy. But to-night this avenging was required. Even Sue's unhappiness was not enough of a blood-offering for all that he hated in Cynthia and resented in her power at Garonlea.

Diana and Sylvester watched him from a doorway in the garlanded ballroom.

"Sylvester, do you think he's possessed?" Diana said. "I've only known one person like this before. Not minding the unhappiness——" It was her mother, iced above all young turbulence of spirits, calm, unconscious, untouchable. Diana could feel her in the room again, commenting coldly on some miserable young girl who had no success, quite unhurt by the dreadful failure of her rare and awful parties. This was Simon. Simon, to-night recreating the atmosphere among his guests so that the moments, long past, lived again with unspent strength.

"I do think it is rather dangerous," Sylvester said. "You knew more about it than I did when you were so frightened."

"Have you been in the library?"

"I can't bear to see all those girls and know I can only dance with one. And I hate to dance."

"Their soured up, awful silence."

"Better than the gay bursts of girlish chatter."

"Sue didn't get many laughs."

309

"Is she now? Listen."

"Yes, listen. Do you hear a difference?"

They went back across the hall and stopped inside the library door, holding their breaths in the change, the warm difference fighting with the chill difficult airs, in the room.

Cynthia was standing with the group on the hearthrug. She looked so of another world and time from them and from the room that Sylvester and Diana hesitated, shocked with surprise in the doorway. She was wearing a black velvet dress of the latest possible mode. It was simple and inevitably clever, flattering her out of two stone of weight and ten years of age. Among all the puffed bosoms and spread hips and frilled shoulders she looked like a very clean, black fish. She looked lovely and clear-lined and active. Beyond any of them she suggested an extremity of luxury and cleanliness. She wore her pearls and diamond bracelets with that sort of bold carelessness that no one could impart to jewels in the days of lockets or in the days of cameos. Her hair was brushed, curling away from her ears. She looked only like herself. Where she was, there was the place to be. There only was strength and laughter and warmth and the reality of present time and the true sense that the past was dead. All these girls who had been cold, stiff and blasted in the hell of Simon's party, were better now—Cynthia with her extraordinary, almost male power of flattering women, had restored in them a sense of their own interest and glamour. It was strange and beautiful, this radiance flowing from Cynthia to the girls and outwards in lessening circles through the terrible room. And it was incomparably stronger than all the chill ghostly currents of the night.

Very soon a servant brought a bottle of wine.

"Bring us another bottle, Thomas," Cynthia said, somehow including him in the lightness and life of her group. "One won't be nearly enough for us. . . . Jean, darling, but you hate champagne. Scotch for Miss Arkwright, Thomas. A double one. Oh, girls, we need a drink up, don't we? You mustn't blame me for this celebration. It's not my idea of fun. I'm all for lots of chaps and sport. Even at my age——"

"She'll steal the whole party," Sylvester whispered to Diana. "You'll see."

"I want her to, Sylvester. Even if it ruins our plans."

"Our plans? Oh, what the Hell? I want her to win it too. Genius putting on that dress——"

"And coming down when it was all touching bottom——"

Sue came flying in in her brown silk.

"Mamma! How marvellous! You are better?"

Cynthia laughed and laughed. "My pet! My poor little McGrath! You do look awful. I hope you won't catch diseases from Granny's old clothes—they never went to the cleaners. Have a nice glass of wine, darling, and take off that monstrous little hat. It gives me back my headache to look at it."

"Oh, Mamma, come and improve the party—you always can." It was odd that Sue who left things half finished, not caring, not even completing her own jokes, should mind so much all this constriction and unhappiness.

"All right, Darling, we'll do our best——"

On the band's platform she looked like a cabaret turn from another life. A preview of Gertrude Lawrence, ageless, enchanting, years before her time. All round the room eyes turned to her. She only saw one person's—Simon's. It was the first time she knew fully the truth of his distrust and hatred. For that was what she saw and she felt towards it exactly as she had felt so long ago towards Lady Charlotte and her bitter powerless hatred. But Simon's was not powerless. Perhaps only for to-night she would be able to defeat it. To-night she felt endlessly strong.

"Listen," she spoke half-apologetically, half alluringly to the waiting listening faces. "The Bar is open—in the schoolroom."

It was nothing and everything. It snatched Simon's party back from reality to masquerade. She was real. She was Cynthia. She knew the horror going on and she would not endure it. She could not suffer its continuing.

XLIV

A BAR—who ever heard of a bar at such a party?
Cynthia was quite right. It was the break up of Simon's
party and the starting point of hers. The schoolroom
empty and pale, nothing there but a lot of bottles
and glasses and the flower piece strong on the wall.
Soon it was stuffed and crowded with people having
a drink, people feeling better already. Girls finding
one another gay, amusing. Not spies and observers of
unsuccess. Cynthia everywhere, with this one, with
that one, giving all that was herself. Taking Simon's
party away from him and from Garonlea and from the
power of the past. Minute by minute changing it
back into to-night. Soon the sedate misery in the
ballroom was broken up. They played musical chairs
and hunt the slipper, and when they were a little
drunk, charades in which one of the mousiest and
dreariest girls was such a success that even Sylvester
was impressed. The party stopped trying to be a ball.
People played Poker and Bridge. If they did want to
dance the band played a little soft music, but most of
the band was playing games soon. Rumour had it
that there were oysters somewhere. And it was true.
All the oysters that you could eat. Really enough
oysters were being opened in the billiard room.
Oysters and champagne and stout. One of the band
turned out to be a truly funny comedian. Encouraged
by Cynthia his cracks grew better and better.

Quite pale with disgust, Violet retired with Muriel
and Simon to the library before the end of that turn.

The library was no longer real, no longer an outpost of dignified restraint. Some one had popped Sue's little hat on one of the black boys. There were forests of empty glasses on all the little tables, plates of oyster shells on the shawled piano. Most curious insult, in the middle of the floor sat a white chamber pot that had figured a little unduly in one of the charades. Violet and Muriel shied violently at the object.

"Well, dear," Violet said smoothly, coming out of her shy on a placid curve, "it's been a wonderful evening, hasn't it, Muriel? Some of the dresses were as pretty as possible. Perhaps now it's time we old people went to bed. What do you say, Muriel? Ah, here's Enid! Are you ready for bed too, Enid?

Simon moved over to the door with them, for all the world as though he would light their candles and watch them up the staircase. There were no candle-sticks but there were good-night kisses.

"Good-night, Simon dear."

"Good-night, Simon dear."

"Good-night, Simon dear."

They went upstairs one behind the other, wearing the dresses that belonged to their youth. They had never escaped their youth. They would never all their lives be free of it.

Simon went back to the library. The gentle ac-companying blur of the piano came through the door from the drawing-room and the husky half-heard voice of the comedian singing soft Bo-Hoo songs now, cowboy songs, cleanest sentiment, whistling bits of them. There were bursts of applause, wild true enjoyment. It was all an excessive emotional success.

Simon picked up the chamber pot and stood with it held absently in his hand listening to his party shaken into ruins around him. His power was gone. The purging fire was dead. The disease he had thought to burn out raged more strongly for its checking. He felt angry and hurt, but principally very tired in his defeat.

Sue came in and found him standing like that.

"Oh, Simon."

"Darling, I've been beastly to you all night. I do deserve this."

"You haven't, Simon. It was my fault. I couldn't play up."

"Where shall we put this? It's such a nuisance to me."

"Behind the curtain, perhaps."

They put it behind the curtain together.

"And your hat, Sue? She took it off the black boy's turbaned head and put it back on her own.

"The thing is, Simon—Have you had some oysters and Black Velvet yet?"

"No, Sue, I've been too proud. Only that awful cup and a chicken pâté."

"I haven't either. Loyalty to you, I expect. Shall we?"

"Well, we might, I suppose. It will help me to take my defeat like a chap."

They left the squalid library hand in hand.

On the party went and on. No one knew that this was Cynthia's last party at Garonlea, this party that she had snatched so wildly from Simon and from Garonlea, from failure to the highest note of success. Only Cynthia knew and she could not own even to her inmost self all that this told her. To-night her determination to defeat Garonlea again was the only thing. She gave herself to it. For to-night, only for to-night. Never had there been such a spectacular moment as this with the house restored and marshalled

315

against her. Never while she lived had Lady Charlotte's power meant what it did to Cynthia, seeing it in Simon's eyes to-night. She saw defeat. She fought for a fitting exit. A death in music. It would never be Cynthia to creep away from Garonlea in weak surrender. She would go out at her own will and time. Not at her child's wish. Not at the insolent insistence of the house and its ghosts.

So great was her effort against Garonlea that the other part of her defeat slackened in urgency. Reuben and Cecil lost their importance as symbols of a more terrifying defeat. For the greater part of the night she did not see them. It was a dream, a panic, her youth and beauty slipping. She felt ageless and secure. Power could not leave her. All night she felt its flow through her, into her, from her. She could never be defeated.

At last the party ended. Every good-bye was a sort of curtain and Cynthia took them all. The thanks, the praise, the love. All the strange emotions that came out of Simon's party had been changed into these things which must always be hers. Always to the end. Was now the end? She found herself standing alone in the hall. The last good-bye over. The sound of cars fainter down the valley.

Cynthia went out through the open door, saying to the tired servant, "I'll shut it when I come in, Thomas. Good-night. You were all marvellous. You looked after us too well." Out on the dewed heavy grass she looked back at the house; pale, immense, sprawling in the night, its creepers like great deep stains spreading on the walls. She walked across to where Diana had planted one of her groups of lilies near trees. Branches pieced the night sky, caging the scent of lilies, that terrible wild scent at night, and she stood there looking back at the house, holding on to her feeling of triumph, her sense of domination. No, she did not have to hold on. It was flowing out of her, she was bathed in it still. She felt beyond triumphing. She was maintaining a sort of ecstasy of power. She had really won. Now if she chose to give back all to-morrow—to-night, she did it out of her own power.

In another minute she would go back into the house. Find Diana—darling Diana, how dreadful to-night must have been to her—and say, "what would you like most? I know—Rathglass. We'll go back. I know about Simon now. I'm not going to sour him up any more."

For another minute she stood in the lilies' scent. It was like drinking the air to stand there, heavy sharp exciting drug of lilies. Another minute she would wait there, looking at the house, conscious of all the forces that she had overcome.

If she had not waited. Ah, if she had not seen them. They came round the wide flagged turn from the terraces, Reuben and Cecil. They walked slowly, their arms round each other, pausing, whispering, head bent, head raised. It was as if the strain of parting must be beyond them. Outside the light of the doorway they embraced. They parted and again embraced. Cynthia watched a mime of good-nights that must be said. Farewells to love in ardour and despair for even such a short good-bye.

She watched till they had gone, standing trembling and cold now in the lilies' bitter rich breath. She saw them still, Cecil in that long dark overcoat lost against him in the darkness, she knew it was good-night they had said, leaving each other till to-morrow. Cecil was too ignorant, too prudish—— Not that—— Too young—that was it—that was the terror and the truth. He would not sleep with her lightly. She was too young. It was all she had. It was everything.

Back to reality a moment came flying. Not softened by time but with its own twist and edge complete. It came back, the first sight she had of Garonlea, immense, pale, river misted in the depth of its valley . . . feeling as though she receded from Desmond. As though impossibly they were parted—their warmth and understanding a cold forgotten breath. Their youth that could never end finished. No more Love. All the meaning of that moment came to her now beyond any meaning it had held then, in its quick passing. He lifted her down from the high dog-cart; decorous, solicitous, tremendously proud. Together terribly emotional. Embraces impossible. Love not spoken then.

And now, all tears shed, Passion unspent, still strong in her. She stood again outside the house, cold in the luxurious animal scent of the lilies and a voice within her was crying and protesting, "Nothing they tell you about Life is true. Nothing. Especially nothing they tell you about old age."

Cold she was now. The night was ending. Mist from the river was taking great shapes like white and silver cows under the dun sky. Cynthia dragged her smart little coat across her breast, and bowed her head and went back across the heavy grass and the wide gravel sweep to the house. She always walked strongly and beautifully. This morning she was like a person walking in a storm or a dream, blown along. She did not remember or trouble to shut the door or put out the light. Through the halls, up the wide stairs, her head still bent, her coat dragged about her for comfort.

Diana was in her room, cosy as though she had

just got out of bed in her fancy dress flannel dressing-
gown and the plaits swinging awkward and rather
stuffy. As though she had just got out of bed twenty-
five years ago. She was slipping a hot-water bottle
down between Cynthia's sheets. There was a spirit
lamp burning on a low table near the fire, clear special
smell and blue flame under a bright little pot.
A tray with two glasses and spoons. A new neat tin of
something. How comforting, just for the moment in
the world's whole emptiness. Could she tell Diana?
Could she say to her, "I'm lost." Am I outside life for
ever? No more love for me? Could you say this to
Diana who had lived so vestal a life? Cold caves of
life with no love ever and all that it means in tears and
dreams. This was her only interpretation of Love, this
warmth and care; hot-water bottles, spirit lamps,
hot milk. No more. But far more than that. Thank
God, one need not speak, need never explain oneself,
apologise, admit defeat or demand praise from her.
She was love itself. With her there was ease and peace.
The drama of saying, "We're going back to Rathglass."
slipped from Cynthia almost, not quite, because she had
to say it, looking over her shoulder from the fire.

"Darling, I know you're right. It's not all because
it's so lovely for me," Diana breathed faster. "I can't
believe it's true." She went to the window, pulling
back the curtain, looking out into the early metal
light as if she could see Rathglass through the mist.
As if she could see into the kitchen as it had been, the
warm air wavering with hot sweet smells. In other
rooms the furniture used daily their own again. . . .
Should she leave her favourite flower vases behind at
Garonlea? Certainly not. . . . The thought of purple
flowers and red flowers was like wine in her mouth.
The toil and peace of having your hands in earth you

319

loved in the place where you lived came back to her, fortifying in her a curious strength for age. No ends of old dreams to startle her into a moment's wild regret. Each present day and hour complete in its own strength.

She saw Cynthia by the fire and knew that somehow, out of this evening, defeat as they had planned had come to her. But she had found and admitted it to herself. No one else knew. It had been as they had planned. But beyond their schemes as she was beyond them. Out of their reckoning. She looked haggard and heavy now, wandering about, the make-up off her face, so that now her skin was clean and old, much older. She had taken off her little coat too and the tight jewelled shoulder-straps of her black dress bit into the solid flesh of her shoulders. She murmured something to herself, putting a Cachet Faivre in a glass of water to melt. Her high-heeled shoes, rather cruel shoes, were off and her black dress seemed yards longer without them. Her figure had lost its balanced lengthy poise. She was taking the earrings out of her small high-set ears and unclasping her pearls and stripping the rings from her faintly swelling fingers. Diana watched her laying earrings and bracelets and pearls and rings in a heap beside Desmond's photograph on her dressing-table. That expressionless picture of a handsome young man in uniform. You could see he had wide-set eyes and a sweet and generous mouth. Otherwise the likeness had no particular character. Cynthia was so accustomed to this picture that she did not see it any more. As they say—"Time is a great Healer."

THE END

VIRAGO MODERN CLASSICS
&
CLASSIC NON-FICTION

❦

The first Virago Modern Classic, *Frost in May* by Antonia White, was published in 1978. It launched a list dedicated to the celebration of women writers and to the rediscovery and reprinting of their works. Its aim was, and is, to demonstrate the existence of a female tradition in fiction, and to broaden the sometimes narrow definition of a 'classic' which has often led to the neglect of interesting novels and short stories. Published with new introductions by some of today's best writers, the books are chosen for many reasons: they may be great works of fiction; they may be wonderful period pieces; they may reveal particular aspects of women's lives; they may be classics of comedy or storytelling.

The companion series, Virago Classic Non-Fiction, includes diaries, letters, literary criticism, and biographies – often by and about authors published in the Virago Modern Classics.

'Good news for everyone writing and reading today'
– *Hilary Mantel*

'A continuingly magnificent imprint' – *Joanna Trollope*

'The Virago Modern Classics have reshaped literary history and enriched the reading of us all. No library is complete without them' – *Margaret Drabble*

VIRAGO MODERN CLASSICS
&
CLASSIC NON-FICTION

Some of the authors included in these two series –

Elizabeth von Arnim, Dorothy Baker, Pat Barker, Nina Bawden,
Nicola Beauman, Sybille Bedford, Jane Bowles, Kay Boyle,
Vera Brittain, Leonora Carrington, Angela Carter, Willa Cather,
Colette, Ivy Compton-Burnett, E.M. Delafield, Maureen Duffy,
Elaine Dundy, Nell Dunn, Emily Eden, George Egerton,
George Eliot, Miles Franklin, Mrs Gaskell,
Charlotte Perkins Gilman, George Gissing,
Victoria Glendinning, Radclyffe Hall, Shirley Hazzard,
Dorothy Hewett, Mary Hocking, Alice Hoffman,
Winifred Holtby, Janette Turner Hospital, Zora Neale Hurston,
Elizabeth Jenkins, F. Tennyson Jesse, Molly Keane,
Margaret Laurence, Maura Laverty, Rosamond Lehmann,
Rose Macaulay, Shena Mackay, Olivia Manning, Paule Marshall,
F.M. Mayor, Anaïs Nin, Kate O'Brien, Olivia, Grace Paley,
Mollie Panter-Downes, Dawn Powell, Dorothy Richardson,
E. Arnot Robertson, Jacqueline Rose, Vita Sackville-West,
Elaine Showalter, May Sinclair, Agnes Smedley, Dodie Smith,
Stevie Smith, Nancy Spain, Christina Stead, Carolyn Steedman,
Gertrude Stein, Jan Struther, Han Suyin, Elizabeth Taylor,
Sylvia Townsend Warner, Mary Webb, Eudora Welty,
Mae West, Rebecca West, Edith Wharton, Antonia White,
Christa Wolf, Virginia Woolf, E.H. Young

☐	Devoted Ladies	Molly Keane	£6.99
☐	Full House	Molly Keane	£6.99
☐	The Knight of Cheerful Countenance	Molly Keane	£5.99
☐	Mad Puppetstown	Molly Keane	£5.99
☐	Taking Chances	Molly Keane	£6.99
☐	Two Days in Aragon	Molly Keane	£5.99

Virago now offers an exciting range of quality titles by both established and new authors. All of the books in this series are available from:

Little, Brown and Company (UK),
P.O. Box 11,
Falmouth,
Cornwall TR10 9EN.
Telephone No: 01326 317200
Fax No: 01326 317444
E-mail: books@barni.avel.co.uk

Payments can be made as follows: cheque, postal order (payable to Little, Brown and Company) or by credit cards, Visa/Access. Do not send cash or currency. UK customers and B.F.P.O. please allow £1.00 for postage and packing for the first book, plus 50p for the second book, plus 30p for each additional book up to a maximum charge of £3.00 (7 books plus).

Overseas customers including Ireland, please allow £2.00 for the first book plus £1.00 for the second book, plus 50p for each additional book.

NAME (Block Letters) ....

....

ADDRESS ....

....

....

☐ I enclose my remittance for ....
☐ I wish to pay by Access/Visa Card

Number ☐☐☐☐☐☐☐☐☐☐☐☐☐☐☐☐

Card Expiry Date ☐☐☐☐